Genetic Engineering

Principles and Methods

Volume 15

GENETIC ENGINEERING
Principles and Methods

Genetic Engineering

Principles and Methods

Volume 15

Edited by

Jane K. Setlow

Brookhaven National Laboratory
Upton, New York

Plenum Press · New York and London

The Library of Congress cataloged the first volume of this title as follows:

Genetic engineering: principles and methods, v. 1-
New York, Plenum Press [1979–
v. ill. 26 cm.
Editors: 1979– J. K. Setlow and A. Hollaender.
Key title: Genetic engineering, ISSN 0196-3716.

1. Genetic engineering—Collected works. I. Setlow, Jane K. II. Hollaender, Alexander, date.
QH442.G454 575.1 79-644807
MARC-S

ISBN 0-306-44526-3

Printed in the United States of America

CONTENTS OF EARLIER VOLUMES

VOLUME 4 (1982)

VOLUME 5 (1983)

VOLUME 6 (1984)

VOLUME 7 (1985)

VOLUME 13 (1991)

VOLUME 14 (1992)

PREFACE TO VOLUME 1

This volume is the first of a series concerning a new technology which is revolutionizing the study of Biology, perhaps as profoundly as the discovery of the gene. As pointed out in the introductory chapter, we look forward to the future impact of the technology, but we cannot see where it might take us. The purpose of these volumes is to follow closely the explosion of new techniques and information that is occurring as a result of the newly-acquired ability to make particular kinds of precise cuts in DNA molecules. Thus we are particularly committed to rapid publication.

Jane K. Setlow

ACKNOWLEDGMENT

Again June Martino is warmly thanked by the Editor for her final processing of the manuscripts. She does a superb job.

CONTENTS

APPLICATION OF COMPUTATIONAL NEURAL NETWORKS TO THE PREDICTION OF PROTEIN STRUCTURAL FEATURES

Stephen R. Holbrook

Structural Biology Division
Lawrence Berkeley Laboratory
Berkeley, CA 94720

INTRODUCTION

The importance of a protein's structure to understanding its function, interaction with other biomolecules, stability and even its possible redesign is apparent to the current generation of biochemists and molecular biologists. Unfortunately, while everyone would like detailed structural information about the protein of his or her interest, physical methods of structure determination, X-ray crystallography and multidimensional NMR can not realistically fulfill this need. Theoretical methods have so far failed to demonstrate their power in protein structure solution. Currently, the best prospect for satisfying the need for protein structural information appears to utilize the large database of three-dimensional protein structures which have been determined experimentally in correlation with the even larger database of protein sequences to find common features relating sequence to structure and conserved features of protein structures themselves. Computer simulated neural networks are a powerful, flexible and easy-to-use tool for extracting empirical relationships between sequence and structure. In this review the use of neural networks for the prediction of protein structural features will be discussed from a practical point of view. The capabilities and drawbacks of this approach will be highlighted and a summary given of what has been accomplished so far.

COMPUTATIONAL NEURAL NETWORKS

A description of the theory and practice of computational neural networks has been given at various levels of sophistication (1,2,3). The method has been presented to molecular biologists in several papers and reviews (4,5,6,7). Only

Genetic Engineering, Vol. 15, Edited by J.K. Setlow
Plenum Press, New York, 1993

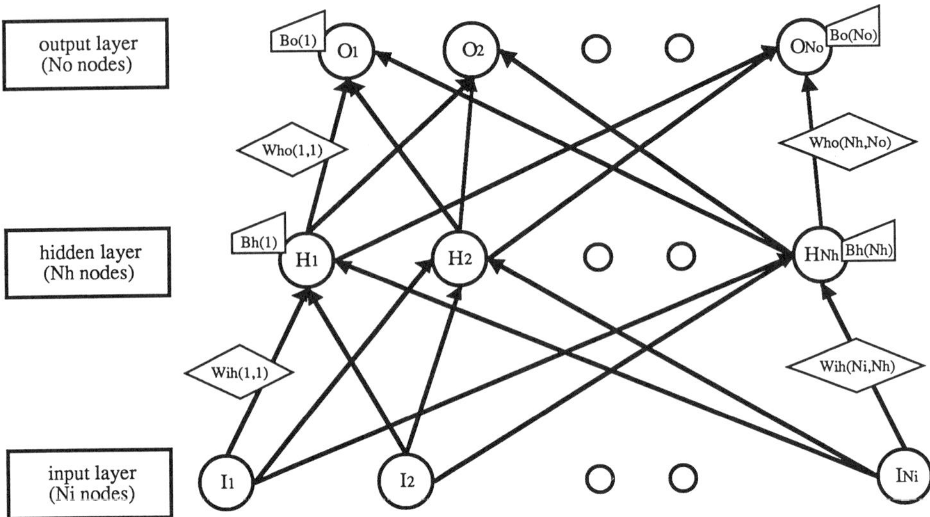

Figure 1. Schematic drawing of a general feed-forward three-layer neural network. Each computational node is represented by a disk. Small circles represent any number of nodes similar to those actually shown. The optimal number of these nodes is problem dependent. Each node in the hidden and output layers is associated with a characteristic *bias* B_h or B_o, as illustrated by the trapezoidal boxes. The links between nodes are represented by lines with arrows pointing in the direction of signal transfer. Each link is associated with a variable weight as indicated in the diamond-shaped box. The values of the input nodes are provided by the user and the values of the hidden and output nodes are determined by their own bias and the accumulated signal from the previous layer.

a general discussion will be given here to refresh the reader and point out some important aspects of the architecture which will be discussed in the text.

There are many types of computational neural networks, most of which will not be treated in this review. All of these consist of simple computational units, called nodes, which are constructed either in hardware or more commonly in software. These individual computational units are capable of performing simple mathematical operations such as summation, multiplication and exponentiation. These nodes typically accept input from other nodes, perform an integration (summation), calculate a response, or activity, usually by a non-linear function, i.e., a sigmoid function, and then pass the calculated response on to other nodes via computational links. These links are associated with weights which are variable parameters. The information describing the relationship between the input and output data is encoded in these linkage weights. These weights are adjusted in an iterative manner so as to minimize the difference between the values of calculated and actual output. This minimization has been done by several methods including backpropagation (8), conjugate gradient minimization (9) and simulated annealing (10).

This discussion will focus on the use of supervised, feed-forward neural networks of the type shown in Figure 1. These networks are usually composed of two or three layers of nodes, the first being the input layer, the second the "hidden" layer and the third being the output layer. Once a hypothesis is formed that certain input parameters are related in some way to the output parameters, the input data must be coded into a numerical pattern for presentation to the network. The input layer, then, is simply a set of real or integer values, one per node, which describes the data hypothesized to be related to the output features of interest. Each node of the input layer is computationally linked to each node of the subsequent layer and each link has a weight associated with it which is initially assigned a random value and then adjusted as the network training proceeds. The activity of each node in the output layer is the accumulated value of the weighted inputs from all the nodes of the previous layer. The activity of each node in the output layer is the prediction of the network. During training these output activities are compared to the known values and the difference minimized by adjustment of the linkage weights throughout the network. When the middle, or hidden, layer is omitted the network is called a perceptron and is expected to perform much like multiple linear regression methods. Including a hidden layer allows the network to extract higher order features, ignored by linear models, which may be very significant in formulating an input-output mapping.

Database Design

Training set. The most important aspect in using a neural network for prediction or analysis of protein structural features is assembly of a list of known examples to be used to train the network. As the network is exposed to each of these examples, the weights on the links are adjusted so as to minimize the difference between the predicted and actual results (output). Protein structural information is found in coordinate form in the Brookhaven Protein Data Bank (11). While this database does include certain other information such as sequence, disulfide bonding, temperature factors and the author's secondary structure assignments, many derived parameters such as torsion angles, secondary structure, energies and surface accessibilities must be calculated by external programs such as the DSSP program of Kabsch and Sander (12). The protein sequence databases, PIR (13) or Swiss-Prot (14), are much larger and in addition to sequence also contain other relevant information such as references, disulfide bonding, phosphorylation and glycosylation sites. Several points should be stressed in deciding which databases to use and which examples from the complete database to include in the training set. These include the reliability of the information (i.e., structural resolution and refinement, method of sequence determination), maximization of the number of examples (perhaps requiring the use of more than one database) and minimization of homology between examples in order to maximize learning. Sufficient nonhomologous examples should also be retained independent from training in order to perform testing as discussed below.

Testing. An accurate assessment of the performance of a neural network is vital to acceptance or rejection of the hypothesis on which it is based. In order to assess network performance, a set of testing examples is required which supply the same input information and for which the desired output information is known. These may be obtained by sampling of the complete database in various ways. A sufficient number of testing examples is necessary so that the performance of the network can be statistically evaluated.

Independent set. In most studies to date a testing data set has been used which is completely different from the training set, that is, there is little or no homology between the testing set examples and those of the training set and none of the testing set examples has been used for training. The testing set should be as large as possible while not significantly affecting training by siphoning off potential training examples. A testing set of 100 to 500 will provide reasonable statistics. If possible it is useful to construct two separate testing sets. A comparison of the network performance on the two sets will provide an estimate of the accuracy of the performance estimate. Also, one testing set may be used to determine when to stop training for maximal performance, while the second remains completely independent to judge the final network accuracy.

Jackknife method. If the total amount of data of known output is insufficient to allow two or even one reasonably large independent testing set, one may consider testing by the jackknife method. In this approach, a single database is constructed for training and testing. Before testing, one example (or a small set of examples) is removed and set aside while the network is trained. After training, this example is tested and returned to the database. Then the next example is selected, removed, the network trained again from scratch, and the second example tested. This process is repeated until all examples in the database have been tested independent of their presence in training. The performance of each of these examples is then averaged and a standard deviation obtained. This is, then, an objective evaluation of network performance. While this method retains the largest number of training examples, one must be careful that none of the examples within the database is homologous, since they are all eventually used for testing while their homologues remain in the training set.

Data representation. The form of the input data to be presented to the network is also a vital consideration. The number of input nodes should be kept to a minimum so as to minimize the number of variable weights. This can often be done by using a different representation of the amino acid sequence. For example, the amino acids can be represented by a set of 20 binary integers or by any of their properties such as hydrophobicity (15), volume, charge, helix propensity, or any combination of these such as the vectors extracted by Kidera (16). Likewise, the overall amino acid sequence can be represented by a sequence window of arbitrary size, the amino acid percent composition, the composition of hydrophobic, neutral and hydrophilic residues, the composition of dipeptides or tripeptides or any combination of these. Other types of information such as chain length (or molecular weight), presence of cofactors such as heme or metals, number of disulfides, or number of subunits may also be important for the network to make accurate predictions of the structural feature being studied.

Advantages

Computational neural networks are able to reproduce any continuous function given sufficient hidden layers and nodes per layer. Thus, they provide a general framework for analysis of information compiled in a database. While many other mathematical techniques may also be used for this purpose, neural networks are a convenient, general method which can be applied without a preconceived model. Some of the applications of neural networks to structural biology are classification of structural types, extraction of rules relating sequence to structure, encoding of data in a pattern of reduced dimension and prediction of structural parameters based on previously observed patterns. The power of neural networks lies in their ability to generalize from known example with which they have been trained to unknown (different) examples which they have never observed.

Pitfalls

Computational requirements. Training of computational neural networks can be extremely intensive, depending on the size of the network as measured by the number of weighted links to be adjusted, the number of training examples and the method of training used. The number of links is easily calculated as: Weights = $(N_i{*}N_h) + (N_h{*}N_o) + N_b$ where N_i is the number of input nodes, N_o is the number of output nodes, N_h is the number of hidden nodes and N_b is the number of biases. The size of the network affects not only the computational time, but usually the memory required by the network, although this may depend on program architecture. The number of training examples should be as large as possible without duplication. Computational time is linear in the number of examples. Adjustment of weights by backpropagation is faster for each iterative cycle, but takes many more cycles to converge than the slower conjugate gradient method. Overall, the conjugate gradient method is faster and more robust. Still, the large number of variable weights leads to a severe problem with location of the global minimum among the multiple local minima. Simulated annealing followed by conjugate gradient minimization widens the range of convergence; however, the computational time is greatly increased. A more commonly used approach is to perform multiple conjugate gradient minimizations starting from different initial points and select the trained network with the lowest minimum.

Memorization (generalization). As in any multivariate analysis, when the number of variable parameters exceeds the number of independent observations the data can be fit to any degree of accuracy. However, in such an underdetermined case there is no generalization of the results, i.e., the fit is exact to the

Table 1
Prediction of Protein Structural Features with Neural Networks

Structural Feature	Network Learns	Database	Accuracy	Weights
Secondary Structure				
α,β, Coil (6)	Local sequence windows	18105	62.7%	819
α,β, Coil (6) Smoothed	Local sequence windows	18105	64.3%	819 + 117
α,β, Coil (21)	Local sequence windows	8315	63.2%	718
α,β, Coil (29) (in all α, all β, α/β classes	Local sequence windows	3864, 5628, 6181	79%, 70%, 64%	819
β-Turns (28) (type I, II, nonspecific, non-turn)	Sequences of β-turns	1265 (100 turns)	56.2%[a] (.43) 53.8%[b] (.41)	320[a] 672[b]
% Composition (α,β,coil) (9)	Number & % amino acids & Heme presence/ absence	104	95%	192[c]
Tertiary Structure				
Distances (32)	Large sequence window	28,200	N.A.	375,900
Torsions (5)	Local sequence window	7,070	N.A.	616
Solvent Accessibility				
Two-state (buried/exposed) (36)	Local sequence windows	3581	72% (.44)	378
Buried/ intermed./-exposed (36)	Local sequence windows	3581	52.0%[a] 54.2%[b]	441[a] 1500[b]
Disulfide Bonding				
Cys bonding state (7)	Local sequence windows	659	80% (.61)	422
Fold Classification				
Immunoglobin Recognition (33)	Sequences in four β-strands	N.A.	92.7%*	832
ATP-binding Proteins (34)	Sequences in conserved region	349	78%	340
Multiple Protein Classes (35)	Number & % amino acids	290, 353[#]	76%, 60%[#]	125

Correlation coefficients in Accuracy column are given in parentheses.
[a]Perceptron network, [b]Hidden layer network, [c]First network of a "tandem" pair
*Does not include false positives
[#]First network distinguishes between four protein classes, second network reflects results for choosing between four protein folds and all other fold types.

specific pattern or distribution of data points, but there is no attempt to extract the most important features of the data. In the case of neural networks the variables are the weights on the links between nodes and the biases of each node. The sum of these numbers should be less than the number of independent examples on which the network is trained. The definition of independent examples implies a total lack of correlation between them, when the more common case is that the examples, though different, are similar.

Interpretation of results. In addition to performing a predictive role, we would like to analyze the basis on which neural networks make their decisions. What kind of information do they learn in training? To do this we must analyze the weights associated with the computational links. This is relatively straightforward for perceptron networks, with larger weights contributing more to the output node with which they are associated. However, in networks containing a hidden layer this analysis can be difficult and in some cases not feasible. Many of the examples of applications discussed below lack a complete analysis of the weights and interpretation in terms of structural chemistry. Others have made an effort in this regard, resulting in interesting scientific insight.

APPLICATIONS

A summary of the current applications of computational neural networks to protein structure prediction is shown in Table 1. The variation in training set size, prediction accuracy and network parameters (weights) provides an indication of how well a network may be expected to perform on a specific protein. A small number of network weights, together with a large training set, should provide the best predictions for a protein which is non-homologous with those in the training set. A large testing set, several consistent and independent testing sets and the use of testing procedures such as the jackknife method are indications that a high prediction accuracy may actually be valid. Following, are listed the various aspects of protein structure prediction to which neural networks have been applied with discussion of the databases, network design, performance on training and testing and examples of predictions which have been made using these networks.

In order for the reader to gain insight into the capabilities and drawbacks of network prediction of protein structure, perceptron networks have been used to predict a variety of features of a specific protein: the c-H-*ras* oncogene protein (17). The crystal structure of this protein contained 178 residues including a nucleotide binding domain which ligands a guanosine diphosphate molecule. A schematic drawing of the protein is given in Figure 2.

Secondary Structure

Three-state model. Initial interest in the use of computational neural networks for predicting protein structural features came in the area of secondary structure prediction. The prediction of secondary structure has long

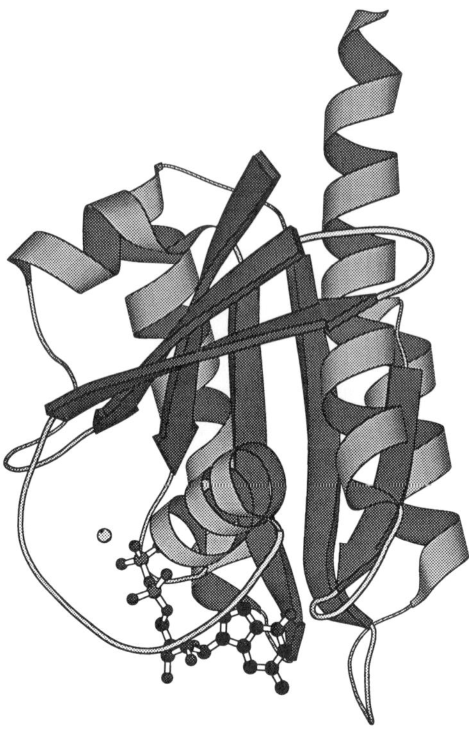

Figure 2. Structure of the c-H-*ras* oncogene protein catalytic domain (17). The barrels represent regions of α-helices and the broad arrows β-strands. The bound GDP molecule is indicated. The illustration was made with the computer program MOLSCRIPT (37).

been approached empirically, with the use of the secondary structures observed in proteins to formulate statistical and rule based algorithms (18–20). It was hoped that the capabilities of neural networks to extract relevant information without the necessity of a preconceived model would allow an improvement in prediction accuracy over the ~60% correct rate previously obtained for a three-state (helix, sheet, coil) model. Qian and Sejnowski (6), Holley and Karplus (21) and Bohr (22) utilized networks of similar, but different, architecture which input a window of local amino acid sequence around the residue of interest to classify the secondary structure of the central residue as helix, strand or coil. The networks, although giving marginally improved results over previous studies, were still limited by the availability of only local sequence information without context in the overall structure. A comparison of a neural network to Bayesian statistical methods (23), which assume that the probability of an amino acid occurring in each position of the protein is independent of amino acids occurring elsewhere, showed no significant differences. While surprising, this is in agreement with the lack of significant improvement in network performance with addition of hidden nodes (6), when compared to a perceptron, and appears to be characteristic of the dependence of secondary structure on local sequence.

While the average overall prediction rate was 63 to 65% correct, it was shown that more accurate predictions were possible for those residues which were most strongly predicted (21). This appears to be a general feature and advantage of predictions made with the use of neural networks; a correlation exists between prediction strength (output activity) and prediction accuracy. It was also shown that prediction accuracy for the aminoterminal region was more accurate, possibly due to the lack of influence of the rest of the structure on its folding. An application of neural networks to the prediction of the secondary structures of the HIV proteins p17, gp41 and gp120 has been presented and the results compared to traditional approaches (24). The neural network approach has been incorporated in at least one microcomputer program for prediction of secondary structure (25).

As an illustration of the performance of neural networks in the prediction of protein secondary structure, the Qian and Sejnowski net (6) was used to predict the secondary structure of the catalytic domain of the c-H-*ras* oncogene protein. The results are summarized in Table 2 and compared to predictions made using the method of Chou and Fasman (26). While almost all of the secondary structural elements are predicted in the proper location by the network, the lengths of the helices and strands are underestimated, resulting in a lower proportion of these elements than are actually observed in the crystal structure. On the other hand, the Chou-Fasman algorithm is less accurate in qualitative location of the helices and strands, but more closely predicts their size for the ones it does locate properly. Since it is now possible to predict accurately the overall proportions of secondary structure from the amino acid composition (see below), it should be possible to improve the problem of overall over- or underprediction by the network by adjustment of the threshold values used for predicting helix, strand or coil so as to best match the calculated overall percentages.

Turns. In addition to α-helices and β-strands, β-turns form a third secondary structural element. Although all β-turns are localized over four residues, there are 7 to 8 different classes defined by hydrogen bonding and conformation. Two-thirds of all β-turns are either Type I or II, which close the turn by a hydrogen bond between the carbonyl oxygen of residue i and the amino of residue i + 3. A neural network has been designed to distinguish between Type I, Type II, and non-specific β-turns and non-turns (27,28). The network was trained by exposure to the four residue amino acid sequences involved in the various types of turns as well as a sampling of four residue sequences in non-turn regions. This network thus consists of 80 input nodes (20 per amino acid), 8 nodes in the hidden layer and 4 output nodes, one for each of the three classes of β-turns identified and one for non-turns. The method is significantly more accurate than prediction of turns with the Chou-Fasman technique. When no hidden nodes are used (perceptron), 26.1% (correlation 0.183) of the predictions are correct while addition of 8 hidden nodes results in a prediction accuracy of 28.7% (correlation 0.204). If a prediction is considered correct when the turn location is displaced by one or two residues,

Table 2
Comparison of Actual and Predicted Secondary Structure in the c-H-*ras* Oncogene Catalytic Domain.

Secondary Structure[1]	Secondary Structure[2]	Q&S Prediction[3]	C-F Prediction[4]
1–9 Strand	2–9 Strand	5–9 Strand	4–9 Strand
15–26 Helix	16–25 Helix	19–24 Helix	*17–29 Strand*
38–44 Strand	38–46 Strand	43–46 Strand	40–46 Strand
51–57 Strand	49–57 Strand	*Not Predicted*	50–56 Strand
69–75 Helix	69–72 Helix	62–71 Helix	65–70 Helix
77–84 Strand	77–83 Strand	79–83 Strand	77–87 Strand
87–104 Helix	87–103 Helix	93–101 Helix	87–102 Helix
110–117 Strand	111–116 Strand	111–114 Strand	109–114 Strand
126–137 Helix	127–137 Helix	121–132 Helix	117–136 Helix
140–144 Strand	141–143 Strand	140–143 Strand	*142–156 Helix*
151–171 Helix	152–168 Helix	155–167 Helix	159–171 Helix
Proportions %H,S,C			
36%H, 30%S, 34%C	35%H, 25%S, 40%C	30%H, 12%S, 58%C	40%H, 29%S, 31%C

[1]Analysis of the secondary structure of the catalytic domain of the c-H-*ras* oncogene protein p21 (17).
[2]Automated assignment of secondary structure by the computer program DSSP (12).
[3]Neural network prediction with the two-layer net of Qian and Sejnowski (6).
[4]Secondary structure prediction with the method of Chou and Fasman (26).

then a perceptron network is 56.2% (correlation 0.434) accurate while one with eight hidden nodes is only 53.8% correct (correlation 0.411).

When this method is applied to the p21 ras protein, a single β-turn is predicted. It is strongly predicted to be centered as S106. This is near the end of a helix in the three-dimensional structure and judged by the DSSP program is consistent with the presence of a β-turn. Other turns identified by DSSP and not predicted by this network occur in regions of nucleotide (GDP) binding.

In protein classes. Kneller–Cohen Langridge (29) attempted to improve on the results of Qian and Sejnowski and others in prediction of protein secondary structure by training separate networks on different protein classes. Specifically, they compiled a database of all α proteins, all β proteins and α/β proteins. For comparison, a general database (differing somewhat from that of Qian and Sejnowski) was compiled and networks were trained and tested on it. The network architecture was very similar to that of Qian and Sejnowski (6).

A testing subset of this general database was predicted with an average accuracy of 63% and correlation coefficients of $C_\alpha = 0.35$, $C_\beta = 0.30$, $C_{coil} = 0.41$, similar to the results obtained in other studies. The all-α, all-β and α/β databases contained 22, 24 and 20 proteins respectively and were trained and tested with the use of a jackknife procedure. The all-α class was predicted with an average accuracy of 79%, all-β proteins at 70% and α/β proteins at 64%. The results for all-α and all-β proteins were much better than the Qian and Sejnowski results on the same proteins. The authors also tested the effect of adding nodes describing helix and strand hydrophobic moment, charge pairing at i and i + 4 residues, and average hydrophobicity over a window. None of these additions made significant improvements, although the slight improvement of helix prediction when helical moment was used prompted them to include this node in their all-α predictions.

Composition. A correlation between amino acid composition and secondary structure content was noted by Davies (30) as early as 1964 as well as Krigbaum and Knutton (31) in 1973. This relationship was quantitated with the use of multiple linear regression. The average error in prediction of secondary structure content with this regression analysis was 12.8% for helix and 10.6% for strand, somewhat worse than the results obtained by merely counting residues predicted to be in helix and strand by secondary structure prediction programs (6). Muskal and Kim recently reported (9) a significant improvement in prediction of secondary structure content using neural networks. They used a "tandem" network approach in which the first network was supplied information of a protein's amino acid composition, molecular weight, and heme presence/absence as a string of 22 real numbers (input nodes), and the network was trained to produce two real numbers describing helix and strand content at its output nodes. A network with eight hidden nodes was found to provide the best prediction; however, since the number of variable weights was greater than the number of training examples (see Table 1), memorization was a problem and it was important to identify at what point during training the network was in a state of generalization, i.e., best able to perform extrapolation and interpolation. A second (tandem) network was designed to determine when to stop the training process in the first network. This second network accepted the hidden node activities from the first network as input and was provided output as to whether or not the first network was in a state of generalization. Predictions made on an independent testing set with this tandem network was approximately 95% accurate in both helix and strand prediction.

This network was applied to prediction of the secondary structure composition of the c-H-*ras* catalytic domain from its amino acid composition. The network predicts Helix as 39%, Strand 21% and Coil 40%, in excellent agreement with the observed values of Helix 25%, Strand 25% and Coil 40% (Table 2). The predicted composition is also much more accurate than that from secondary structure prediction (Table 2) and can thus be used to improve those results.

Tertiary Structure

Distance matrix. A common representation of the secondary and tertiary folding of a protein is given by a binary distance matrix (dotplot) in which the protein sequence is plotted along both the vertical and horizontal axes and points are placed on the graph to indicate where two C_α positions are within a specified distance in the three-dimensional structure. Bohr et al. (32) have attempted to use a neural network to incorporate this information from many proteins and used the trained network to make predictions concerning the secondary and tertiary structure of unknown proteins. Their network used an extremely large number of units and weights, much greater than the number of training examples used (see Table 1). The authors demonstrated the ability of the network trained on 13 proteases to generate a distance matrix quite similar (at least for interactions between residues nearby in sequence) to that of rat trypsin from its sequence. Subsequent steepest descent minimization starting with a different trypsin, but subject to the network distance constraints converged to an rms difference in C_α atoms of 3.0Å. This result may be dependent on starting model and is clearly biased by the presence of trypsins in the training set. Nevertheless, this paper points a direction for eventual prediction of protein tertiary structure.

Torsion angles. An alternate way of representing tertiary structure (as well as secondary structure) is by the variable conformation or torsion angles, phi and psi of each residue in the polypeptide backbone. Helical and strand regions have well-defined ranges of these torsion angles and an accurate prediction of all torsion angles would specify the three-dimensional structure of the protein. However, since any error in these variables would be cumulative and change the relative orientation of distant regions of the proteins by a lever-arm effect, these predictions must be highly accurate in order to be valuable for tertiary structure prediction. An attempt to predict the phi-psi angles in proteins from neural networks has been described (5), with the use of an architecture similar to that used in secondary structure prediction. A local sequence window of seven residues was used as input together with four output nodes defining ranges of phi-psi angles in some of these networks. In other networks there is a more finely partitioned output grid. As expected, predictions made with this method show good correspondence to the actual structure on the secondary structure level, but the overall fold of the protein is incorrect. This is likely due to several factors: accumulation of small errors and the lever effect, the presence of one or a few large errors, or poor prediction of non-helical and strand regions for which the various conformations are underrepresented.

Folding patterns and motifs. Three groups have recently directed their efforts toward prediction of protein folding patterns or classes from amino acid sequence. Two of these approaches use local sequence windows as input, while the third uses the overall amino acid composition of the protein to make the classification. These predictions, together with secondary structure information, may enable construction of actual three-dimensional models where experimental structures are not available.

Immunoglobins. Immunoglobin domains are formed by β-sheets bound by disulfide bonds. Proteins belonging to the immunoglobin superfamily may contain many of these domains. It has been shown that the β-strands B,C,E and F are the most highly conserved segments of the domain. Bengio and Pouliot (33) have trained a network accepting a window of 5 consecutive amino acids as input to determine if that sequence is compatible with one of these four β-strand motifs. When each of the four β-strands is recognized above a threshold, in the correct order and with reasonable separation, the protein is identified as containing an immunoglobin domain. An input representation of 20 nodes per amino acid was coupled with a hidden layer of 8 nodes and an output layer of 4 nodes corresponding to the 4 conserved strands. The network was trained on only 30 proteins containing immunoglobin domains, but the effective number was increased by including "pseudo sequences" generated by known variable residues. This network recognized 98% of the immunoglobin test sequences with only 7% false positives.

ATP-binding motifs. Hirst and Sternberg (34) have trained a network to distinguish the ATP-binding motif based on a window of 17 residues around the consensus sequence of the phosphate binding motif. This perceptron network correctly classified 78% of the sequences tested, much better than a simple consensus sequence search, but roughly equivalent to the performance of a sophisticated statistical method (34).

Classification of Folding Patterns

Recently, Dubchak and co-workers (35) have extended the approach which Muskal and Kim originally applied to prediction of secondary structure composition, to the discrimination of protein folding patterns. In this method, there are 21 real-valued input nodes representing the amino acid percent composition of the protein as a whole, or that of the domain of interest. A series of networks was constructed, trained and tested with from one to four outputs representing the four different protein folding patterns predicted by the authors. The network with a single output node is designed to distinguish a particular fold from all other folds, while networks with two, three, or four output nodes are useful in distinguishing between either one, two, or three other folds and the fold of interest. The folding classes tested in this work and the accuracy of their prediction in a single output network were the 4α-helical bundle (81%), the eight-stranded parallel α/β barrel (85%), the nucleotide binding or Rossmann fold (91%) and the immunoglobin fold (78%). Although examination of the percent amino acid composition of the various classes revealed no statistical differences in single amino acid percentage with either all proteins in the database or another folding class, the networks were able, by formation of combinations of these small differences, to distinguish between folding types. Analysis of the network weights revealed which amino acids are important in formation of the specific folding patterns.

A critical problem in classification of protein folding types is the number of protein examples of known three-dimensional structure. Dubchak et al. (35) have overcome this problem by using the much larger protein sequence database. Proteins of the same family, showing high sequence homology with proteins of known structure, were assumed to have the same overall folding pattern and were used to increase greatly the number of input examples for use in network training and testing.

A prediction of the folding class of the c-H-*ras* oncogene was made as a test of these networks. No ras or ras-related proteins were included in network training. The networks clearly predicted this protein as a member of the class containing the nucleotide binding fold in agreement with the crystal structure.

Disulfide bonding. Cysteine performs special functional and structural roles in proteins. Functionally, it may act as a ligand for metals or other cofactors or it may serve as an active chemical moiety such as a nucleophile in thiol proteases. Structurally, disulfide bonds formed by pairs of cysteines provide stability to small proteins, polypeptides, and extended loops in larger proteins. In addition, intermolecular disulfides may stabilize oligomer formation.

Cysteine bonding state. As an initial step in the prediction of disulfide bonding patterns neural networks were trained to learn the difference between a cysteine involved in a disulfide bond and that existing as a free sulfhydryl (7). A database of proteins of known three-dimensional structure, containing both cysteine and cystine, was selected from the Brookhaven Protein Data Bank. The local amino acid sequence surrounding (but not including) the cysteine of interest was used as input to the network on the hypothesis that a cysteine's propensity and/or aversion for disulfide bond formation depends on its neighbors in sequence. A network trained with a sequence window of ten residues, five to either side of the cysteine, was able to predict the presence of a disulfide bond in 81.4% of the cases tested and free (non-disulfide bonded) cysteines in 80% of the cases tested. Analysis of the variable parameters (weights) in the network provides information concerning the make-up of the sequence surrounding disulfide-bonded cysteines versus those not involved in S-S bonding. For example, the presence of another Cys at a position + or - three amino acids from the cysteine of interest is not favorable for disulfide bond formation, whereas a second Cys one residue away is. Also, polar and/or charged flanking residues were observed to favor disulfide bond formation while hydrophobic neighbors tended to oppose disulfide formation. This observation may be rationalized by the tendency of a hydrophobic sequence to be buried and thus inaccessible for formation of disulfide bonding.

This method was applied to cysteine residues in the c-H-*ras* p21 protein. Of the three cysteines in the catalytic domain, two were clearly predicted to be not involved in disulfide bonding. The third was marginally predicted to be part of a disulfide bond. However, since two cysteines are the minimum required, it is predicted by implication that there are no intramolecular disulfides formed in this protein. This is in agreement with findings from the crystal structure determination.

Recognizing partners. If only a single disulfide is present in a protein, then assignment of disulfide bonding state to each Cys will specify which two cysteines form the S-S pair. However, in many proteins there exist multiple disulfide bonds which place severe constraints on the tertiary structure. Assignment of the disulfide bonding partners then is an important, and difficult, task. For example, in the case of the intensely sweet protein, thaumatin, there are eight disulfide bonds representing over two million possible pairing combinations. A network, using as input the local sequence around cysteines and separation of the cysteines, has been trained (Holbrook, Dubchak and Kim, unpublished results) to predict which cystine residues are disulfide bonded to which other cystines. While the accuracy of this network is only around 70% in correctly recognizing Cys-Cys partners, pairs predicted with higher network activities have been shown to be predicted more accurately. Thus, it may be possible for a given protein to specify with a high degree of certainty some of the disulfide bonding pairs and conversely which cystines definitely are not paired.

Surface Accessibility

Knowledge of which protein residues are exposed to solvent and which are buried is an important criterion in evaluating possibilities for interactions with other macromolecules and ligands, antigenic determinants and potential mutation sites. The exposure to solvent, or surface accessibility, can be directly calculated if the three-dimensional structure of a protein is known (12). In the absence of a known structure, the intrinsic hydrophobicity of the individual amino acids has been used to infer exposure, i.e., hydrophilic residues are exposed and hydrophobic are buried. This approach is unreliable due to the wide variation in the hydrophobicity scales (15), especially the disagreement on classification of the more neutral residues. A database of 20 high resolution protein crystal structures was used (36) to train neural networks to predict: 1) whether a residue is buried or exposed (binary model), 2) whether a residue is completely buried, partially exposed, or completely exposed (ternary model) and 3) the actual exposure of the residues (continuous model). The input to the network is the amino acid whose exposure is in question and its flanking sequence on both sides. All amino acids are represented by activating one of 21 binary input nodes, one for each of the 20 possible amino acids and one node for positions beyond the amino- or carboxyterminus. The surface accessibility is calculated as a fraction of that observed in the central residue of extended tripeptides, which is taken as fully exposed. In the binary model residues of less than 20% fractional exposure were considered buried and greater than 20% exposed. In the ternary model buried was defined as 0 to 5% exposure, intermediate as 5 to 40% exposure and fully accessible as greater than 40%.

For the binary model the most accurate results on an independent set of testing proteins was 72% correct prediction using a window size of nine residues (the central amino acid and four on either side). The influence of the surrounding residues was shown to be quite small with 70% prediction obtained using only the central residue. There does appear to be some influence, however, since when the central residue itself is omitted from input, a correlation of 0.10 and prediction accuracy of 55.3% is still obtained. The networks trained for binary model prediction were not improved by addition of a hidden layer, implying a lack of correlation between amino acids in the input window. An interesting finding was that residues near the N-terminus are predicted more accurately than the overall rate. The ten aminoterminal residues were predicted correctly as buried or exposed in 84% of the cases. This effect was also observed in secondary structure prediction and is attributed to the fact that this region is synthesized prior to the rest of the protein and must therefore assume a stable structure in the absence of intramolecular packing interactions.

The networks constructed to predict a ternary model were 52% correct in their predictions on a testing set for a maximal window size of seven residues. A residue predicted to be fully exposed was actually found to be fully or partially exposed over 89% of the time, while a residue predicted to be buried was found fully or partially buried in 95% of the cases. Again, little improvement was obtained by including a hidden layer of nodes in the network.

A network trained to have a single output node corresponding to the real value of the fractional accessibility and using a window of 9 amino acid residues was able to predict the exposure of residues in a testing set with a correlation coefficient of 0.508 and average deviation between observed and calculated exposure of 18%.

An important feature often observed with neural networks is that the strength of the prediction is correlated with the accuracy of prediction. In the case of the binary networks, for example, the half of residues which are the most strongly predicted (highest activities), are predicted correctly in over 80% of the cases. A large improvement in performance was also noted with the use of average predictions over several members of a protein family (36).

It is possible to gain some insight into the factors affecting surface accessibility by analysis of the weights in the neural networks. Although the primary factor affecting exposure is the identity of the central amino acid itself as judged by its high weight, the flanking sequence is more influential in the case of neutral or amphiphilic residues such as proline and glycine. In the binary model, hydrophobic residues two or three amino acids away from the central residues favor its burial. Analysis of the weights in the ternary model show a greater influence of surrounding residues, implying that the surrounding sequence may affect the degree to which a residue is buried or exposed rather than qualitatively determining whether a residue is buried or exposed. Large residues such as Trp, Tyr and His tend to be partially accessible and in general neighboring hydrophilic residues favor exposure of the central residue.

To illustrate this method, surface accessibilities were calculated for the residues of the c-H-*ras* protein with the use of the binary model and compared to the fractional accessibilities determined from the crystal structure. Overall, the predicted residue accessibility was 77% correct; 63 residues were correctly predicted as exposed and 68 as buried; there were only 18 false positives and 22 accessible residues which were not predicted. Figure 3 compares the predicted and actual accessibility. Many incorrectly predicted residues occur in the regions of nucleotide binding; this may be due to a different conformation assumed on nucleotide ligation.

FUTURE PROSPECTS

While the future of protein structure prediction in general and the use of neural networks as a tool for identification of empirical relationships appear bright, several obstacles must be overcome before further significant progress will result. First, there is still a severe restriction on the size of the database of

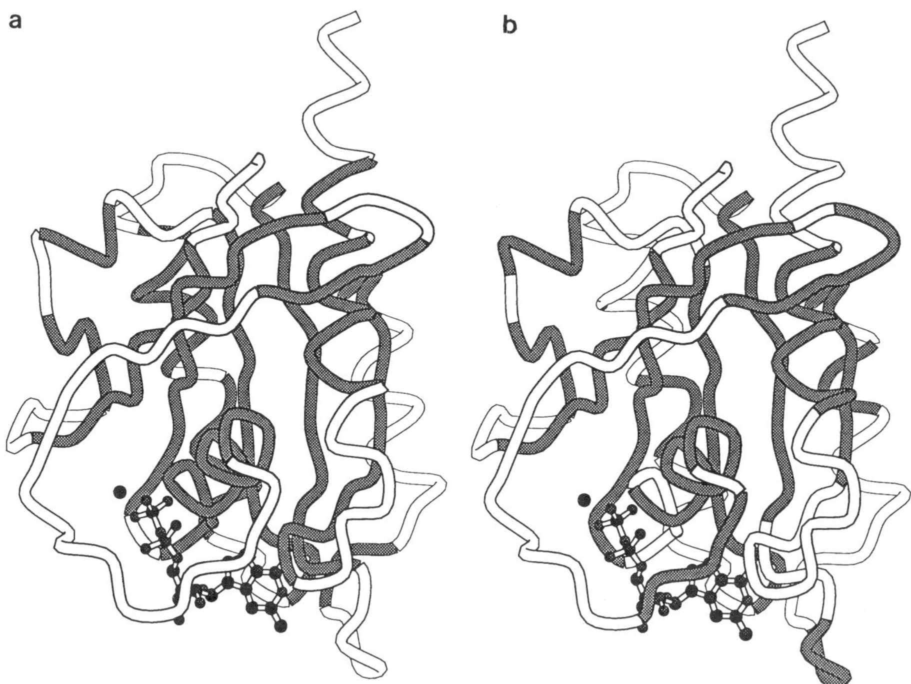

Figure 3. The predicted surface accessibility of the residues is shown on the model of the c-H-*ras* oncogene protein catalytic domain. The C_α positions of the protein are traced by a tube and the GDP and tightly bound Mg^{++} ion are shown in ball and stick form. The figures were made with MOLSCRIPT (37). a) The solvent accessible (greater than 20% exposure) residues as calculated from the X-ray crystallographic coordinates using the DSSP program (12) are unshaded, while the buried residues are shaded. Smoothing was applied for clarity by disregarding single residue breaks in a run of buried or exposed residues. b) Predictions of the residue accessibility based on sequence with the use of the network designed by Holbrook et al. (36). Shaded residues are predicted to be buried. Smoothing was again applied.

known structures. With the rapid growth in structure determination by both NMR and crystallographic methods this will be overcome in time. Meanwhile, it may be possible to use the much larger sequence databases, as some have already done, to enhance the number of known examples where at least the general features of proteins are known. Second, the studies to date have established statistical relationships between sequence and structure, many with only a moderate predictive value. Not only must effort be exerted to identify methods to improve the accuracy of the predictions, but the reasons for lack of accuracy should be identified. For example, cases of incorrect prediction should be analyzed with zeal equal to that of the analysis of those which are successful. Finally, even with powerful mathematical tools such as neural networks, it is up to the innovation and insight of the researcher to formulate a hypothesis to be tested. It seems unlikely that a direct link will be found between sequence and three-dimensional structure of most proteins. Therefore, an intelligent choice of the features to be determined and the parameters upon which they may depend will be the key to approaching an overall solution of the protein folding problem.

We can expect that ever more complex problems will soon be solvable by the approaches outlined above. These may include identification of protein modification sites, prediction of specific residue-residue tertiary interactions and supersecondary structure prediction. These studies must increasingly focus not only on the prediction of aspects of protein structure, but must also begin to analyze the chemical, biological and evolutionary basis of these empirical relationships. A deeper understanding of the principles of protein folding should be our ultimate goal.

REFERENCES

1 Hertz, J., Krogh, A. and Palmer, R.G. (1991) Introduction to the Theory of Neural Computation (Addison-Wesley Publ. Co., Santa Fe, NM).
2 Rumelhart, D.E., McClelland, J.L. and the PDP Research Group (1986) Parallel Distributed Processing: Explorations in the Microstructure of Cognition (MIT Press, Cambridge, MA).
3 Crick, F. (1989) Nature 337, 129–132.
4 Holbrook, S.R., Muskal, S.M. and Kim, S.-H. (1992) in Artificial Intelligence and Molecular Biology (Hunter, L., ed.) pp. 161–194, AAAI Press, Menlo Park, CA.
5 Vanhala, J. and Clementi, E. (1991) in Modern Techniques in Computational Chemistry: MOTECC-91 (Clementi, E., ed.) pp. 991–1015, ESCOM Science Publishers B.V., Leiden, The Netherlands.
6 Qian, N. and Sejnowski, T.J. (1988) J. Mol. Biol. 202, 865–884.
7 Muskal, S.M., Holbrook, S.R. and Kim, S.H. (1990) Protein Eng. 3, 667–672.
8 Rumelhart, D.E., Hinton, G.E. and Williams, R.J. (1986) Nature, 323, 533–536.
9 Muskal, S.M. and Kim, S.-H. (1992) J. Mol. Biol. (in press).
10 Engel, J. (1988) Complex Systems 2, 641–648.
11 Bernstein, F.C., Koetzle, T.F., Williams, G.J.B., Meyer, E.F.J., Brice, M.D., Rodgers, J.R., Kennard, O., Shimanouchi, T. and Tasumi, M. (1977) J. Mol. Biol. 112, 535–542.
12 Kabsch, W. and Sander, C. (1983) Biopolymers 22, 2577–2637.
13 George, D.G., Barker, W.C. and Hunt, L.T. (1986) Nucl. Acids Res. 14, 11–15.
14 Bairoch, A. and Boeckmann, B. (1991) Nucl. Acids Res. 19, Suppl.: 2247–2249.
15 Eisenberg, D. (1984) Annu. Rev. Biochemistry. 53, 595–623.
16 Kidera, A., Konishi, Y., Oka, M., Ooi, T. and Scheraga, H.A. (1985) J. Protein Chem. 4, 23–55.

17 Tong, L., de Vos, A.M., Milburn, M.V. and Kim, S.-H. (1991) J. Mol. Biol. 217, 503–516.
18 Chou, P.Y. and Fasman, G.D. (1974) Biochemistry 13, 211–222.
19 Garnier, J., Osguthorpe, D.J. and Robson, B. (1978) J. Mol. Biol. 120, 97–120.
20 Lim, V.I. (1974) J. Mol. Biol. 88, 873–894.
21 Holley, L.H. and Karplus, M. (1989) Proc. Nat. Acad. Sci. U.S.A. 86, 152–156.
22 Bohr, H., Bohr, J., Brunak, S., Cotterill, R.M., Lautrup, B., Norskov, L., Olsen, O.H. and Petersen, S.B. (1988) FEBS Lett. 241, 223–238.
23 Stolorz, P., Laepedes, A. and Xia, Y. (1992) J. Mol. Biol. 225, 363–377.
24 Andreassen, H., Bohr, H., Bohr, J., Brunak, S., Bugge, T., Cotterill, R.M.G., Jacobsen, C. Kusk, P., Lautrup, B., Petersen, S.B., Saermark, T. and Ulrich, K. (1990) J. AIDS 3, 615–622.
25 Pascarella, S. and Bossa, F. (1989) CABIOS 5, 319–320.
26 Chou, P.Y. and Fasman, G.D. (1974) Biochemistry. 13, 222–244.
27 McGregor, M.J., Flores, T.P. and Sternberg, M.J. (1989) Protein Eng. 2, 521–526.
28 McGregor, M.J., Flores, T.P. and Sternberg, M.J.E. (1990) Protein Eng. 3, 459–460.
29 Kneller, D.G., Cohen, F.E. and Langridge, R. (1990) J. Mol. Biol. 214, 171–182.
30 Davies, D. (1964) J. Mol. Biol. 9, 605–609.
31 Krigbaum, W.R. and Knutton, S.P. (1973) Proc. Nat. Acad. Sci. U.S.A. 70, 2809–2813.
32 Bohr, H., Bohr, J., Brunak, S., Cotterill, R.M.J., Fredholm, H., Lautrup, B. and Petersen, S.B. (1990) FEBS Lett. 261, 43–46.
33 Bengio, Y. and Pouliot, Y. (1990) CABIOS 6, 319–324.
34 Hirst, J.D. and Sternberg, M.J.E. (1991) Protein Eng. 4, 615–623.
35 Dubchak, I., Holbrook, S.R. and Kim, S-H. (1992) Proteins (in press).
36 Holbrook, S.R., Muskal, S.M. and Kim, S.H. (1990) Protein Eng. 3, 659–665.
37 Kraulis, P.J. (1991) J. Appl. Cryst. 24, 946–950.

HUMAN CELLULAR PROTEIN PATTERNS AND THEIR LINK TO GENOME DNA MAPPING AND SEQUENCING DATA: TOWARDS AN INTEGRATED APPROACH TO THE STUDY OF GENE EXPRESSION

Julio E. Celis,* Hanne H. Rasmussen,* Henrik Leffers,* Peder Madsen,* Bent Honoré,* Kurt Dejgaard,* Paul Gromov,* Eydfinnur Olsen,* Hans J. Hoffmann,* Morten Nielsen,* Borbala Gesser,* Magda Puype,† Josef Van Damme† and Jôel Vandekerckhove†

*Institute of Medical Biochemistry and Danish Center for Human Genome Research, Aarhus University, DK-8000 Aarhus C, Denmark
†Laboratory of Physiological Chemistry, University Ghent, Belgium

INTRODUCTION

The haploid human genome consists of 3×10^9 base pairs of DNA distributed in 23 distinct chromosomes. Current estimates indicate that there are about 50,000 to 100,000 genes with perhaps 2,000 to 4,000 genes in each chromosome (Figure 1) (1). Fortunately, only a fraction of the total number of genes is expressed in a distinct cell type at any given time, with perhaps no more than 5,000 different expressed proteins together with their modified variants per cell. Of these, about 70 to 80% may represent household proteins that are shared by all cell types and that are expressed in variable amounts. Assuming that there are at least 250 different cell types in the adult human body (2), each differing from the rest in about 300 to 400 proteins, one ends up with a total number of polypeptides that is reasonably close to the putative number of genes.

To date, only a small proportion of the total set of proteins from any mammalian organism has been identified and in particular, little is known about the protein composition of differentiated cell types. Proteins orchestrate most of the cell functions and it is therefore of importance to identify and functionally characterize as many as possible as well as to reveal those related to disease. Furthermore, given the concerted effort to map and sequence the entire human genome, it is urgent to develop strategies to integrate protein and

Genetic Engineering, Vol. 15, Edited by J.K. Setlow
Plenum Press, New York, 1993

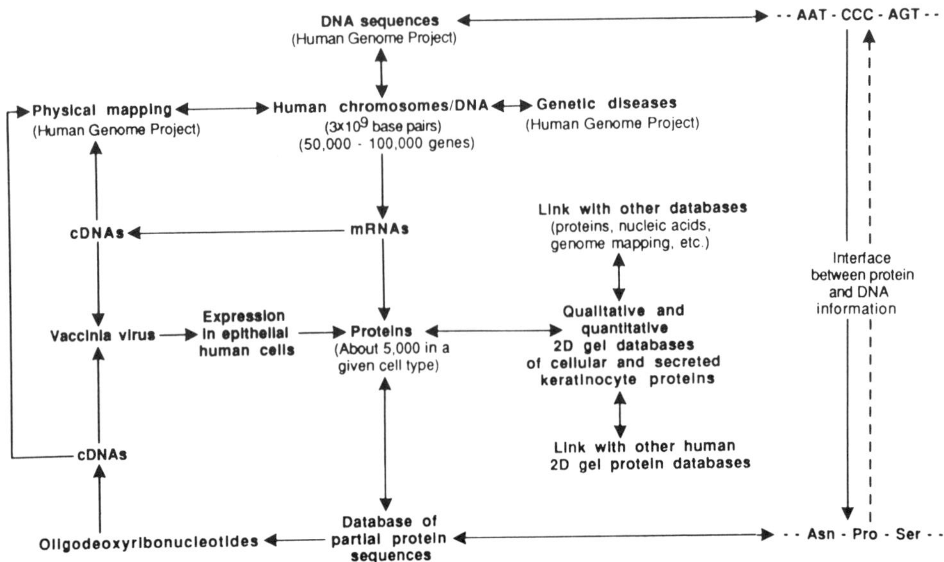

Figure 1. Interface between comprehensive two-dimensional gel protein databases, databases of partial protein sequences and the human genome mapping and sequencing program.

DNA information and to generate comprehensive approaches to the study of the human cell (Figure 1) (3,4).

For the past 15 years, high resolution two-dimensional gel electrophoresis has been the technique of choice to determine the protein composition of a given cell type and for monitoring changes in gene activity through the quantitative and qualitative analysis of the thousands of proteins that orchestrate various cellular functions (3–10 and references therein). The technique which was originally described by O'Farrell (5) separates proteins both in terms of their isoelectric point (pI) and molecular weight. Usually, one chooses a condition of interest and lets the cell reveal the global protein behavioral response as all detected proteins can be analyzed both qualitatively and quantitatively in relation to each other. At present, most available two-dimensional gel techniques (regular gel format) can resolve between 1,000 and 2,000 proteins from a given mammalian cell type, a number that corresponds to about 2 million base pairs of coded DNA. Less abundant proteins can be detected by the analysis of partially purified cellular fractions.

Two-dimensional gel electrophoresis has been widely applied to analysis of cellular protein patterns from bacteria to mammalian cells (3–10, and references therein). In spite of much work, however, information gathered from these studies has not reached the scientific community in its fullness because of lack of standardized gel systems (reagents, equipment, etc.) and the lack of means for storing and communicating protein information. Only recently, thanks to the development of appropriate computer software, has it been possible to scan gels, assign numbers to individual proteins and store the wealth of information in quantitative and qualitative comprehensive two-dimensional gel protein databases (3,4,9,10,18–33), i.e., those containing information about the various properties (physical, chemical, biological, biochemical, physiological, genetic, immunological, architectural, etc.) of all the proteins that can be detected in a given cell type (Figure 1). Such integrated two-dimensional gel protein databases offer an easy and standardized medium in which to store and communicate protein information and provide a unique framework in which to focus a multidisciplinary approach to the study of the cell. Once a protein is identified in the database, all of the information accumulated can be easily retrieved and made available to the researcher. In the long run, protein databases are expected to foster a wide variety of biological information that may be instrumental to researchers working in many areas of biology—among others, cancer and oncogene studies, differentiation, development, drug development and testing, genetic variation and diagnosis of genetic and clinical diseases (Figure 1).

Systematic two-dimensional gel protein analysis has gained a new dimension with the advent of techniques to microsequence major proteins recorded in the databases (34–53 and references therein). Partial protein sequences can be used to search for protein identity as well as to prepare specific DNA probes for cloning as yet uncharacterized proteins (Figure 1). As these sequences can be stored in the database, they offer a unique opportunity to link protein information with the existing or forthcoming DNA sequence data coming from the human genome project (Figure 1).

A few two-dimensional gel protein databases that are accessible in a computer form have been published *in extenso:* these correspond to the protein-gene database of *Escherichia coli* K-12 developed by Neidhardt and colleagues (18,27,33), the rat REF 52 database established by Garrels and co-workers at Cold Spring Harbor (22,26), the human plasma protein database established by the Andersons (30,54) and a few human databases (transformed amnion cells (19,24,28), normal embryonal lung MRC-5 fibroblasts (21,25), keratinocytes (23,29) and peripheral blood mononuclear cells (19) developed in Aarhus). In addition, there are several smaller cellular databases being established in human (normal human diploid fibroblasts, lymphocytes, leukocytes, leukemic cells) mouse (NIH/3T3 cells, T-lymphocytes), *Aplysia,* yeast (*Saccharomyces cerevisae*), plants (wheat, barley, sorghum) and *Euglena.* Databases of tissue proteins, (brain, whole mouse, liver) and body fluid proteins (plasma proteins, cerebrospinal fluid, urine and milk) are being established in several laboratories. The reader is directed to the review by Celis et al. (9) for details and references concerning these databases. Given space limitations and to keep this review in focus, we will mainly concentrate on the computerized analysis of human cellular two-dimensional gel patterns, and in particular on the steps involved in establishing comprehensive two-dimensional gel databases that will link protein and DNA information.

MAKING AND MANAGING A COMPREHENSIVE TWO-DIMENSIONAL GEL DATABASE OF HUMAN CELLULAR PROTEINS

The first step in making a comprehensive two-dimensional gel protein database is to prepare a synthetic image (digital form of the gel image) of the gel (fluorogram, Coomassie blue or silver-stained gel) to be used as a standard or master reference. This can be done with laser scanners, charge couple device (CCD) array scanners, television cameras, rotating drum scanners and multiwire chambers (17). Computerized analysis systems for spot detection, quantitation, pattern matching and data handling (access and retrieval of information, database making) have been described in the literature (ELSIE (55), GELLAB (15), HERMeS (56), MELANIE (14), QUEST (13), TYCHO (12) and CREAM (57)) and some are available commercially (PDQUEST, Protein Databases Inc., Huntington, N.Y.; KEPLER, Large Scale Biology, Rockville, MD; Visage, BioImage Corporation, Ann Arbor, MI; Gemini, Joyce Locbl, Gatcshcad; Microscan 1000 Technology Resources Inc., Nashville, TN, MasterScan™, Billerica, MA and CREAM, KEM-EN-TEC, Copenhagen). Unfortunately, most of these systems are incompatible with one another and their advantages and disadvantages have been discussed by Miller (17).

In our workstation in Aarhus, fluorograms are scanned with a Molecular Dynamics laser scanner and the data are analyzed with the use of the PDQUEST II software (Protein Databases Inc.) (13) running on a SPARK station computer 4100 FC-8-P3 from SUN Microsystems, Inc. The scanner measures intensity in the range of 0 to 2.0 absorbance. A typical scan of a 17 cm x 17 cm fluorogram takes about 2 min. Steps in image analysis include: initial smoothing, background subtraction, final smoothing, spot detection and fitting of ideal Guassian distributions to spot centers. Spot intensity is calculated as the integration of a fitted Gaussian. If calibration strips containing individual segments of known amount of radioactivity are used, it is possible to merge multiple exposures of the sample image into a single data image of greater dynamic range. Once the synthetic image is created it can be stored on disk and displayed directly on the monitor. Functions that can be used to edit the images include: cancel (for example to erase scratches that may have been interpreted as spots by the computer; cancel streaks or low dpm spots), combine (sometimes a spot may be resolved into several closely packed spots), restore, uncombine and add spot to the gel. The editing process is time consuming—about 1 to 1 1/2 day per image. Figure 2A shows a standard synthetic image (IEF, isoelectric focusing) of a fluorogram of [^{35}S]-methionine labelled cellular proteins from human keratinocytes (master database) (29). Images can be displayed either in black and white (resembling the original fluorograms) or in color depending on the need. As shown in Figure 2B, each polypeptide is assigned a number by the computer, which facilitates the entry and retrieval of qualitative and quantitative information for any given spot in the gel (29). The standard image can be matched automatically by the computer to other standard or reference gels provided a few landmark spots

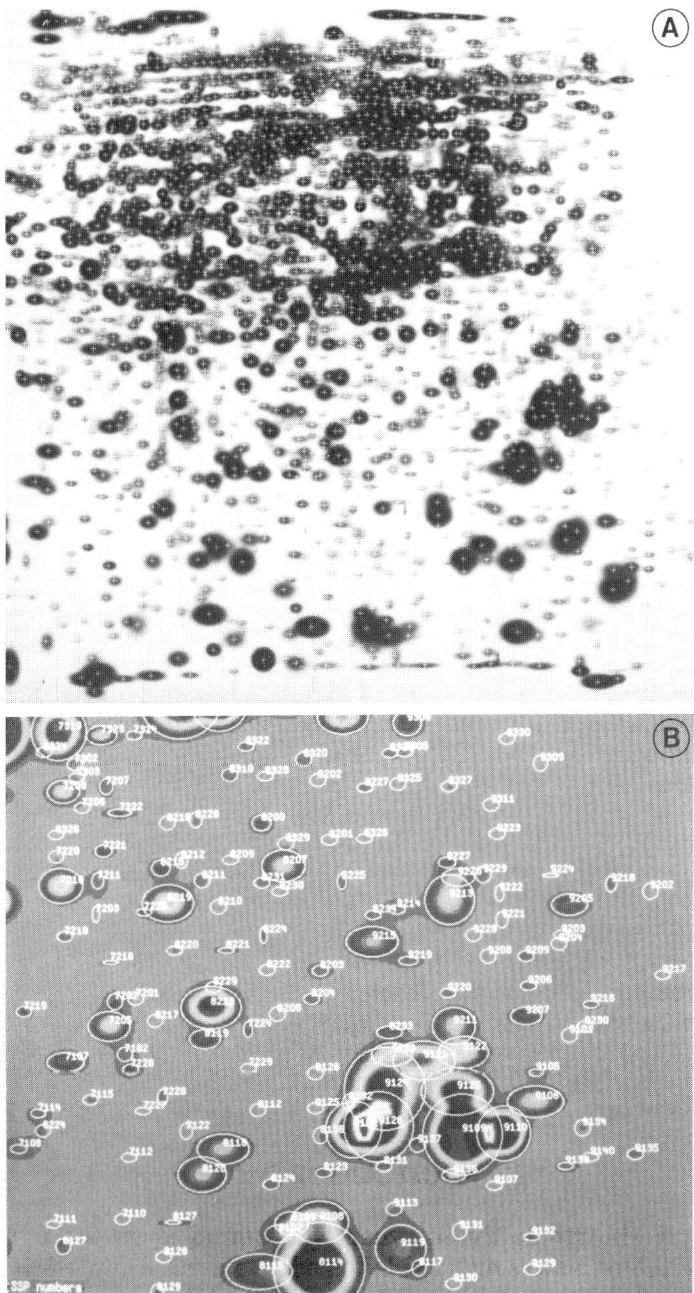

Figure 2 A) Synthetic image (master image) of a fraction of an IEF fluorogram of [^{35}S]-methionine-labelled proteins from human keratinocyte proteins. B) Image showing numbers assigned by the computer to each spot.

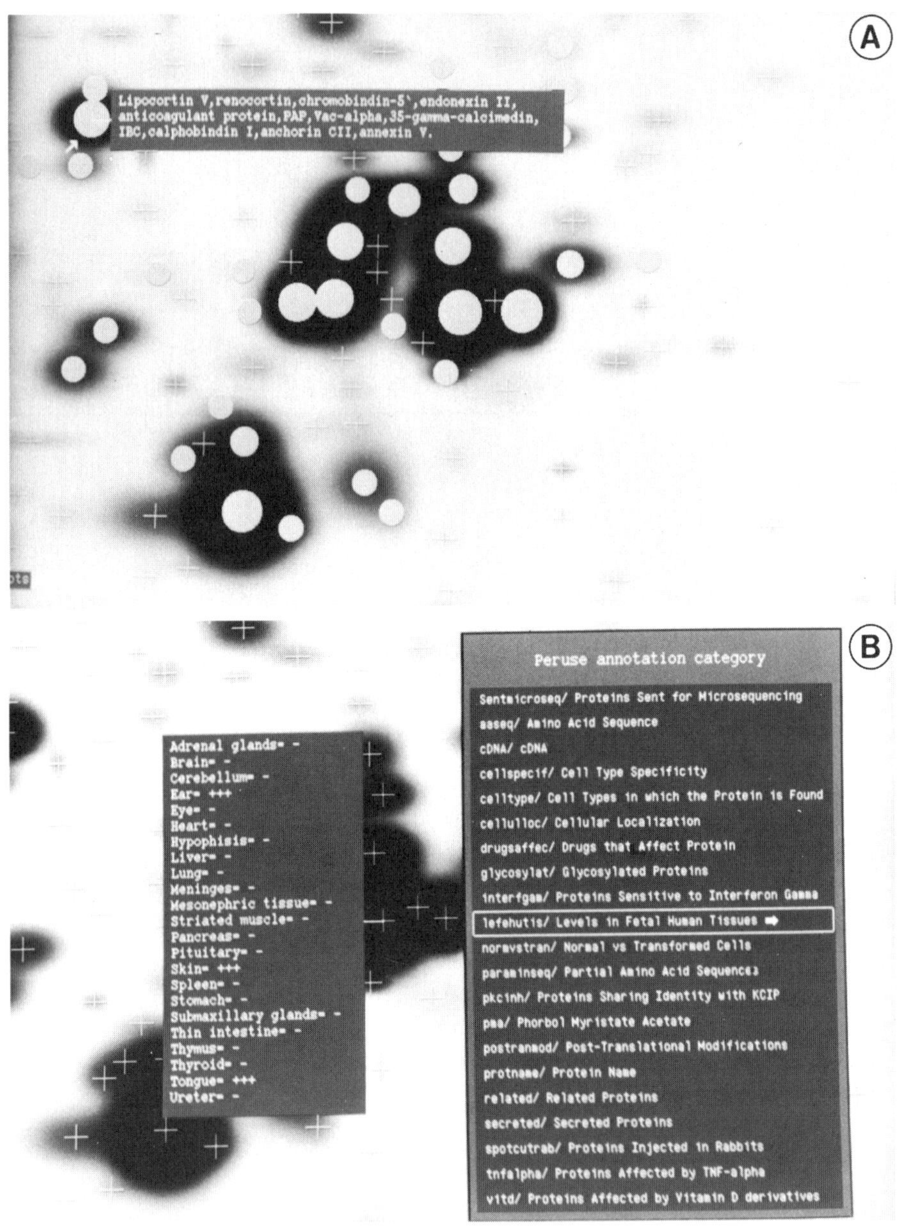

Figure 3 A) Synthetic image of a fraction of an IEF fluorogram showing proteins (flagged with white) that contain information under the entry *protein name.* Only the information contained in one of the spots is displayed. About 600 proteins contain information in this category. B) As A) but using the function *peruse annotation for spot* to inquire about categories and information available for the epithelial marker stratifin.

are given manually as reference to initiate the process. It should be stressed that proteins are matched according to their gel position and, therefore, additional ways to verify their relatedness are needed before one can take full advantage of these data.

Once a standard map of a given protein sample is made, one can enter qualitative annotations to establish a reference database. Our current master two-dimensional gel database of human keratinocyte cellular proteins (update 1992; this database is updated yearly in Electrophoresis) lists 2,980 polypeptides having pI's ranging from 4 to 13 and molecular weights between 8.5 and 230 kD. The most abundant protein in the database corresponds to total actin (about 90 million molecules per cell) while the less abundant of the recorded polypeptides are present in around 5,000 molecules per cell. Some annotation categories that we are using to establish the master keratinocyte database include: 1) protein identification (co-migration with purified proteins, two-dimensional gel immunoblotting, microsequencing, cDNA expression in the vaccinia virus expression system); 2) amounts (total amounts and levels of synthesis); 3) subcellular localization (nuclear, cytoskeletal, membrane, membrane receptors, specific organelles, etc.); 4) antibody against protein; 5) post-translational modifications (phosphorylation, glycosylation, myristoylation, etc.); 6) microsequencing; 7) cell cycle specificity (specific variations in levels of synthesis and amount); 8) regulatory behavior (effect of hormones, growth factors, cytokines, heat shock, etc.); 9) rate of synthesis in normal and transformed cells (proliferation sensitive proteins, cell cycle specific proteins, differentiation markers, oncogenes, components of the pathways that control cell proliferation); 10) function (mainly from co-migration with proteins of known function); 11) sets of proteins that are coordinately regulated (hierarchy of controls, differential gene expression in various stages of differentiation, etc.); 12) cDNAs (cDNAs cloned in the laboratory); 13) proteins that are specific to a given disease (for example systematic comparison of protein patterns from normal and psoriatic keratinocytes); 14) expression and exploitation of transfected cDNAs; 15) pathways (signal transduction, metabolic, others); 16) gene localization (genetic and physical) and 17) effect of microinjected antibody on patterns of protein synthesis.

Information entered for any spot in a given annotation category can be easily retrieved by asking the computer to display the information on the screen. For example, Figure 3A shows a synthetic image of an IEF gel (master keratinocyte database) displaying the information contained under the entry *protein name.* All the spots flagged with white in Figure 3A contain information in this category. By using the function *show annotation for spot* it is possible to display the annotation for a particular protein. Alternatively, one can use the function *peruse annotations for spot* directly to ask the computer to list all the entries available for a particular protein (in this case stratifin). By clicking the mouse in a given entry (presence in fetal human tissues) it is possible to take a quick look at the information in that particular entry (Figure 3B). Table 1 lists entries available for the epithelial marker stratifin.

One of the short-term goals we have set in building the master database of human keratinocyte proteins (29) has been the identification of as many proteins as possible. This is currently being done by one or a combination of the following procedures: 1) co-migration with known proteins, 2) 2-D gel immunoblotting with the use of specific antibodies, 3) microsequencing of Coomassie brilliant blue stained human proteins recovered from dried 2-D gels and 4) expression of known cDNAs in the vaccinia virus expression system. So far, we have received about 900 antibodies from laboratories all over the world and these are being systematically tested by 2-D gel immunoblotting for antigen determination. Similarly, purified proteins and organelles provided by several laboratories have greatly aided the identification of proteins. We routinely request antibodies and protein samples and promise the donors to make available all of the information we may have accumulated on that particular protein. In general, the process of data collection and dissemination is laborious and requires substantial manpower. To date, about 600 proteins (~20% of the total number of proteins recorded) have been identified in the keratinocyte database.

As mentioned earlier, one distinct advantage of two-dimensional gel electrophoresis is the possibility of studying quantitative variations in cellular protein patterns that may lead to the identification of groups of proteins that are expressed coordinately during a given biological process. Quantitation, however, is not an easy task as reflected by the lack of published data on global cellular protein patterns. We believe that this is partly due to difficulties in obtaining sets of gels that are suitable for computer analysis (streaking, material remaining at the origin, etc.) as well as to limitations (laborious editing time, need of calibration strips to merge images, limited dynamic range, etc.) in the computer analysis systems available at the moment. Perhaps the most advanced quantitative studies published so far with computer analysis have been carried out by Garrels and co-workers (22,26). In particular, these investigators have established a quantitative rat protein database (22,26) designed to study growth control (proliferation, growth inhibitors and stimulation) and transformation in well-defined groups of cell lines obtained by transformation of rat REF52 cells with SV40, adenovirus and the Kirsten murine sarcoma virus. These studies have revealed clusters of proteins induced or repressed during growth to confluence as well as groups of transformation-sensitive proteins that respond in a differential fashion to transformation by DNA and RNA viruses. A most interesting feature of this quantitative database is the discovery of a group of co-regulated proteins that show similar expression patterns as the cell cycle-regulated DNA replication protein known as proliferating cell nuclear antigen (PCNA)/cyclin (58).

In our human databases, most quantitations have been carried out by estimating the radioactivity contained in the polypeptides by direct counting of the gel pieces in a scintillation counter (24,25). Up to 700 proteins can be cut out through appropriate exposed films in a period of time comparable to that required for editing a synthetic image. Manual quantitation of this large number of spots is difficult, however, without the assistance of a master

Table 1
Some Entries for Stratifin in the Human Keratinocyte Two-dimensional Gel Protein Database

Entries for Stratifin (IEF SSP 9109)	Information Entered
1. Protein name	Stratifin
2. Apparent molecular weight (Mr)	30.0 kD
3. Isoelectric point (pI)	4.4
4. Method (or methods) of identification	Microsequencing, cDNA cloning and expression
5. Antibody against protein	Polyclonal (rabbit) J.E. Celis and B. Basse, Aarhus
6. Cellular localization	Cellular and secreted
7. Partial amino acid sequence	YEDMAAF (19-25), NLLSVAYK (42-49), VFYLK (118-122), YLAEVATGDD (130-139) LGLALNFSVFXY (170-181), XYEIANSPEE(A)I (180-191), DNLTL (225-229), (T/A)ADNAGEEG (231-239)
8. cDNA sequence	Known. H. Leffers et al. (60). (Mr = 27,773 D, pI = 4.5 from translated sequence.)
9. Levels in fetal human tissues	Adrenal glands = -; brain = -; cerebellum = -; ear = +++; eye = -; heart = -; hypophysis = -; liver = -; lung = -; meninges = -; mesonephric tissue = -; striated muscle = -; pancreas = -; skin = +++; spleen = -; stomach = -; submandibular gland = -; small intestine = -; thymus = -; thyroid gland = -; tongue = +++; ureter = -.
10. Levels in quiescent, proliferating and transformed MRC-5 fibroblasts	Q (quiescent) = 1.3; P (proliferating) = 1.0; T (SV40 transformed) = 0.1
11. Cell specificity	Only in cultured epithelial cells
12. Modification	Glycosylated
13. Phorbol esters	Down regulated

reference image and a numbering system that can be used to identify the spots. Using this approach, we have recorded quantitative changes in the relative abundance of 592 [^{35}S]-methionine-labelled proteins synthesized by quiescent, proliferating, and SV40-transformed human embryonic lung MRC-5 fibroblasts (25). At present, we are investigating the possibility of using the phosphoimager to obtain quantitative data. Our studies as well as those of Garrels and co-workers (22,26) may in the long run help define patterns of gene expression that are characteristic of the transformed state.

MICROSEQUENCING HAS ADDED A NEW DIMENSION TO COMPREHENSIVE TWO-DIMENSIONAL GEL DATABASES: A DIRECT LINK BETWEEN PROTEINS AND GENES

The development of highly-sensitive amino acid gas-phase or liquid-phase sequenators (24), together with the establishment of efficient protein and peptide sample preparation methods, has opened the possibility to perform a

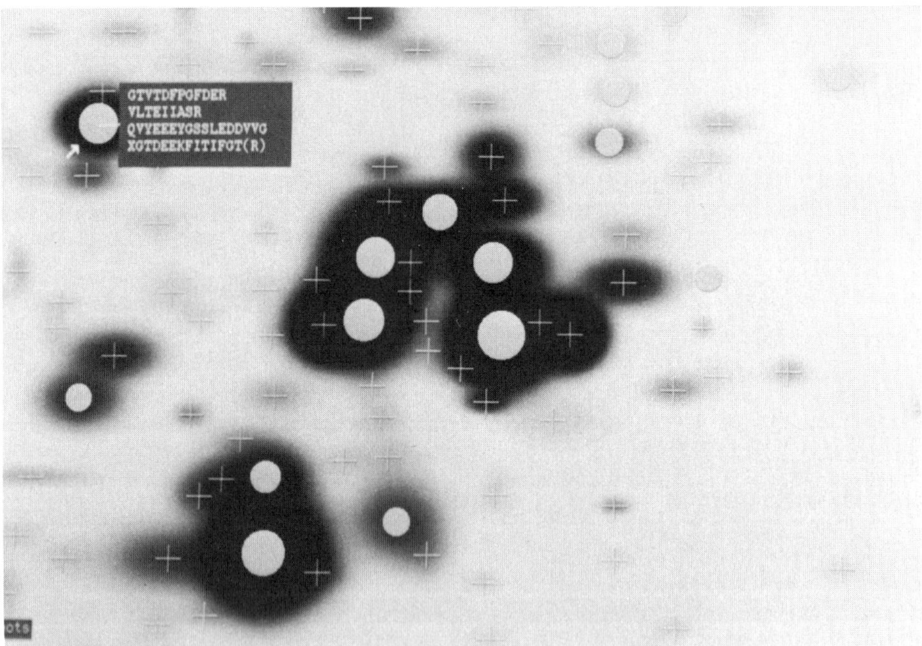

Figure 4. Synthetic image of a fraction of an IEF fluorogram showing polypeptides that contain information under the category of *partial amino acid sequences.* Only the information contained in one of the spots is displayed. About 250 proteins contain information in this category.

systematic sequence analysis of proteins resolved by two-dimensional gel electrophoresis. Indeed, generated pieces of protein sequences can be used to search for protein identity (comparison with available sequences stored in databanks), as well as for preparing specific DNA probes for cloning of as yet uncharacterized proteins (Figure 1). In addition, partial protein sequences can be stored in two-dimensional gel databases (Figure 4; all the proteins flagged with white contain information in this category) and offer a unique link between proteins and genes (Figure 1).

In the early 1970s, gel electrophoresis was used to purify proteins for sequencing purposes (reviewed by Weber and Osborn in (35)). Proteins were recovered by diffusion and sequenced by the manual dansyl-Edman degradation at the nanomole level. This technique was further refined by using electro-elution to recover proteins and by miniaturizing the system (36). This method has been used extensively, but showed increasing drawbacks (low yields, protein samples contaminated by free amino acids, and NH_2-terminal blocking) as the amounts of handled protein gradually became smaller (e.g., at the picomole level).

The introduction of protein-electroblotting procedures (37–42) and chemically inert membranes made it possible to sequence the immobilized proteins directly without additional manipulations, generally yielding NH_2-terminal sequences containing 10 to 40 residues. A major difficulty encountered in this procedure is the occurrence of frequent artifactual blockage of the proteins. In addition to this primarily technical problem, many proteins are blocked *in vivo* by acylation or by a pyrrolidone carboxylic acid cap.

The problem of partial or complete NH_2-terminal blockage can be circumvented by generating internal amino acid sequences. This is achieved by fragmenting the protein present in the gel (gel *in situ* cleavage), or by cleaving it while bound to the membrane (membrane *in situ* cleavage) (43–45). This method has been described for Ponceau red-stained proteins on nitrocellulose blots (44), for Amido-black-stained Immobilon-bound proteins, and for fluorescamine-detected proteins on glass-fiber membranes (45). The proteases used (trypsin, chymotrypsin, V8 or pepsin) cleave at multiple sites, generating small peptides which elute from the blot into the digestion buffer from which they are purified by reverse-phase high performance liquid chromatography (HPLC) before being sequenced individually. As membrane-immobilized proteins are not homogeneously digested, but rather show protease sensitivity next to resistant regions, the number of peptides generated is much lower than expected from the number of potential cleavage sties. Consequently, HPLC peptide chromatograms are less complex and most peptides can be recovered in pure form.

As only limited amounts of a protein mixture can be loaded on a two-dimensional gel, proteins of interest are often obtained in yields insufficient for the currently available sequencing technology. More material can be obtained by enriching for a certain subcellular fraction (purified cell organelles) or by exploiting affinity (dyes, metals, drugs, etc.) or hydrophobic properties of proteins before gel analysis. All of the sequencing results accumulated so far in the human protein databases (24,29,53) (an example is shown in Figure 4)

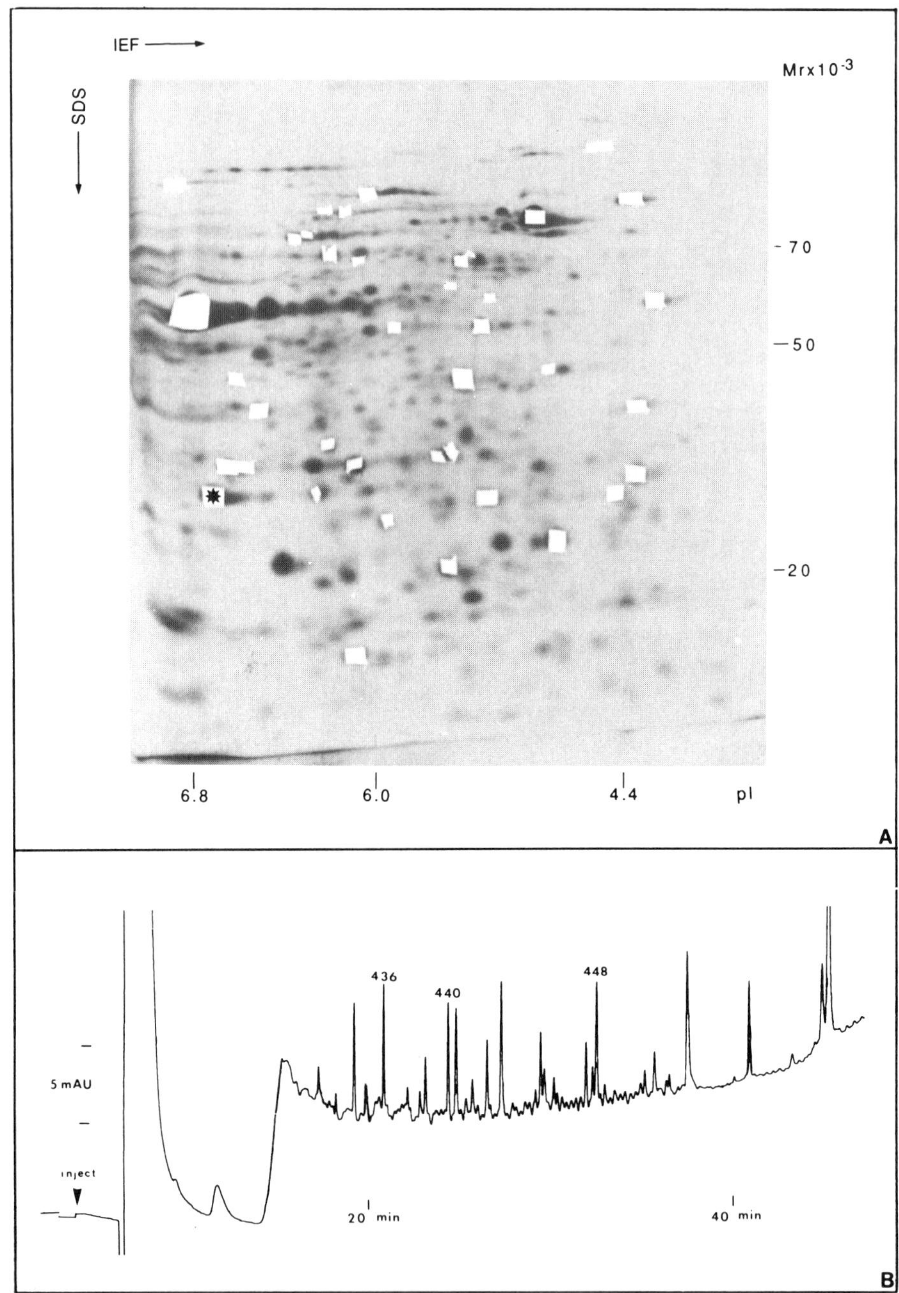

Figure 5. The various steps involved in microsequencing are illustrated by the analysis of protein IEF SSP 2119 (24). (A) Representative Coomassie brilliant blue stained gel of partially purified Molt-4 proteins (about 1 mg of total protein was loaded) used to cut protein IEF SSP 2119 (indicated with an asterisk). Spots from 6 gels were used in this particular analysis. (B) HPLC chromatogram of peptides generated by *in situ* digestion of electroblotted protein IEF SSP 2119. The column was a C18 normal-bore. (C) Phenylthio-

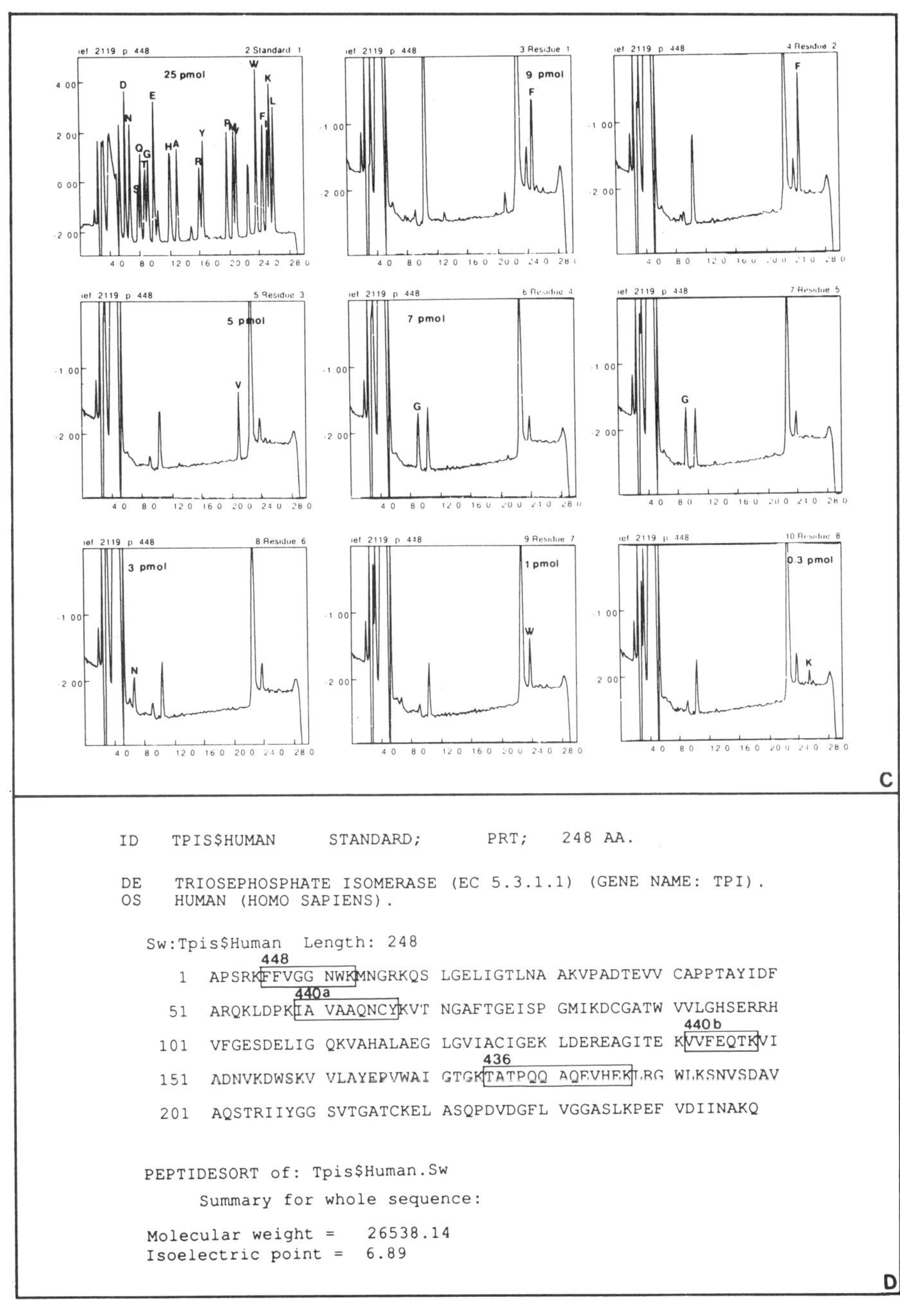

ID TPIS$HUMAN STANDARD; PRT; 248 AA.

DE TRIOSEPHOSPHATE ISOMERASE (EC 5.3.1.1) (GENE NAME: TPI).
OS HUMAN (HOMO SAPIENS).

Sw:Tpis$Human Length: 248

```
          448
  1  APSRKFFVGG NWKMNGRKQS LGELIGTLNA AKVPADTEVV CAPPTAYIDF
            440a
 51  ARQKLDPKIA VAAQNCYKVT NGAFTGEISP GMIKDCGATW VVLGHSERRH
                                                 440b
101  VFGESDELIG QKVAHALAEG LGVIACIGEK LDEREAGITE KVVFEQTKVI
                               436
151  ADNVKDWSKV VLAYEPVWAI GTGKTATPQQ AQEVHEKLRG WLKSNVSDAV
201  AQSTRIIYGG SVTGATCKEL ASQPDVDGFL VGGASLKPEF VDIINAKQ
```

PEPTIDESORT of: Tpis$Human.Sw
Summary for whole sequence:

Molecular weight = 26538.14
Isoelectric point = 6.89

hydantoin (PTH) chromatograms from amino acid sequencing of the tryptic peptide 448. Data were collected on-line with the 610A data system. (D) Amino acid sequences from three tryptic peptides were used for homology searches that gave total homology to human triosephosphate isomerase. The molecular weight and pI calculated from the published sequence matched completely the gel coordinates recorded in the master AMA two-dimensional gel protein database (24).

have been obtained from the analysis of protein spots collected from two-dimensional gels that had been stained with Coomassie brilliant blue according to standard procedures and dried for storage. Proteins are recovered from the collected gel pieces by elution-concentration gel electrophoresis. Details of this technique have been reported previously (52,53) and a brief outline is given below (see also Figure 5).

Combined gel pieces are allowed to swell in gel sample buffer (a total volume of 1.5 ml). The gel pieces combined with the supernatant are then loaded into the slot of an elution-concentration gel. In this way the protein is efficiently eluted from the gel pieces and concentrated from a large volume into a narrow spot as a result of both a vertical stacking and horizontal contraction of the protein band. The highly concentrated (about 5 mm^2) protein spot is then electroblotted on polyvinyldiene difluoride (PVDF)-membranes, stained with Amido black and *in situ* digested with trypsin. The peptides generated during digestion elute from the membrane and are separated by normal or narrow bore reverse-phase HPLC and collected individually for sequence analysis. In our hands, this approach is routinely applied to gel-purified proteins available in few μgs. The various steps involved in microsequencing are illustrated in Figure 5 (analysis of protein IEF SSP 2119).

Using this and previous procedures (47,49,52,53), we have so far microsequenced 250 protein spots collected from two-dimensional gels (24,47,49,52,53 and unpublished observations), a number we expect to increase by at least 100 per year. Of the microsequenced proteins about 50% corresponded to unknown proteins.

GENERAL CONSIDERATIONS AND PERSPECTIVES

One of the major advantages in developing computer-accessible human 2-D gel databases in which most of the known proteins are identified is the wealth of new proteins that will become amenable to experimentation both at the biochemical and molecular biology level. Accordingly, a high-priority goal in our program has been to establish a database of partial sequences of unknown proteins that may be interfaced with the forthcoming DNA sequence information from the Human Genome Project (Figure 1). Besides aiding the identification of genes, these sequences can be used to prepare oligodeoxyribonucleotides which in turn are used to clone the corresponding cDNAs (Figure 1). Complete cDNAs can then be sequenced and used for physical mapping in collaborative studies. Many proteins, however, are members of families, and therefore it is necessary to determine which protein is actually coded for by a given cDNA. This has been achieved by expressing the cDNA in human epithelial amnion (AMA) cells with the use of a vaccinia virus vector that turns off the host cells' protein synthesis while producing large quantities of the proteins coded in the virus genome (59–61 and references therein) (Figure 6). The protein coded by the cDNA is then identified by superimposing the autoradiogram of the proteins synthesized by the recombinant virus-infected cells with the silver-stained gel showing the "background" of unlabelled cellular

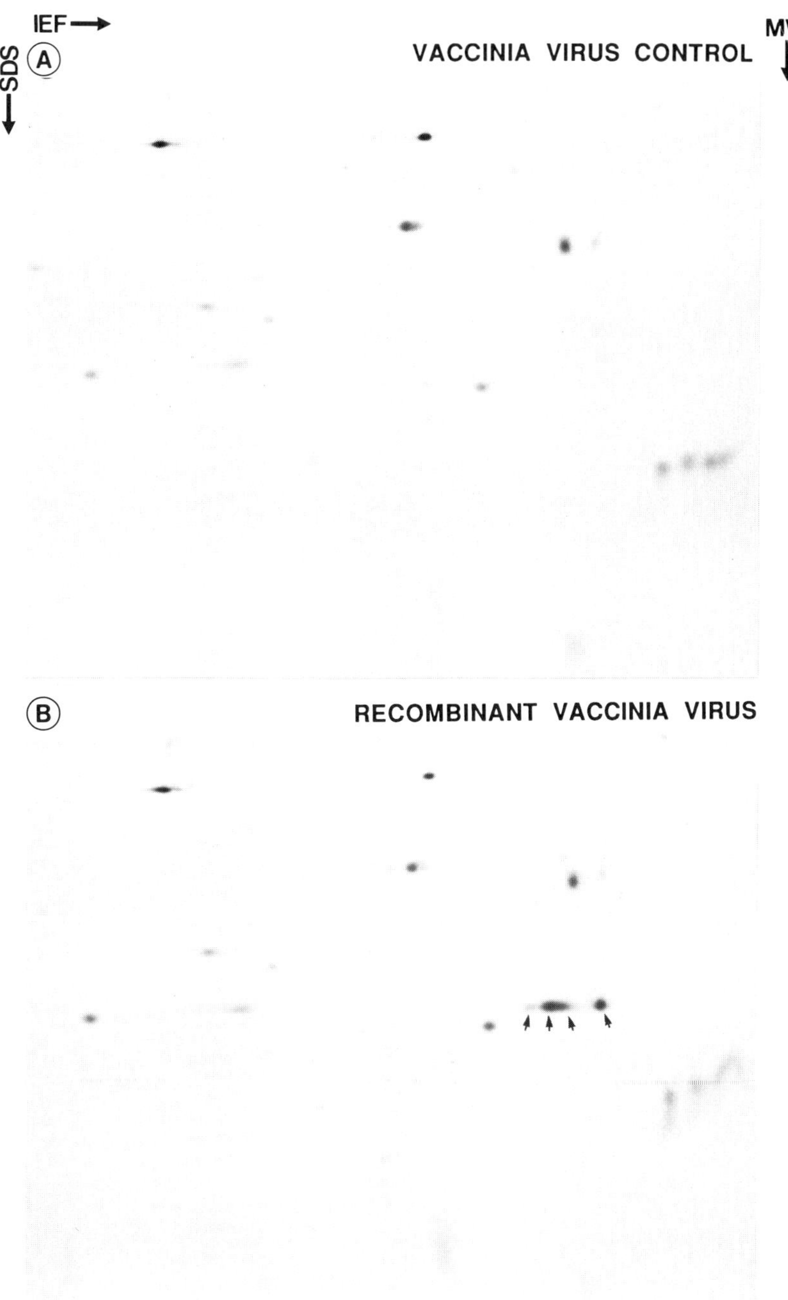

Figure 6. IEF autoradiograms of [^{35}S]-methionine-labelled proteins synthesized by AMA cells infected with (A) control vaccinia virus and (B) vaccinia virus carrying the 8106 clone (60).

proteins. The method also allows us to identify other cloned proteins in the protein databases by either asking researchers to send their cDNA clones to us or by cloning the cDNAs ourselves using the PCR method. So far we have cloned and sequenced about 15 specific cDNAs.

An alternative approach to increase the number of cDNAs ("non-specific" cDNAs, Figure 1) that will be available for physical mapping is currently being pursued by H. Leffers. The procedure is as follows: we pick cDNA clones at random from *Not* I-poly [dT] primed libraries, sequence both ends of the insert, analyze the insert size by gel electrophoresis, check if they are full length by Northern blotting, recombine the full-length cDNAs into vaccinia virus and determine which polypeptide they code for using two-dimensional gel electrophoresis. The main obstacle in the random approach, however, is the large amount of DNA sequencing required. We recently started a collaborative project with W. Ansorge's group at the European Molecular Biology Laboratory (EMBL; sponsored by the European Community Human Genome Analysis Program) to sequence some of the cDNAs using automated DNA sequencing machines.

SUMMARY

Analysis of cellular protein patterns by computer-aided two-dimensional gel electrophoresis together with recent advances in protein sequence analysis and expression systems have made possible the establishment of comprehensive two-dimensional gel protein databases that may link protein and DNA mapping and sequence information and that offer an integrated approach to the study of gene expression. With the integrated approach offered by two-dimensional gel protein databases it is now possible to reveal phenotype-specific protein(s), to microsequence them, to search for homology with previous identified proteins, to clone the cDNAs, to assign partial protein sequences to genes for which the full DNA sequence and the chromosome location are known, and to study the regulatory properties and function of groups of proteins that are coordinately expressed in a given biological process. Comprehensive two-dimensional gel protein databases will provide an integrated picture of the expression levels and properties of the thousands of protein components of organelles, pathways, and cytoskeletal systems, both under physiological and abnormal conditions, and are expected to lead to the identification of new regulatory networks.

So far, about 20% (600 out of 2,980) of the total number of proteins recorded in the human keratinocyte protein database have been identified and we are actively gathering qualitative and quantitative biological data on all resolved proteins. Given the current improvements on microsequencing as well as the availability of specific antibodies, it seems feasible to expect that most known keratinocyte proteins will be identified in the very near future. This feast will reveal a wealth of new proteins that will become amenable to experimentation both at the biochemical and molecular biology level.

REFERENCES

1 Human Genome II Conference, San Diego, 1990.
2 Geneser, F. (1986) Textbook of Histology, Munksgaard, Copenhagen.
3 Special issue: Two-dimensional Gel Protein Databases (1990) Electrophoresis 11, 989–1168.
4 Celis, J.E., Rasmussen, H.H., Leffers, H., Madsen, P., Honoré, B., Gesser, B., Dejgaard, K. and Vandekerckhove, J. (1991) FASEB J. 5, 2200–2208.
5 O'Farrell, P.H. (1975) J. Biol. Chem. 250, 4007–4021.
6 O'Farrell, P.Z., Goodman, H.M. and O'Farrell, P.H. (1977) Cell 12, 1133–1142.
7 Special Issue: Two-dimensional Gel Electrophoresis (1982) Clin. Chem. 28, 737–1092.
8 Celis, J.E. and Bravo, R. (eds.) (1984) Two-Dimensional Gel Electrophoresis of Proteins: Methods and Applications. Academic Press, New York, NY.
9 Celis, J.E., Madsen, P., Gesser, B., Kwee, S., Nielsen, H.V., Rasmussen, H.H., Honoré, B., Leffers, H., Ratz, G.P., Basse, B., Lauridsen, J.B. and Celis, A. (1989) In Advances in Electrophoresis 3 (Chrambach, C., Dunn, M.J. and Radola, B.J., eds.) VCH, Weinheim.
10 Special Issue: Two-dimensional Gel Protein Databases (Celis, J.E., ed.) (1990) Electrophoresis 11, 989–1166.
11 Garrels, J.I. (1983) Methods Enzymol. 100, 411–423.
12 Anderson, N.L., Hofmann, J.P., Gemmell, A. and Taylor, J. (1984) Clin. Chem. 30, 2031–2036.
13 Garrels, J.I., Farrar, J.T. and Burwell, C.B. (1984) in Two-dimensional Gel Electrophoresis of Proteins. Methods and Applications (Celis, J.E. and Bravo, R., eds.) pp. 37–91, Academic Press, New York, NY.
14 Vincens, P. and Tarroux, P. (1988) Internat. J. Biochem. 20, 499–509.
15 Appel, R., Hochstrasser, D., Roch, C., Funk, M., Muller, A.F. and Pellegrini, C. (1988) Electrophoresis 9, 136–142.
16 Lemkin, P.F. and Lester, E.P. (1989) Electrophoresis 10, 122–139.
17 Miller, M.J. (1989) in Advances in Electrophoresis 3 (Chrambach, C., Dunn, M.J. and Radola, B.J., eds.) pp. 182–217, VCH, Weinheim.
18 Philips, T.D., Vaughn, V., Bloch, P.L. and Neidhardt, F.C. (1987) in *Escherichia coli* and *Salmonella typhimurium,* Cellular and Molecular Biology, Gene-Protein Index of *Escherichia coli* K-12, 2nd ed. (Neidhardt, F.C., Ingraham, J.I., Low, K.B., Magasanik, B., Schaechter, M. and Umbarger, H.E., eds.) pp. 919–966, American Society for Microbiology, Washington, DC.
19 Celis, J.E., Ratz, G.P., Celis, A., Madsen, P., Gesser, B., Kwee, S., Nielsen, H.V., Yde, H., Lauridsen, J.B. and Basse, B. (1988) Leukemia 2, 561–601.
20 Special Issue: Protein Databases in Two-dimensional Electrophoresis (Celis, J.E., ed.) (1989) Electrophoresis 10, 73–164.
21 Celis, J.E., Ratz, G.P., Madsen, P., Gesser, B., Lauridsen, J.B., Brogaard Hansen, K.P., Kwee, S., Rasmussen, H.H., Nielsen, H.V., Crüger, D.,

Basse, B., Leffers, H., Honoré, B., Mϕller, O. and Celis, A. (1989) Electrophoresis 10, 76–115.
22 Garrels, J.I. and Franza, B.R. (1989) J. Biol. Chem. 264, 5283–5298.
23 Celis, J.E., Crüger, D., Kiil, J., Dejgaard, K., Lauridsen, J.B., Ratz, G.P., Basse, B., Celis, A., Rasmussen, H.H., Bauw, G. and Vandekerckhove, J. (1990) Electrophoresis 11, 242–254.
24 Celis, J.E., Gesser, B., Rasmussen, H.H., Madsen, P., Leffers, H., Dejgaard, K., Honoré, B., Olsen, E., Ratz, G., Lauridsen, J.B., Basse, B., Mouritzen, S., Hellerup, M., Andersen, A., Walbum, E., Celis, A., Bauw, G., Puype, M., Van Damme, J. and Vandekerckhove, J. (1990) Electrophoresis 11, 989–1071.
25 Celis, J.E., Dejgaard, K., Madsen, P., Leffers, H., Gesser, B., Honoré, B., Rasmussen, H.H., Olsen, E., Lauridsen, J.B., Ratz, G., Mouritzen, S., Basse, B., Hellerup, M., Celis, A., Puype, M., Van Damme, J. and Vandekerckhove, J. (1990) Electrophoresis 11, 1072–1113.
26 Garrels, J.I., Franza, B.R., Chang, C. and Latter, G. (1990) Electrophoresis 11, 1114–1130.
27 VanBogelen, R.A., Hutton, M.E. and Neidhardt, F.C. (1990) Electrophoresis 11, 1131–1166.
28 Celi, J.E., Leffers, H., Rasmussen, H.H., Madsen, P., Honoré, B., Gesser, B., Dejgaard, K., Olsen, E., Ratz, G.P., Lauridsen, J.B., Basse, B., Andersen, A.H., Walbum, E., Brandstrup, B., Celis, A., Puype, M., Van Damme, J. and Vandekerckhove, J. (1991) Electrophoresis 12, 765–801.
29 Celis, J.E., Madsen, P., Rasmussen, H.H., Leffers, H., Honoré, B., Gesser, B., Dejgaard, K., Olsen, E., Magnusson, N., Kiil, J., Celis, A., Lauridsen, J.B., Basse, B., Ratz, G.P., Andersen, A.H., Walbum, E., Brandstrup, B., Pedersen, P.S., Brandt, N.J., Puype, M., Van Damme, J. and Vandekerckhove, J. (1991) Electrophoresis 12, 802–872.
30 Anderson, N.L. and Anderson, N.G. (1991) Electrophoresis 12, 883–906.
31 Anderson, N.L., Esquer-Blasco, R., Hofmann, J.-P. and Anderson, N.G. (1991) Electrophoresis 12, 907–930.
32 Wirth, P.J., Luo, L., Fujimoto, Y., Bisgaard, H.C. and Olson, A.D. (1991) Electrophoresis 12, 931–954.
33 VanBogelen, R.A. and Neidhardt, F.C. (1991) Electrophoresis 12, 955 –994.
34 Hewick, R.M., Hunkapiller, M.W., Hood, L.E. and Dreyer, W.J. (1981) J. Biol. Chem. 256, 7990–7997.
35 Weber, K., and Osborn, M. (1985) in The Proteins and Sodium Dodecyl Sulfate, Molecular Weight Determination on Polyacrylamide Gels and Related Procedures. (Neurath, H. et al., eds.) Vol. 1, pp. 179–223, Academic Press, New York, NY.
36 Hunkapiller, M.W., Lujan, E., Ostrander, F. and Hood, L.E. (1983) Methods Enzymol. 91, 227–236.
37 Vandekerckhove, J., Bauw, G., Puype, M., Van Damme, J. and Van Montagu, M. (1985) Eur. J. Biochem. 152, 9–19.
38 Aebersold, R.H., Teplow, D.B., Hood, L.E. and Kent, S.B.H. (1986) J. Biol. Chem. 261, 4229–4238.

39 Bauw, G., De Loose, M., Inzé, D., Van Montagu, M. and Vandekerckhove, J. (1987) Proc. Nat. Acad. Sci. U.S.A. 84, 4806–4810.
40 Matsudaira, P. (1987) J. Biol. Chem. 262, 10035–10038.
41 Eckerskorn, C., Mewes, W., Goretzki, H. and Lottspeich, F. (1988) Eur. J. Biochem. 176, 509–519.
42 Moose, M., Jr., Nguyen, N.Y. and Liu, T.-Y. (1988) J. Biol. Chem. 263, 6005–6008.
43 Kennedy, T.E., Gawinowicz, M.A., Barzilai, A., Kandel, E.R. and Sweatt, J.D. (1988) Proc. Nat. Acad. Sci. U.S.A. 85, 7008–7012.
44 Aebersold, R.H., Leavitt, J., Saavedra, R.A., Hood, L.E. and Kent, S.B.H. (1987) Proc. Nat. Acad. Sci. U.S.A. 84, 6970–6974.
45 Bauw, G., Van den Bulcke, M., Van Damme, J., Puype, M., Van Montagu, M. and Vandekerckhove, J. (1988) J. Prot. Chem. 7, 194–196.
46 Celis, J.E., Ratz, G.P., Madsen, P., Gesser, B., Lauridsen, J.B., Kwee, S., Rasmussen, H.H., Nielsen, H.V., Crüger, D., Basse, B., Leffers, H., Honoré, B., Mφller, O., Celis, A., Vandekerckhove, J., Bauw, G., Van Damme, J., Puype, M. and Van den Bulcke, M. (1989) FEBS Lett. 244, 247–254.
47 Bauw, G., Van Damme, J., Puype, M., Vandekerckhove, J., Gesser, B., Ratz, G.P., Lauridsen, J.B. and Celis, J.E. (1989) Proc. Nat. Acad. Sci. U.S.A. 86, 7701–7705.
48 Aebersold, R. and Leavitt, J. (1990) Electrophoresis 11, 517–527.
49 Bauw, G., Rasmussen, H.H., Van den Bulcke, M., Van Damme, J., Puype, M., Gesser, B., Celis, J.E. and Vandekerckhove, J. (1990) Electrophoresis 11, 528–536.
50 Tempst, P., Link, A.J., Riviere, L.R., Fleming, M. and Elicone, C. (1990) Electrophoresis 11, 537–553.
51 Eckerskorn, C. and Lottspeich, F. (1990) Electrophoresis 11, 554–561.
52 Rasmussen, H.H., Van Damme, J., Bauw, G., Puype, M., Gesser, B., Celis, J.E. and Vandekerckhove, J. (1991) in Methods in Protein Sequence Analysis (Jörnvall, H., Höög, J.-O. and Gustavsson, A.-M., eds.) pp. 103–114, Birkhäuser Verlag, Basel.
53 Rasmussen, H.H., Van Damme, J., Puype, M., Gesser, B., Celis, J.E. and Vandekerckhove, J. (1991) Electrophoresis 12, 873–882.
54 Anderson, N.G. and Anderson, N.L. (1979) Behring. Inst. Mitt. 63, 169–210.
55 Olson, A.D. and Miller, M.J. (1988) Anal. Biochem. 169, 49–70.
56 Vincens, P., Paris, N., Pujol, J.L., Gaboriaud, C., Rabilloud, T., Pennetier, J., Matherat, P. and Tarroux, P. (1986) Electrophoresis 7, 347–356.
57 Hagerup, M., Conradsen, K., Bφgh-Hansen, T.C., Bouchelouche, P. and Schafer-Nielsen, C. (1988) Electrophoresis '88, Sixth Meeting of the International Electrophoresis Society July 4–7, Copenhagen, pp. 369–373.
58 Celis, J.E., Madsen, P., Celis, A., Nielsen, H.V. and Gesser, B. (1987) FEBS Lett. 220, 1–7.
59 Moss, B. (1991) Science 252, 1662–1667.
60 Leffers, H., Madsen, P., Honoré, B., Rasmussen, H.H., Andersen, A.H., Walbum, E., Vandekerckhove, J. and Celis, J.E. (1992) (unpublished data).

61 Madsen, P., Rasmussen, H.H., Leffers, H., Honoré, B., Dejgaard, K., Olsen, E., Kiil, J., Walbum, E., Andersen, A., Basse, B., Lauridsen, J., Ratz, G., Celis, A., Vandekerckhove, J. and Celis, J.E. (1991) J. Invest. Dermatol. 97, 701–712.

REGULATION OF TRANSLATION IN PLANTS

Avihai Danon, Christopher B. Yohn and Stephen P. Mayfield

Department of Cell Biology
The Scripps Research Institute
10666 N. Torrey Pines Road
La Jolla, CA 92037

INTRODUCTION

Protein synthesis in plants takes place in two distinct compartments, the cytoplasm and organelles. Translation in the cytoplasm of plants exhibits the characteristics of a typical eukaryotic system such as m7G capped mRNAs, polyadenylated 3′ ends and cap-dependent initiation. On the other hand, translation in organelles such as the chloroplast shows homologies to translation in prokaryotes. Chloroplastic mRNAs are not polyadenylated, exhibit internal initiation of translation and contain sequences which are reminiscent of Shine-Dalgarno sequences. The symbiotic state of the chloroplast within the eukaryotic cell necessitates the close cooperation of gene expression between these two compartments. This cooperation of gene expression may require the participation of nuclear-encoded factors in chloroplast translation resulting in the utilization of eukaryotic factors in a prokaryotic-like system.

Translational regulation can be divided into two major groups: the regulation of the general translation within the cell and the regulation of translation of specific mRNAs. The overall rate of protein synthesis is synchronized with the growth rate of the organism while the regulation of translation of specific mRNAs is controlled in response to unique requirements such as adjustment to specific environmental or developmental changes. Mechanistically, regulation of translation can take place during two key processes, initiation and elongation.

Genetic Engineering, Vol. 15, Edited by J.K. Setlow
Plenum Press, New York, 1993

INITIATION IN EUKARYOTES

Initiation of translation can essentially be broken down into two steps: recognition and selection of the mRNA by the ribosome, and recognition of the initiation codon. Several aspects of the mRNA structure can influence these two recognition steps: 1) the m7G cap, 2) the sequence context of the AUG, 3) the position of the initiator codon relative to other AUGs, 4) length of the 5′ untranslated region (UTR) and 5) mRNA secondary structure surrounding the AUG (1). The predominant mechanism of translational initiation in the cytoplasm of eukaryotes is the ribosome scanning model. The scanning model (2) predicts that the 40S ribosomal subunit complexed with Met-tRNA$_i^{Met}$ and a set of initiation factors recognizes the 5′ end of capped mRNAs. The complex migrates linearly down the mRNA and the first AUG that is in the appropriate context is selected. The first peptide bond is formed once the 60S ribosomal subunit joins this complex at the AUG.

The first recognition event, that of the 40S ribosomal subunit for the mRNA, is mediated through the m7G cap. The mRNA m7G cap interacts with a 25 kD cap binding protein (CBP), eIF-4E (3), which associates with two other proteins, eIF-4A and p220, to complete the eIF-4F complex (4). These proteins may already be, or subsequently become, associated with the 40S ribosomal subunit and position the complex on the 5′ end of the mRNA. Another initiation factor that plays a role in forming the pre-initiation complex is eIF-3. In wheat germ, the binding of eIF-3 to the mRNA increases upon formation of an eIF-4F/mRNA complex. The interaction of eIF-3 with the cap may help in positioning the ribosome (5). The cap recognition sequence is one point where mRNA translation can be regulated. The binding of CBP can be inhibited by either RNA secondary structure or specific proteins which bind the mRNA and compete with CBP binding.

While the scanning mechanism can account for the majority of eukaryotic translation initiation, some evidence is available for alternatives to the scanning model. A set of eukaryotic mRNAs appears to initiate translation by a cap- and 5′ end-independent internal ribosome-binding mechanism. These transcripts, of which the best studied are picornavirus mRNAs, fail to follow the rules set forth by the scanning model (6). The genome of a picornavirus consists of an uncapped RNA with a very long open reading frame and a long, highly structured 5′ UTR. Additionally, there are several upstream AUGs preceding small ORFs which do not seem to be translated, suggesting a mechanism which uses internal initiation. Experiments with dicistronic messages have provided evidence to support this model (7,8). In such experiments, two protein coding cistrons are generated on one mRNA. Normally, when the scanning mechanism is in use, the second cistron is translated much less efficiently. However, when picornavirus leader regions are placed in front of the second ORF, the protein product from this cistron can be detected earlier and at a higher level, implying that ribosomes are bypassing the first cistron and initiating directly on the downstream ORF. The internal ribosome entry sites (IRES) that have been identified show similarities to the prokaryotic Shine-Dalgarno sequence, although, unlike prokaryotic mRNAs, the positioning of this cis-acting element relative to other sequences (e.g. the AUG) is strictly fixed (9).

Initiation factors can affect internal ribosome binding as well. eIF-4F stimulates translation from a second initiator codon of middle component RNA (M-RNA) of cowpea mosaic virus (a member of the plant comoviruses). Insertion of the putative IRES from this M-RNA between a translation inhibiting stem-loop and the initiator codon in the ornithine decarboxylase (ODC) mRNA confers a marked increase in ODC translation, supporting the function of the IRES as an initiation site. Additionally, the stimulatory affect of eIF-4F was observed in translation of ODC, showing that this initiation factor is able to stimulate internal initiation on this eukaryotic IRES (10). In experiments with capped bicistronic mRNAs, eIF-4F and eIF-4B have selectively enhanced translation of the second ORF relative to the first. Further, these factors facilitated translation from both ORFs in uncapped bicistronic messages (11). The transactivator (TAV) protein from cauliflower mosaic virus (CaMV) is able to activate downstream ORFs in the viral genome. While the ribosome seems to enter at the cap site, the TAV probably plays a role in reinitiation at the downstream AUG or in forming a "train" in which a 43S ribosome is physically coupled to the 80S ribosome translating the upstream ORF for delivery to the downstream AUG, allowing for a higher rate of reinitiation (12).

Another site of translational regulation may reside in the 5′ untranslated region (UTR) of the mRNA, and is influenced by the stability of the secondary RNA structure. A high level of stable secondary structure within the UTR can reduce the level of initiation and thus influence the translation rate of specific mRNAs (1). One can postulate the existence of mRNA specific proteins which recognize and act to melt this secondary RNA structure to allow increased initiation of translation of specific mRNAs (13). Helicase activity (melting of secondary structure) has been described for eIF-4A, and this activity is conserved between eukaryotic species as diverse as mammals, yeast and plants (14). In wheat germ, eIF-4B stimulates the helicase activity of eIF-4F and eIF-4A in combination (14). The 5′ UTR of mRNAs may also include short open reading frames (uORFs) that are located upstream to the main coding region. The 5′ UTR of the translationally regulated GCN4 mRNA in yeast includes 4 uORFs that are used, under nonlimiting levels of amino acids, prematurely to terminate translation before the ribosomes approach the GCN4 ORF (15). Under amino acid starvation conditions, modification of eIF-2α by a specific kinase, GCN2, causes a slower rate of reinitiation which allows a higher percentage of ribosomes to bypass uORFs 2–4 and reinitiate at the GCN4 ORF (15). Whether additional mRNA specific factors are interacting with the phosphorylated eIF-2α or whether the eIF-2α itself shows mRNA specificity, it is interesting that the phosphorylation of this general initiation factor affects the translation of a specific mRNA.

The poly(A) tail of eukaryotic mRNAs also has a role in translational regulation. Poly(A)$^+$ mRNAs have higher translational rates when compared to poly(A)$^-$ mRNAs in wheat germ cell-free extracts (16). A poly(A) binding protein may facilitate the binding of an initiation factor or ribosomal subunit at the mRNA 5′ end (17). The 3′ UTR of eukaryotic mRNAs may also participate in message-specific translational repression. A soluble factor(s) binding to the 3′ UTR of creatine kinase B mRNA has been shown to repress translation (18). This interaction probably acts at a step near or at the end of initiation.

ELONGATION IN EUKARYOTES

Other proteins regulate translation after the initiation events, the most abundant of which are the elongation factors. eEF-1α is an essential enzyme for general protein translation, promoting the GTP-dependent binding of aminoacyl-tRNA to ribosomes during elongation. Modulation of eEF-1α affects general translation levels, and has been shown to correlate with high protein synthesis rates in tobacco and tomato, both developmentally and in response to growth hormones (19,20). The activity of elongation factor 1α is regulated by autocatalytic phosphorylation (21). The phosphorylation of eEF-1β in wheat embryos has been shown to cause stimulation of elongation *in vitro* (22). There are several ways that changes in elongation rate can affect gene expression besides the obvious overall increase or decrease in protein synthesis (23). Inhibition of the overall elongation rate will stimulate the relative translation from mRNAs with low initiation rates, as elongation becomes the rate-limiting step for protein synthesis. Elongation rate changes can also affect mRNA stability, perhaps simply by physical protection of the mRNA by stalled ribosomes. Transient inhibition of elongation can eliminate proteins that have very short half-lives. If such a protein is a repressor of translation of a specific mRNA, elongation inhibition could de-repress the translation of this specific mRNA.

TRANSLATION IN PROKARYOTES

Prokaryotic translational initiation involves the same two recognition steps as in eukaryotes. The similarities continue to some degree, with some recognition events being analogous between the two systems. In both systems, the selection of the initiator codon utilizes codon:anticodon pairing, and selection of the initiator tRNA is mediated by an initiation factor (IF-3 in prokaryotes and probably eIF-2β in eukaryotes). Additionally, S1 binding to mRNA and Shine-Dalgarno recognition in prokaryotes may mirror some functions of the cap binding complex (24). Finally, EF-Tu, the prokaryotic functional analog of eEF-1α, has a GTPase activity and is involved in the elongation cycle of translation (25).

There are two major differences between the eukaryotic and prokaryotic mechanisms of translation initiation. Eukaryotes use RNA helicase and ATP to aid in the search for the initiating AUG, while prokaryotes do not utilize ATP to aid in the search for the initiating AUG. Most eukaryotic mRNAs do not contain an obvious Shine-Dalgarno homologue and do not have any nucleotides beyond the hairpin stem at the 3′ end of 18S rRNA, where this homologue is found in prokaryotes (24). These components of prokaryotic translation allow for internal translation initiation and polycistronic mRNAs which are rare in eukaryotes.

TRANSLATION IN THE CHLOROPLAST

Translation in the chloroplast resembles prokaryotic, but some similarities to translation in eukaryotes also exist. The 16S rRNA from *Zea mays* chloroplast shows strong homology to the ribosomal RNA from *E. coli*, while yeast 18S rRNA shows only very few homologous regions when compared to the sequence from maize (26). Additionally, 23S rRNA from *E. coli* is highly homologous to *Zea mays* chloroplastic 23S rRNA when the maize chloroplast 4.5S rRNA is included as an equivalent of the *E. coli* 23S rRNA $3'$ end (27). The $5'$ leader of the mRNA encoding the large subunit of ribulose bisphosphate carboxylase (LS of RuBPCase) contains a perfectly positioned GGAGG, a Shine-Dalgarno equivalent sequence, upstream of the initiator AUG (28). A similar sequence occurs in other chloroplastic genes (29). A deletion of the sequence GGAG, identified as a putative ribosome binding site, from the *psbA* mRNA resulted in a complete loss of D1 synthesis in *Chlamydomonas reinhardtii* (Mayfield and Danon, unpublished data). Chloroplast ribosomes have been shown to bind selectively to purine-rich oligonucleotides that are complementary to the $3'$ terminus of small subunit rRNA (30). Even though these sequences are not identical to those in *E. coli*, it seems that the mechanism of recognition is similar to that in bacteria. In *Euglena gracilis* chloroplasts, initiation of polypeptide chain synthesis is *via* an N-formylmethionine as in *E. coli* (31). Barkan (32) has shown that spliced genes, encoded within a chloroplastic polycistronic unit (maize *psb* B gene cluster), are translated whether on mono- or poly-cistronic transcripts and regardless of upstream or downstream sequences. These analogies argue for a similar mechanism of ribosomal subunit assembly and translational initiation in both prokaryotes and chloroplasts.

Elongation regulation in the chloroplast probably occurs by mechanisms similar to prokaryotes as well. The chloroplast EF-Tu from both *E. gracilis* and *C. reinhardtii* is highly homologous to *E. coli* EF-Tu (33,34). These factors are also functionally exchangeable (35) and show immunological homology as well (36).

Similarities to the mechanism of eukaryotic cytoplasmic translational regulation also exist in the chloroplast. The aminoterminus of eEF-1α shows striking homology to the functionally analogous prokaryotic and chloroplastic counterparts, EF-Tu (37). A homologue to eIF-4A has been identified in chloroplasts of several higher plants, including pea, spinach and tobacco (Owttrim and Kuhlemeier, personal communication).

TRANSLATIONAL REGULATION OF CYTOPLASMIC mRNAs

Translational regulation can be exercised *via* processes involved in either initiation of translation or elongation of the nascent polypeptide and both types of translational regulation are used in plants. Research on photosynthetic gene expression has yielded the most extensive evidence for translational regulation in plants. Translational regulation has been shown to be a major component in the expression of photosynthetic genes in the chloroplast, while nuclear-encoded photosynthetic genes appear to be largely regulated at the transcriptional level. However, evidence for an additional component of translational regulation has been found in the expression of several light-regulated nuclear-encoded genes.

Light-regulated translational control has been detected in a variety of mRNAs that are translated in the cytoplasm. Transcriptional and translational regulation have been shown to act in concert to control the light-responsive expression of the nuclear *rbcS* gene, coding for the small subunit of ribulose-1,5-bisphosphate carboxylase (RuBPCase) (SSU of RuBPCase). In dark-grown amaranth seedlings the synthesis of SSU of RuBPCase occurs three to five hours after exposure to light and is independent of the level of *rbcS* mRNA (38). This regulation appears to act at the level of initiation as *rbcS* mRNA is not associated with polysomes in the dark (38). A chimeric nitrate reductase (*nit*) gene containing a constitutive promoter revealed a post-transcriptional component of nitrate reductase gene expression. Nitrate reductase mRNA accumulates in these transgenic plants independent of light, whereas the level of NR protein is 3- to 4-fold higher in the light than in the dark, implicating light-controlled translational regulation in NR protein synthesis (39). These data underline the conclusion that elucidating the level of mRNA is necessary but not sufficient for determination of gene expression as an additional level of regulation, such as translational control, may also exist.

In *C. reinhardtii* only light-grown cells accumulate HSP22 protein in the chloroplast following heat stress, although comparable levels of *hs22* mRNA are found in both light- and dark-grown heat-stressed cells (40). The *dsp-22* mRNA accumulates in desiccated plants grown in the dark but its chloroplast-targeted protein product accumulates only in light-grown desiccated plants (41). These data may reveal the existence of a light-regulated translational control of stress-induced proteins that is in addition to the stress-induced transcriptional regulation. Interestingly, *hs22* mRNA extracted from dark-grown cells was found to be inefficiently translated *in vitro* (40). Cat2 catalase (*cat2*) mRNA is rendered untranslatable in the dark as measured by immunoprecipitation of both *in vivo* and *in vitro* labeled proteins in maize (42). Again, mRNAs extracted from dark-grown leaves are not translated into the CAT-2 protein *in vitro* (42). The inefficient level of translation of *hs22* and *cat2* mRNAs in cell-free translation systems indicates that the inhibitory factor was inseparable from the mRNA itself. Although the nature of this inhibition has not been elucidated (40), one possible explanation is the existence of inhibitory *cis*-elements and whose effect could be relieved by a second factor such as a translational activator.

Other types of stress, such as hypoxia and wounding, have been shown to regulate the synthesis of a selective set of proteins in potato tubers (43). The synthesis of wound-specific proteins was inhibited during hypoxia stress while the wound-specific mRNAs were still present in the polysomal fraction, indicating that inhibition of translation may be occurring during polypeptide chain elongation.

Translational regulation of storage protein synthesis has recently been shown in alfalfa embryos (44) and oat seeds (45). Analysis of the distribution of mRNAs encoding storage proteins in the pre-cotyledonal stage of alfalfa seedlings has shown that these mRNAs are present only in the ribonucleoprotein fraction. Upon cotyledon emergence these mRNAs are found in the polysomal fraction and translated *in vivo* , suggesting that the regulation of translation occurs at initiation (44). Modulation of the level of translation of two storage proteins, avenins and globulins, has been detected in oat seeds (45). The mRNAs encoding these two proteins are found to be associated with polysomes at comparable levels and both are synthesized *in vivo* as detected by pulse labeling. However, the synthesis rate of globulins is about nine-fold higher than that of avenins, suggesting that the translation rate of at least one of the mRNAs is modulated during elongation.

TRANSLATIONAL REGULATION IN THE CHLOROPLAST

In the chloroplast, the mRNA encoding the P700 apoprotein of photosystem I is found to be associated with the polysomal fraction in both dark- and light-grown barley plants but the P700 protein is synthesized in leaves only after 1 hour of illumination (46). The induction of translation can also be achieved in plastids isolated from dark-grown leaves and illuminated *in vitro*, demonstrating that the light-induction mechanism resides within the chloroplast (46). The expression of the chloroplast-encoded LS of RuBPCase has been shown to be translationally regulated in peas. Dark- or light-grown plants contain the same amount of *rbcL* mRNA but light-grown plants synthesize more of the LS protein (47). Polysome fractionation has been used to show that the light-mediated expression of the LS of RuBPCase is regulated via initiation of translation in amaranth cotyledons (48).

The aquatic higher plant *Spirodela* can be grown heterotrophically in the dark to allow for the separation of the effect of light from the effect of energy starvation, which cannot be done in most dark-grown higher plants. Under these conditions, *psbA* gene expression is independent of transcript level in *Spirodela*, whereas energy starvation reduces the level of *psbA* mRNA in the dark-grown plants (49), suggesting that light may directly regulate the translation of the *psbA* mRNA in chloroplasts of *Spirodela* . The distribution of *psbA* mRNA between membrane-associated polysomes and ribonucleoprotein particles has been analyzed in spinach (50) and *C. reinhardtii* (51). About half of the *psbA* mRNA is associated with ribosomes and half is detected as ribonucleoprotein particles (50). However, there is no shortage of ribosomes available for the synthesis of the D1 protein, as an excess of monosomes is found in the stroma (50). These results indicate that the limiting factor for the synthesis of the D1 proteins is not the availability of either ribosomes or *psbA* mRNA, but rather that the limiting factor likely involves initiation.

The *psbA* mRNA is constitutively expressed in light- or dark-grown *C. reinhardtii* wild type (wt) and y-1 cells (52). Unlike higher plants, wt cells of *C. reinhardtii* have a constitutive chlorophyll synthesis pathway and thus are fully photosynthetic competent in dark. This feature allows the identification of light-induced gene expression in the absence of light-induced responses that are due to chloroplast development. y-1 is a mutant cell line that lacks the constitutive, but contains the light-induced chlorophyll synthesis pathway mimicking higher plants (53). The D1 protein is synthesized rapidly in illuminated wt or y-1 cells, is synthesized at a very low rate in dark-grown wt cells and is not synthesized at any detectable level in dark-grown y-1 cells. Upon illumination, dark-grown wt cells increase synthesis of the D1 protein several fold, whereas the increase in D1 protein synthesis rate in similarly treated y-1 cells occurs only after a time lag of about an hour (52), suggesting both a light-regulated and developmental regulation of *psbA* mRNA translation. Translational regulation under a developmental program is also evident in root amyloplasts of spinach plants, in that mRNAs for photosynthetic genes are present but are excluded from the polysome fraction relative to mRNAs encoding ribosomal proteins (54). The synthesis of the D1 protein and LS of RuBPCase declines during leaf senescence, while the abundance of the *psbA* and *rbcL* mRNAs remains unchanged (55). Taken together, these results indicate that the expression of *psbA* is translationally regulated under chloroplast development and by environmental signals (light).

REGULATION OF TRANSLATION INITIATION BY PHOSPHORYLATION

Translation initiation and elongation have been shown to be affected by phosphorylation of translation factors. In eukaryotic cells, the phosphorylation state of eIF-4B, eIF-4Fα and the ribosomal protein S6 was positively correlated, and that of eIF-2α negatively correlated, with the level of general protein synthesis (56). Phosphorylation of eIF-2α by a specific kinase GCN2 has been shown to be a key regulatory event in the specific de-repression of GCN4 gene expression in yeast (15). These data demonstrate that the modulation of activity of a translation factor can affect either the general rate of cellular protein synthesis or the translation of a specific mRNA. Evidence for phosphorylation of translation factors has also been found in plants. Changes in the phosphorylation state of two isoforms of eIF-4F have been identified in maize roots in response to hypoxia (57). A five-fold increase in the ratio of phosphorylated to nonphosphorylated forms is detected within 20 minutes following application of hypoxia stress (57). A wounding-specific 32 kD phosphoprotein is associated with polysomes in potato tubers. This protein is kinased only during the first 24 hours following wounding and has been postulated to mediate recognition of wound-specific mRNAs by ribosomes (43).

TRANSLATION REGULATION BY TRANS-SPLICING AND EDITING

Translation of specific proteins can be controlled by an intermolecular trans-splicing of a 5′ end sequence onto an otherwise untranslatable mRNA. Repetitive sequences have been shown to be transferred to and serve as a short leader of actin mRNA in *C. elegans* (58). A high degree of homology among leader sequences which are present in many of the nuclear-encoded mRNAs but which are not present in the corresponding genomic clones has been observed in *Euglena gracilis*, suggesting that the leaders may have been spliced onto the mRNAs (59).

Editing of organelle mRNAs can affect sequences important for translation. Evidence for editing of the ACG codon to an initiator AUG codon in chloroplasts has been obtained for both the tobacco *psbL* mRNA (60) and the maize *rpl2* mRNA (61). Comparison of genomic sequence of the *psbL* gene from ten species revealed that only two, tobacco and spinach, contained the ACG codon and the other eight contained the AUG codon (60), suggesting that editing compensated for the loss of AUG that occurred in tobacco and spinach and may have restored the translatability of the *psbL* transcript.

MECHANISMS OF TRANSLATIONAL REGULATION IN PROKARYOTES AND HOMOLOGIES TO THE CHLOROPLAST

The mechanisms of translational control in prokaryotes predominantly involve occluding (repression of translation) or exposing (enhancement of translation) the Shine-Dalgarno sequence and the initiator codon. Repressor proteins or mRNA secondary structures can obstruct the ribosome binding site (RBS) and prevent efficient translation. Activating mechanisms involve modifying the RNA structures or removal of the repressor protein to allow access to the RBS. Bacteriophage T4 RegA protein binds to target mRNAs near the initiating AUG and blocks ribosome binding. Sequence comparison of *regA* -repressed mRNAs suggests that the AUG is necessary but not sufficient for RegA binding (62). In phage λ, partitioning between and overall repression of the two functional initiator codons of the *S* gene is accomplished by two stem loop structures, one immediately upstream of the first start codon and the other 10 codons downstream. Relaxation or instability at either stem-loop structure allows a complementary change in the ratio of translational initiation at the two start codons (63). The *E. coli* ribosomal protein S4 is a translational repressor regulating expression from the ribosomal genes in the α operon. S4 specifically binds to a pseudoknot in the *rps* M message which also contains the binding site for the ribosome, although the RBS is distinct from the repressor binding site. Site-directed mutagenesis of the mRNA can relieve repression while leaving S4 binding unaffected. Thus, repressor binding is allosterically coupled to ribosome binding, so that binding of S4 causes a conformational change of the mRNA which prohibits ribosome binding (64).

Activation of several prokaryotic genes is controlled by a mechanism involving structure relaxation. RNase III binds to an upstream sequence of the phage λ *cIII* gene to relieve translational repression (65). The Com protein from phage Mu is a translational enhancer that binds specifically to the translation initiation region of the *mom* transcript and causes conformational changes in the secondary structure of this region to relieve repression of translation imposed by this secondary structure (66). Com is a zinc-finger protein which binds its target RNA and alters the secondary structure without requiring ATP (67).

These prokaryotic mechanisms of translational regulation should provide important clues in the elucidation of such mechanisms in organelles like the chloroplast. RNA structures similar to the prokaryotic control elements can be found in chloroplastic transcripts (68), and message-specific translational effector proteins are beginning to be found (69). Unlike prokaryotic organisms, the chloroplast does not contain the complete genetic information that is required for independent metabolism. In addition, to maintain high efficiency, chloroplast gene expression has to be tuned to and interact with nuclear gene expression. These requirements may call for the interaction of eukaryotic factors with this prokaryotic-like system. Nuclear-encoded factors that control translation of specific chloroplastic genes have been genetically identified in *C. reinhardtii* (70–72). Nuclear-encoded activators of the mitochondrial *coxIII* mRNA have been identified in yeast (73). The target site for these translational activators resides within the 5′ UTR of the *coxIII* mRNA (73). Although the identity of these factors has yet to be elucidated, by analogy with other systems we can imagine that they interact with the 5′ UTR of the mRNA to regulate initiation of translation.

MECHANISMS OF LIGHT-ACTIVATED TRANSLATIONAL REGULATION OF CHLOROPLAST GENE EXPRESSION

The most detailed description of the mechanism of translational regulation has been provided by the light-regulated translational control of specific chloroplast mRNAs in *C. reinhardtii*. Genetic analysis has identified several mutants that exhibited impaired translation of specific proteins in the chloroplast (70–72). Several of these mutants have been mapped to the nuclear genome revealing that nuclear-encoded factors participate in the activation of translation of specific mRNAs in the chloroplast. Gene-specific translation mutants of chloroplast origin have also been identified and the mutations found to be alterations to a stem-loop RNA structure located in the 5′ untranslated region of the chloroplastic mRNA. In addition, chloroplast suppressors of one of the nuclear mutants contained an altered 5′ stem-loop RNA structure (71). These data identify 5′ UTR stem-loop RNA structures as the site of interaction with the nuclear-encoded translational activators. These factors, which must be transported into the chloroplast, provide a link between the nucleus and the chloroplast, that can synchronize the light-regulated expression of these two distinct genomes.

Nuclear-encoded translational activators of *psbA* were identified and purified from *C. reinhardtii* cells (69). Four proteins were detected that bind specifically to the 5′ UTR of *psbA* mRNA with a 47 kD protein showing the most extensive contact with the RNA. The 47 kD protein and an additional protein of 60 kD were found to be the minimum set of proteins comprising the *psbA* RNA-binding activity. Dark-grown cells contained only one isoform of the 47 kD protein while light-grown cells contained three additional isoforms (69). It is possible that these additional isoforms, which are due to post-translational modifications, are responsible for the differences in the *psbA* mRNA-binding activity observed between light- and dark-grown cells (69). This increase in *psbA* mRNA-binding activity may enhance *psbA* translation in light-grown cells (52). Post-translationally modulating the *psbA* mRNA binding activity provides the capability for a fast response in protein synthesis to fluctuating environmental conditions. Such fast response in *psbA* mRNA translation is observed in *C. reinhardtii* cells upon illumination (52). The binding site of the RNA-binding protein complex was identified as a stem-loop RNA structure located upstream and adjacent to the initiator codon of the *psbA* mRNA. The ribosome binding site of the *psbA* mRNA is situated in a bulge within this stem-loop (69). It is possible that the *psbA* mRNA-binding protein complex may alter the accessibility of the RBS to incoming ribosome subunits, thus determining the rate of initiation of translation. The abundance of the 47 kD protein is reduced in dark-grown y-1 cells, as determined by Western blotting analysis (69), which may explain the difference in response time of *psbA* translation observed between dark-grown wt and y-1 cells, upon illumination (52).

Light-induced enhancement of translation has been observed in isolated chloroplasts (46), suggesting that the photoreceptor for translational control may be localized in the chloroplast and predicting that the light-induced signal transduction pathway that modifies translation in the chloroplast is self-contained. This signal transduction pathway probably includes a photoreceptor and a protein kinase, as minimum components.

From the above data and by analogy with other translationally regulated systems we can construct a model for the light-activated translational regulation in the chloroplast (Figure 1). In this model, nuclear-encoded translational activator proteins are synthesized in the cytoplasm and post-translationally transported into the chloroplast in inactive forms. A photoreceptor, localized in the chloroplast, initiates a signal transduction pathway which results in the phosphorylation and activation of the mRNA-binding proteins, here identified as a helicase. The binding of these proteins to the 5′ UTR stem-loop controls the degree of exposure of the ribosome binding site. The accessibility of the ribosome binding site to incoming 30S ribosomal subunits would in turn determine the rate of translation initiation. The accessibility of the ribosome binding site to the 30S ribosomal subunit could be regulated in several ways. In the first, as diagrammed in Figure 1, the protein complex binding to the 5′ stem-loop may include initiation factors with helicase activity that utilize ATP to melt RNA secondary structure, there by exposing the RBS and allowing binding of 30S ribosomal subunits. This type of translation initiation would be analogous to internal binding and initiation in eukaryotes utilizing the eIF-4F complex as a helicase (10,11). During binding of the 30S ribosomal subunit, the RNA-binding proteins would dissociate and be recycled according to the illuminated state of the cell.

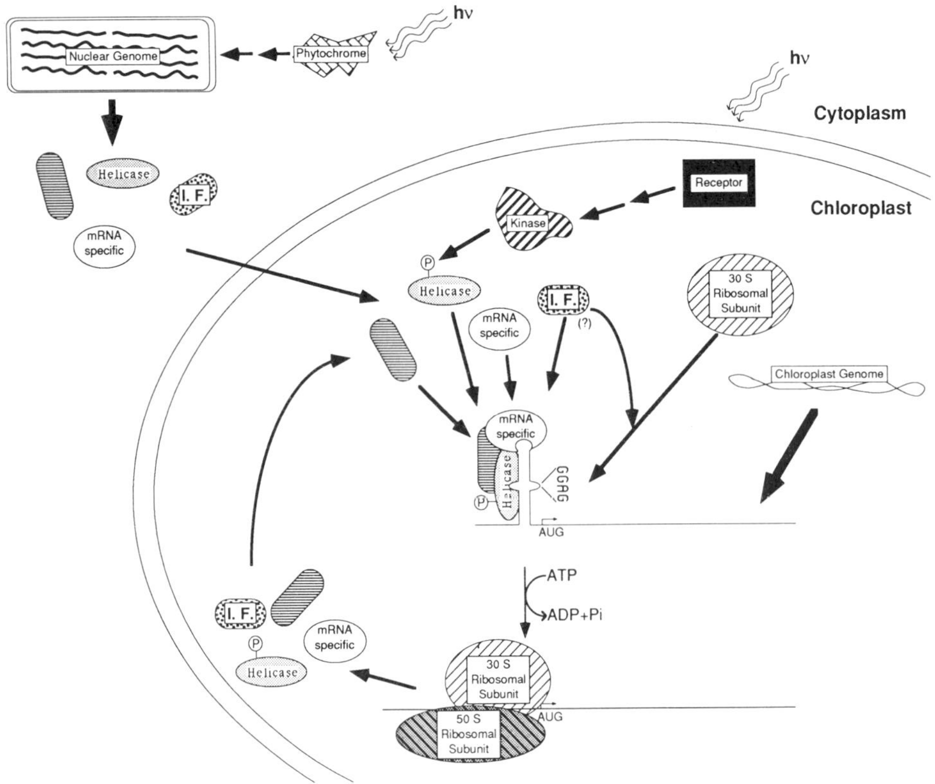

Figure 1. Model of light activation of translation in the chloroplast. I.F. = initiating factor. For details, see text.

Alternatively, the RNA-binding may act in a fashion analogous to the binding of the prokaryotic Com protein in which protein binding allosterically alters the RNA structure to allow binding of 30S ribosomal subunits (67).

According to this model there are two key points where light can influence translation in the chloroplast. The first point is the photoregulation of the transcription of the nuclear-encoded translational activators and the second is the chloroplast-localized light-activated post-translational activation of the mRNA-binding proteins. These two light-activated regulatory pathways provide the means for both fast response in translation of chloroplast mRNAs, tuned to the fluctuating illuminated state of the chloroplast, and long-term regulation, which is responsive to light and to the developmental state of the chloroplast, through transcription and accumulation of the nuclear-encoded translational activators. This model also predicts that chloroplast gene expression can be developmentally regulated by controlling the expression of the nuclear-encoded translational activators. For example, synthesis of the D1 protein can be blocked in non-green plastids, independent of *psbA* mRNA levels, by repressing the expression of its translational activator in these cell types.

CONCLUSIONS

The accumulating data in plants and other systems suggest that translational regulation is an important and apparently prevalent mechanism of gene regulation. Translational regulation provides the means for a fast response in gene expression and, as such, is heavily utilized in the photoregulation of chloroplast gene expression. However, translational regulation is not limited to the chloroplast and has also been detected in the expression of a number of mRNAs that are translated in the cytoplasm (38–45). Thus, measuring transcription and determining the abundance of mRNA is necessary but not sufficient for understanding gene expression, as an additional level of regulation, such as translational control, may also exist.

REFERENCES

1 Kozak, M. (1991) J. Biol. Chem. 266, 19867–19870.
2 Kozak, M. (1989) J. Cell Biol. 108, 229–241.
3 Jaramillo, M., Pelletier, J., Edery, I., Nielson, P.J. and Sonenberg, N. (1991) J. Biol. Chem. 266, 10446–10451.
4 Thach, R.E. (1992) Cell 68, 177–180.
5 Carberry, S.E. and Goss, D.J. (1991) Biochemistry 30, 6977–6982.
6 Pilipenko, E.V., Blinov, V.M., Romanova, L.I., Sinyakov, A.N., Maslova, S.V. and Agol, V.I. (1989) Virology 168, 201–209.
7 Pelletier, J. and Sonenberg, N. (1988) Nature 334, 320–325.
8 Jang, S.K., Davies, M.V., Kaufman, R.J. and Wimmer, E. (1989) J. Virol. 63, 1651–1660.
9 Pilipenko, E.V., Gmyl, A.P., Maslova, S.V., Svitkin, Y.V. Sinyakov, A.N. and Agol, V.I. (1992) Cell 68, 119–131.
10 Thomas, A.A.M., ter Haar, E., Wellink, J. and Voorma, H.O. (1991) J. Virol. 65, 2953–2959.
11 Anthony, D.D. and Merrick, W.C. (1991) J. Biol. Chem. 266, 10218–-10226.
12 Fütterer, J. and Hohn, T. (1991) EMBO J. 10, 3887–3896.
13 Fu, L.N., Ye, R.Q., Browder, L.W. and Johnston, R.N. (1991) Science 251, 807–810.
14 Jaramillo, M., Browning, K., Dever, T.E., Blum, S., Trachsel, H., Merrick, W.C., Ravel, J.M. and Sonenberg, N. (1990) Biochim. Biophys. Acta 1050, 134–139.
15 Lanker, S., Bushman, J.L., Hinnebusch, A.G., Trachsel, H. and Mueller, P.P. (1992) Cell 70, 647–657.
16 Rubin, H.N. and Halim, M.N. (1987) Biochem. Biophys. Res. Commun. 144, 649–656.
17 Munroe, D. and Jacobson, A. (1990) Mol. Cell. Biol. 10, 3441–3455.
18 Ch'ng, J.L.C., Shoemaker, D.L., Schimmel, P. and Holmes, E.W. (1990) Science 248, 1003–1006.
19 Ursin, V.M., Irvine, J.M., Hiatt, W.R. and Shewmaker, C.K. (1991) The Plant Cell 3, 583–591.

20 Pokalsky, A.R., Hiatt, W.R., Ridge, N., Rasmussen, R., Houck, C.M. and Shewmaker, C.K. (1989) Nucl. Acids Res. 17, 4661–4673.
21 Tuhácková, Z., Ullrichová, J. and Hradec, J. (1985) Eur. J. Biochem. 146, 161–166.
22 Ejiri, S. and Honda, H. (1985) Biochem. Biophys. Res. Commun. 128, 53–60.
23 Ryazanov, A.G., Rudkin, B.B. and Spirin, A.S. (1991) FEBS Lett. 285, 170–175.
24 Gold, L. and Hartz, D. (1990) in NATO ASI Series, 49, pp. 433–441. Post-transcriptional Control of Gene Expression (McCarthy J.E.G. and Tuite, M.F., eds.), Springer-Verlag, Berlin, Heidelberg.
25 Miller, D.L. and Weissbach, H. (1977) in Molecular Mechanisms of Protein Biosynthesis (Weissbach, H. and Pestka, S., eds.), pp. 323–373, Academic Press, New York, NY.
26 Schwarz, Z. and Kössel, H. (1980) Nature, Lond. 283, 739–742.
27 Edwards, K. and Kössel, H. (1981) Nucl. Acids Res. 9, 2853–2869.
28 McIntosh, L., Poulson, C. and Bogorad, L. (1980) Nature, Lond. 288, 556–560.
29 Krebbers, E.T., Larrinua, I.M., McIntosh, L. and Bogorad, L. (1982) Nucl. Acids Res. 10, 4985–5002.
30 Steege, D.A., Graves, M.C. and Spremulli, L.L. (1982) J. Biol. Chem. 257, 10430–10439.
31 Schwarz, J.H., Meyer, R., Eisenstadt, J.M. and Brawerman, G. (1967) J. Mol. Biol. 25, 571–574.
32 Barkan, A. (1988) EMBO J. 9, 2637–2644.
33 Montandon, P.-E. and Stutz, E. (1983) Nucl. Acids Res. 11, 5877–5892.
34 Watson, J.C. and Surzycki, S.J. (1982) Proc. Nat. Acad. Sci. U.S.A. 79, 2264–2267.
35 Breitenberger, C.A., Graves, M.C. and Spremulli, L.L. (1979) Arch. Biochem. Biophys. 194, 265–270.
36 Breitenberger, C.A. and Spremulli, L.L. (1980) J. Biol. Chem. 255, 9814–9820.
37 Slobin, L.I., Clark, R.V. and Olson, M.O.J. (1983) Biochemistry 22, 1911–1917.
38 Berry, J.O., Breiding, D.E. and Klessig, D.F. (1990) Plant Cell 2, 795–803.
39 Vincentz, M. and Caboche, M. (1991) EMBO J. 10, 1027–1035.
40 Ish-Shalom, D., Kloppstech, K. and Ohad, I. (1990) EMBO J. 9, 2657–2661.
41 Bartels, D., Hanke, C., Schneider, K., Michel, D. and Salamini, F. (1992) EMBO J. 11, 2771–2778.
42 Skadsen, R.W. and Scandalios, J.G. (1987) Proc. Nat. Acad. Sci. U.S.A. 84, 2785–2789.
43 Crosby, J.S. and Vayda, M.E., (1991) Plant Cell 3, 1013–1023.
44 Pramanik, S.J., Krochko, J.E. and Bewley, J.D. (1992) Plant Physiol. 99, 1590–1596.
45 Boyer, S.K., Shotwell, M.A. and Larkin, B.A. (1992) J. Biol. Chem. 267, 17449–17457.
46 Laing, W., Kreuz, K. and Apel, K. (1988) Planta 176, 269–276.

47 Inamine, G., Nash, B., Weissbach, H. and Brot, N. (1985) Proc. Nat. Acad. Sci. U.S.A. 82, 5690–5694.
48 Berry, J.O., Breiding, D.E. and Klessig, D.F. (1990) Plant Cell 2, 795–803.
49 Fromm, H., Devic, M., Fluhr, R. and Edelman, M. (1985) EMBO J. 4, 291–295.
50 Minami, E.-I., Shinohara, K., Kawakami, N. and Watanabe, A. (1988) Plant Cell Physiol. 29, 1303–1309.
51 Breidenbach, E., Jenni, E., Leu, S. and Boschetti, A. (1988) Plant Cell Physiol. 29, 1–7.
52 Malnoë, P., Mayfield, S.P. and Rochaix, J.-D. (1988) J. Cell Biol. 106, 609–616.
53 Ohad, I., Siekevitz, P. and Palade, G.E. (1967) J. Cell Biol. 35, 553–584.
54 Deng, X.-W. and Gruissem, W. (1988) EMBO J. 7, 3301–3308.
55 Bate, N.J., Rothstein, S.J. and Thompson, J.E. (1991) J. Exp. Bot. 42, 801–811.
56 Duncan, R. and Hershey, J.W.B. (1985) J. Biol. Chem. 260, 5493–5497.
57 Webster, C., Gaut, R.L., Browning, K.S., Ravel, J.M. and Roberts, J.K.M. (1991) J. Biol. Chem. 266, 23341–23346.
58 Krause, M. and Hirsh, D. (1987) Cell 49, 753–761.
59 Tessier, L.-H., Keller, M., Chan, R.L., Fournier, R., Weil, J.-H. and Imbault, P. (1991) EMBO J. 9, 2621–2625.
60 Kudla, J., Igloi, G.L., Metzlaff, M., Hagemann, R. and Kössel, H. (1992) EMBO J. 11, 1099–1103.
61 Hoch, B., Maier, R.M., Appel, K., Iglo, G.L. and Kössel, H. (1991) Nature 353, 178–180.
62 Winter, R.B., Morrisey, L., Gauss, P., Gold, L., Hsu, T. and Karam, J. (1987) Proc. Nat. Acad. Sci. U.S.A. 84, 7822–7826.
63 Bläsi, U., Nam, K., Hartz, D., Gold, L. and Young, R. (1989) EMBO J. 8, 3501–3510.
64 Tang, C.K. and Draper, D.E. (1990) Biochemistry 29, 4434–4439.
65 Altuvia, S., Locker-Giladi, H., Koby, S., Ben-Nun, O. and Oppenheim, A.B. (1987) Proc. Nat. Acad. Sci. U.S.A. 84, 6511–6515.
66 Wulczyn, F.G. and Kahmann, R. (1991) Cell 65, 259–269.
67 Hattman, S., Newman, L., Krishna Murthy, H.M. and Nagaraja, V. (1991) Proc. Nat. Acad. Sci. U.S.A. 88, 10027–10031.
68 Erickson, J.M., Rahire, M. and Rochaix, J.-D. (1984) EMBO J. 3, 2753–2762.
69 Danon, A. and Mayfield, S.P. (1991) EMBO J. 10, 3993–4001.
70 Chua, N.-H. and Bennoun, P. (1975) Proc. Nat. Acad. Sci. U.S.A. 72, 2175–2179.
71 Rochaix, J.-D., Kuchka, M., Mayfield, S.P., Schirmer-Rahire, M., Girard-Bascou, J. and Bennoun, P. (1989) EMBO J. 8, 1013–1021.
72 Kuchka, M.R., Mayfield, S.P. and Rochaix, J.-D. (1988) EMBO J. 7, 319–324.
73 Costanzo, M.C. and Fox, T.D. (1988) Proc. Nat. Acad. Sci. U.S.A. 85, 2677–2681.

ON THE ORIGINS, STRUCTURES AND FUNCTIONS OF RESTRICTION-MODIFICATION ENZYMES

Joseph Heitman

Section of Genetics and Department of Pharmacology
Howard Hughes Medical Institute
Duke University Medical Center, Box 3546
322 Carl Building, Research Drive
Durham, NC 27710

Typically, restriction-modification (RM) systems consist of two enzymes: an endonuclease that recognizes and cleaves a specific DNA sequence, and a methyltransferase that modifies the same sequence to protect the host chromosome from cleavage. These enzymes also play an important role in genetic engineering and provide insight into the basis of sequence-specific DNA-protein interactions. A growing number of type II restriction endonucleases and methyltransferases are being subjected to biochemical and genetic studies which, when combined with ongoing X-ray crystallographic analyses, promise to provide detailed models for mechanisms of DNA recognition and catalysis. Studies on anti-restriction systems, the repair of DNA single- and double-strand breaks, and the roles of DNA lesions in recombination initiation, suggest RM systems may provoke genome rearrangements and assimilate foreign DNA into the host genome and point towards possible evolutionary sources of this interesting group of DNA metabolizing enzymes.

INTRODUCTION

RM systems are elaborated by many species of bacteria as enzymatic defense systems which serve as a bacterial immune system to destroy foreign DNA entering the cell (1). The first observations on the phenomenon of restriction and modification were described by Luria and Human in 1952 (2), and Bertani and Weigle in 1953 (3). Bertani and Weigle (3) observed that stocks of λ phage prepared from *E. coli* strain C grew poorly when propagated on a different strain, K-12. This effect was named restriction. Rare phage that

Genetic Engineering, Vol. 15, Edited by J.K. Setlow
Plenum Press, New York, 1993

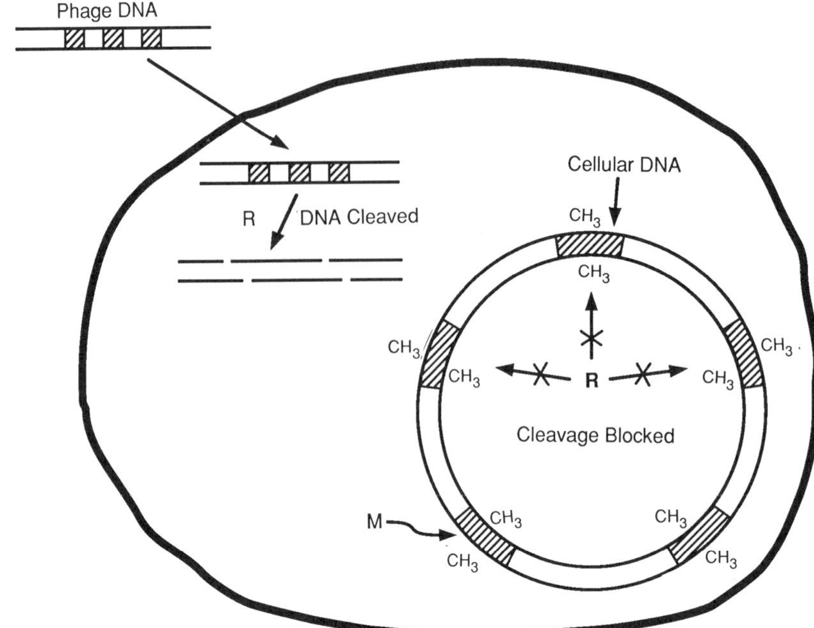

Figure 1. Restriction-modification. Restriction-modification (RM) systems typically consist of an endonuclease (R) and a methyltransferase (M). Hatched boxes represent specific DNA recognition sites modified by methylation or subject to endonuclease cleavage. The host chromosome is protected by methylation whereas unmethylated foreign DNA that enters the cell is restricted.

escaped restriction produced progeny which now grew with equal efficiency on either host. These survivors were not mutants because another cycle of growth on the *E. coli* C host restored sensitivity to restriction. In some way these phage were transiently altered or modified during growth on K-12. It is now known that specific enzymes form the molecular basis underlying this epigenetic phenomenon (1,4). Restriction results from the action of an endonuclease that recognizes a specific DNA sequence and cleaves there or elsewhere (Figure 1). Modification is the work of a DNA methyltransferase which recognizes and methylates the same DNA sequence to protect the host chromosome from endonuclease action. Phage resistant to restriction occur when the methyltransferase modifies each phage-borne recognition site prior to endonuclease action.

Based on the properties of the purified enzymes, restriction-modification systems fall into four classes: type I, type II, type III and the methylation-dependent restriction systems (MDRS) (5–9). Type I systems are the most complex and consist of three subunits (R = restriction, M = modification, and S = specificity), which form a pentameric (R_2M_2S) multifunctional enzyme, exhibit complex cofactor requirements, cleave at random sites distant from the site of recognition, unwind DNA to form a loop between sites of recognition and cleavage, and are inactive after a single round of DNA scission. Type III systems are the most mysterious, and consist of two subunits, R and M, which form a methyltransferase of subunit M alone and a methyltransferase-endonuclease complex containing both subunits, require but do not hydrolyze ATP as a cofactor during DNA scission, and cleave at unique sites but yield partial digests that do not proceed to completion. Lastly, type III enzymes modify only one DNA strand, raising a conundrum, only recently answered, as to how the unmethylated daughter molecule produced after replication is protected. Type II systems are the simplest and many have now been cloned and characterized (6,7,10). Type II endonucleases are typically homodimers distinct from and with no sequence identity to partner monomeric DNA methyltransferases. This difference in subunit arrangement may arise from the nature of the recognition sites. The endonuclease usually binds a palindromic DNA sequence and symmetrically cleaves both strands. The typical methyltransferase substrate is asymmetric hemi-methylated DNA produced by replication. Thus, as monomers, the methyltransferases are well suited to methylate the one remaining strand. Each enzyme requires a single cofactor: Mg^{2+} for scission and S-adenosylmethionine (SAM) for methylation.

Most restriction enzymes cleave unmethylated DNA and are blocked by DNA methylation. The first exception was the type II restriction enzyme *Dpn* I from *Streptococcus pneumoniae,* which cleaves the sequence GATC only when methylated at the N^6 position of both adenines (11). While studying cloned RM systems from bacterial species other than *E. coli,* we and others observed that expression of the methyltransferase genes alone was in some cases detrimental to *E. coli* (12–16). This results from several *E. coli* restriction systems that cleave methylated DNA: Mrr, McrA, and McrBC (8).

Like all weapons, restriction endonucleases are potentially dangerous both for the intended victim(s) (foreign DNA) as well as the aggressor. Host bacteria have evolved or acquired DNA methyltransferases that modify and thereby protect the host chromosome from cleavage. Similarly, some phage, plasmids and bacteria express additional proteins that antagonize restriction system function, and these are called anti-restriction systems (9,17). Studies of anti-restriction reveal that restriction-modification may be coordinated with other cellular processes, such as DNA repair, and provide insights into the origin and function of RM systems.

Since the discovery of RM, many have suggested that in addition to or instead of cellular defense, RM systems function *in vivo* to foster recombination and speed genomic rearrangements (4). Some early disputed observations supported this notion (18). More recent observations have demonstrated that restriction endonucleases can indeed produce DNA double-strand breaks that promote recombination (19–26). In parallel, studies on naturally occurring recombination initiation sites and integration of linear DNA in yeast following transformation have led to and supported the double-strand break repair model of recombination (27). Related studies on the consequences and repair of *in vivo* DNA nicks and double-strand breaks (28,29) suggest that some types of DNA DSBs do not require recombinational repair. Further studies on DNA nick and break repair and recombination initiation by DNA lesions are clearly warranted and may unveil broader roles for RM systems in recombination.

TYPE II RESTRICTION-MODIFICATION SYSTEMS

*Eco*RI

In the recently revised X-ray crystal structure of an *Eco*RI-substrate complex (solved in the absence of Mg^{2+} to prevent scission), the enzyme binds DNA as a symmetric homodimer that projects one extended chain and two α-helices from each monomer into the DNA major groove to contact each purine and pyrimidine of the substrate (30,31). Unlike an earlier model (32), this revised structure is compatible with site-directed mutagenic (33–36) and biochemical studies (37–39) and robustly supports the pyrimidine contact model, initially proposed based on genetic and biochemical studies (28,33,40), in which *Eco*RI contacts the pyrimidine bases of the substrate.

Attempts to alter the substrate specificity of *Eco*RI to engineer new restriction enzymes have so far been unsuccessful. Mutant enzymes with increased star activity which accept both the wild-type substrate (GAATTC) and additional sites (GACTTC,AAATTC) have been isolated and in some cases identify residues that the X-ray crystal structure suggests participate in direct DNA-protein interactions (28,33). Judicious combinations of increased star activity mutations, site-directed alterations of the substrate binding pocket, and molecular modelling should allow us to begin the next round elucidating the basis of *Eco*RI-substrate recognition by attempting to change enzyme specificity. For readers interested in a more thorough discussion of *Eco*RI, several reviews have appeared recently (28,31,41,281).

*Rsr*I

Rhodobacter spheroides elaborates the *Rsr*I restriction-modification system, which recognizes the sequence GAATTC and cleaves between the guanine and adenine (42). Thus, *Rsr*I and *Eco*RI are isoschizomers. The *Rsr*I restriction system has been cloned and sequenced, and both the endonuclease and methyltransferase have been purified (43–46). In contrast to most other isoschizomeric RM systems, the *Eco*RI and *Rsr*I endonucleases are remarkably similar, sharing 50% amino acid identity, whereas the *Eco*RI and *Rsr*I methyltransferases share little or no homology, even though both are N^6-adenine methyltransferases that recognize GAATTC and modify the internal adenines. Moreover, *Rsr*I methyltransferase is dimeric in contrast to the monomeric *Eco*RI methyltransferase (47). Further studies on these four enzymes should provide a wealth of information concerning different protein structures that recognize the same DNA sequence.

Like the *Eco*RI endonuclease, the *Rsr*I endonuclease is a dimer that forms a specific stable substrate-enzyme complex in the absence of the Mg^{2+} cofactor required for catalysis. In these complexes, both *Eco*RI and *Rsr*I footprint 12 basepairs, bend DNA by 50°, and unwind the DNA helix by 25°,

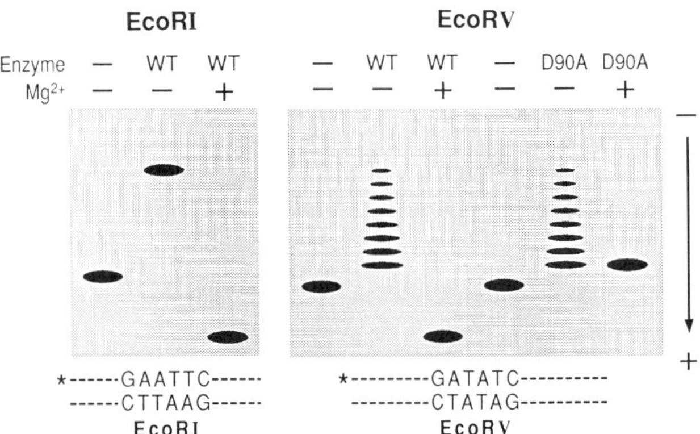

Figure 2. DNA binding by *Eco*RI and *Eco*RV endonucleases. *Eco*RI and *Eco*RV serve as paradigms of two distinct classes of type II restriction endonucleases. This figure presents a schematicized version of non-denaturing agarose gel mobility assays with purified enzymes and substrate-containing DNA. *Eco*RI forms a specific enzyme-substrate complex in the absence of Mg^{2+} whereas *Eco*RV does not. Addition of Mg^{2+} to a DNA-binding proficient catalysis-defective *Eco*RV mutant (D90A) converts the ladder of nonspecific complexes to a single specific enzyme-substrate complex. Based on original data in (53,59).

suggesting that the two substrate-enzyme complexes are remarkably similar (48). Nonetheless, the ideal reaction conditions differ for the two enzymes. *Rsr*I is inherently temperature sensitive and less prone to star activity than *Eco*RI. In addition, studies with oligonucleotides containing nucleotide analogs reveal *Rsr*I is more sensitive than *Eco*RI to substrate functional group losses (49). Thus the two enzymes may differ in substrate recognition, possibly forming different substrate-enzyme transition state complexes. If so, this must involve amino acids other than most of those presently thought to be engaged in substrate recognition (Q115,M137,A138,G140,N141,R145,R200, R203 in *Eco*RI) and catalysis (P90,D91,E111 and K113 in *Eco*RI), because these residues are identical in *Eco*RI and *Rsr*I. Candidates include I197 (M202 in *Rsr*I, residue designations differ because *Rsr*I has 5 additional N-terminal amino acids), E192 (V197 in *Rsr*I), and Y193 (H198 in *Rsr*I), which have been implicated in substrate recognition by crystallographic (31) or genetic findings (28,40). Further studies with *Rsr*I site-directed mutants and *Rsr*I-*Eco*RI hybrid proteins may answer this question.

*Eco*RV

Our understanding of the structure and function of the *Eco*RV restriction endonuclease has advanced rapidly (41,50–59) and *Eco*RV now rivals *Eco*RI as a paradigm for type II restriction enzymes. One might have anticipated *Eco*RV would share many properties with *Eco*RI; however the two enzymes are remarkably different and may represent two distinct classes of type II RM systems. Insightful and up-to-date reviews on *Eco*RV have recently appeared (41,51).

The *Eco*RV RM system recognizes the sequence GATATC. Scission is at the center of the site, producing blunt ends, and methylation is at the N^6 amino groups of the outer adenines. The *Eco*RV RM system has been cloned and sequenced, the proteins overexpressed and purified, and the endonuclease crystallized (60–62). These studies disclose that, unlike *Eco*RI, which binds its substrate with high affinity and specificity in the absence of Mg^{2+}, *Eco*RV binds nonspecific and specific DNA sequences with the same affinity. Thus in gel shift experiments, *Eco*RI forms a unique enzyme-substrate complex whereas *Eco*RV forms largely nonspecific DNA-protein complexes that migrate as a ladder in which decreasing electrophoretic mobility reflects an increasing number of bound proteins per DNA molecule (53,59) (Figure 2). One consequence might have been that *Eco*RV would have less substrate specificity compared to *Eco*RI. However, *Eco*RV and *Eco*RI exhibit comparable levels of substrate discrimination and thus, for *Eco*RV, recognition must occur during Mg^{2+} binding or catalysis. *Eco*RV bound to nonspecific DNA has lower affinity for Mg^{2+} compared to *Eco*RV bound to substrate (50), suggesting that the high-affinity Mg^{2+} binding site only assembles in the substrate-enzyme complex. In turn, Mg^{2+} binding may allosterically activate DNA scission by *Eco*RV.

Both X-ray crystal structures and site-directed mutant enzymes provide a more detailed view of *Eco*RV function. Winkler and co-workers have solved high resolution X-ray crystal structures of *Eco*RV endonuclease free in solution, and in complex with nonspecific and specific DNA sequences (41) (Winkler, personal communication). *Eco*RV is a symmetric homodimer and each monomer bears a surface-exposed loop that fits into the DNA major groove. The structures of free and DNA-bound *Eco*RV differ, implying conformational changes during binding. While DNA in the nonspecific *Eco*RV-DNA complex is largely B-form, DNA in the specific *Eco*RV-DNA complex is drastically bent. Winkler has suggested that this DNA distortion, which only transpires at the specific DNA recognition sequences, underlies the remarkable sequence specificity of this and other restriction enzymes. In essence, only the substrate can undergo conformational changes required to draw the DNA backbone into the jaws of the enzyme. The *Eco*RV and *Eco*RI enzyme-substrate X-ray structures further unveil a set of four conserved residues (P73,D74,D90,K92 in *Eco*RV and P90,D91,E111,K113 in *Eco*RI) deployed at similar positions about the scissile phosphodiester bonds. Genetic and biochemical analysis of mutant enzymes supports the view that these residues constitute the enzyme active sites (41,51,57,58,63,64). For more in-depth discussion, readers are referred to two excellent reviews exploring the X-ray crystal structures of *Eco*RV endonuclease and *Eco*RV-DNA complexes (41) and detailed biochemical and genetic studies on *Eco*RV substrate recognition and catalysis (51).

*Bam*HI

The *Bam*HI type II RM system is expressed by *Bacillus amyloliquefaciens* H and recognizes the sequence GGATCC. The genes encoding the *Bam*HI enzymes have been cloned and sequenced, and the *Bam*HI endonuclease has been overexpressed, crystallized and subjected to genetic analysis (65–68). *Bam*HI endonuclease mutants that lack or have greatly reduced DNA cleavage activity were isolated based on ability to survive in the absence of *Bam*HI methyltransferase expression (67). Three such mutant proteins (D94N, E77K and E113K) have little or no cleavage activity but retain DNA binding of wild-type (D94N) or enhanced affinity (E77K,E113K), implying the acidic residues D94, E77 and E113 are essential for catalysis but dispensable for binding. This phenotype is reminiscent of the E111 mutants of the *Eco*RI endonuclease, which bind DNA with normal or greatly increased affinity but are largely defective in cleavage (63,64). The revised *Eco*RI-substrate crystal structure reveals E111 near the scissile phosphodiester bond to chelate the essential Mg^{2+} cofactor in the enzyme active site (31). E111 is also one of four conserved residues which may comprise the active sites of both *Eco*RI and *Eco*RV (57).

The E77K *Bam*HI mutant protein has been purified to homogeneity. *In vitro* the mutant enzyme has nondetectable cleavage activity with Mg^{2+} as cofactor, but weak ($< 10^2$ U/mg) *Bam*HI activity with Mn^{2+} as cofactor (67). In contrast, wild-type *Bam*HI is ten-fold less active when Mn^{2+} replaces Mg^{2+} as cofactor. Thus the cofactor preference of the E77K mutant is reversed compared to wild-type, suggesting the mutation may alter the cofactor binding pocket. E77 (and also D94 or E113) may therefore perform a similar role in *Bam*HI as residue E111 in *Eco*RI.

Expression of the E77K *Bam*HI mutant *in vivo* partially induces the SOS DNA repair response, reflecting weak DNA cleavage activity (67). The mutant enzyme retains specificity for *Bam*HI sites because induction is blocked by coexpression of *Bam*HI methyltransferase. Xu and Schildkraut isolated second site suppressors of the E77K mutation by screening for increased induction of an SOS::*lacZ* fusion (68). Two suppressors, R76K and P79T, enhance the activity of the respective double mutants (R76K E77K and E77K P79T) *in vivo* (as assayed by SOS induction) and by the purified mutant proteins *in vitro* (5×10^3 U/mg compared to $< 10^2$ U/mg for E77K and 10^6 U/mg for wild-type *Bam*HI). The double mutant enzymes retain *Bam*HI substrate specificity because *Bam*HI methyltransferase blocks SOS induction *in vivo* and DNA cleavage *in vitro*.

Interestingly, these double mutant proteins now prefer Mg^{2+} over Mn^{2+} as cofactor, albeit with somewhat less preference than the wild-type enzyme (two-fold versus ten-fold). This suggests that the cofactor binding pocket is altered in the E77K mutant to better accommodate Mn^{2+} and that the nearby R76K or P79T suppressor mutations may increase activity by restoring the cofactor binding pocket configuration.

*Taq*I

The *Taq*I restriction-modification system is produced by the thermophilic bacteria *Thermus aquaticus* and recognizes the tetranucleotide TCGA (69). The *Taq*I system has been cloned, sequenced, and the endonuclease overexpressed and crystallized (70–72). A set of insertion mutants has been described which should prove informative once the crystal structure is solved (73,74). Biophysical analysis reveals that *Taq*I requires Mg^{2+} and high temperature (50° C) to form a tight substrate-enzyme complex (75). Similarly, both *Eco*RV (53,59) and *Pae*R7 (76) fail to form a tight specific enzyme-DNA complex in the absence of a Mg^{2+} cofactor. In contrast, *Eco*RI and *Rsr*I form specific enzyme-substrate complexes in the absence of Mg^{2+}. Thus type II endonucleases may fall into two distinct classes, those that require cofactor for substrate recognition and those that do not. The names SEL (specificity early and late) and SLO (specificity late only) have been suggested (75). As an alternative, I suggest type IId (for cofactor dependent specific binding) and type IIi (for cofactor independent specific binding), similar to the already existing type IIs designation for type II enzymes with split recognition and cleavage sites.

Zebala et al. have recently described extensive studies on the interaction of *Taq*I with modified oligonucleotides (77), revealing that *Taq*I is remarkably tolerant of functional group changes. K_{cat}, and not K_m, was largely altered by these functional group changes, arguing that catalysis is slowed but binding is not perturbed. In addition, *Taq*I makes no apparent contacts with substrate pyrimidines, in contrast to both *Eco*RI and *Eco*RV, even though the *Taq*I recognition site is four instead of six base pairs and one might have expected the enzyme would need more contacts to extract sufficient sequence information. These findings suggest that *Taq*I, and probably also *Eco*RV and *Pae*R7, recognize substrate in a fundamentally different way compared to *Eco*RI or *Rsr*I. The *Taq*I class of enzymes appears to form substrate-specific complexes only in the presence of the Mg^{2+} cofactor whereas *Eco*RI forms a specific enzyme-substrate complex in its absence. It has been suggested that additional conformational changes that contribute to substrate recognition may occur in *Eco*RI after formation of the initial substrate-enzyme complex (31,40,63,64), and such changes could be provoked by Mg^{2+} binding. In the case of *Taq*I and *Eco*RV, allosteric activation models may be even more appropriate since the cofactor stimulates sequence-specific DNA binding.

Recently the genes encoding a *Taq*I isoschizomer, *Tth*HB8I, have been cloned from *Thermus thermophilus* HB8 (78). In contrast to other isoschizomer enzyme pairs but like *Eco*RI and *Rsr*I, the *Taq*I and *Tth*HB8I endonculeases share 77% identity. In contrast to the *Eco*RI and *Rsr*I RM systems, the *Taq*I and *Tth*HB8I methyltransferases are also remarkably similar and share 79% identity. Further studies on the *Taq*I and *Tth*HB8I RM systems should contribute to our understanding of how related enzymes discriminate DNA sequences, in this case the tetranucleotide TCGA.

*Fok*I

*Fok*I RM system is a member of the type IIs family of enzymes, which are distinct from type II enzymes and cleave at a defined distance outside their recognition site, irrespective of the sequence at the cleavage site. This property suggested that the DNA recognition and cleavage domains of the type IIs endonucleases might be distinct, and that one might be able to construct endonucleases of novel specificities by fusing a different DNA recognition domain to the cleavage domain. While this has not yet been realized, two important steps have been taken. First, a large number of *Fok*I endonuclease mutants have been isolated with the use of SOS::*lacZ* fusion strains to test for *in vivo* endonuclease activity (Waugh and Sauer, personal communication). Several mutant enzymes have increased propensity to cleave hemi-methylated *Fok*I sites, suggesting they either increase DNA binding affinity or specifically alter substrate discrimination. Second, proteolytic fragments of the *Fok*I endonuclease have defined an N-terminal DNA binding domain and a C-terminal fragment with nonspecific DNA cleavage activity (79). The next step will be to fuse different DNA binding domains to the C-terminal cleavage domain and screen for active endonucleases with altered substrate specificity.

*Pvu*II

The *Pvu*II type II RM system is produced by the bacteria *Proteus vulgaris* and recognizes the sequence CAGCTG (80). Cleavage occurs at the center of the site to yield blunt ends, and methylation is at the central cytosines. The *Pvu*II methyltransferase is one of an atypical group of cytosine methyltransferases which modify the N^4 amino group of cytosine (81). Modification by *Pvu*II confers sensitivity to the McrBC methylation-dependent restriction system (12,14).

The *Pvu*II restriction-modification system has been cloned (14), sequenced (82,83), and the endonuclease overproduced and crystallized. Remarkably, a third gene, *C*, was found to lie between those encoding the *Pvu*II endonuclease and methyltransferase (83). The *C* gene product displays striking similarity to the lambda phage cI repressor; disruption of the *C* gene blocks restriction but not methylation by the *Pvu*II RM system, and the cloned *C* gene complements the *C* gene disruption when provided *in trans*. Thus the *C* gene product is likely to be a DNA-binding protein that regulates *Pvu*II endonuclease gene expression, possibly to ensure integrity of the host genome following RM system transfer into a naive recipient cell. This model makes the testable prediction that the wild-type $R^+C^+M^+$ *Pvu*II RM system should transform cells with higher efficiency compared to a $R^+C^-M^+$ mutant.

Similar *C* genes are found in association with at least three other restriction systems: *Bam*HI, *Eco*RV and *Sma*I. Three of these RM systems have a similar gene arrangement (*RCM*, where *RC* is transcribed leftward and *M* rightward) and the fourth is *CRM* (where *CR* is transcribed rightward and *M* leftward). Thus in each case, the *C* gene lies upstream of the *R* gene and both are transcribed in the same direction.

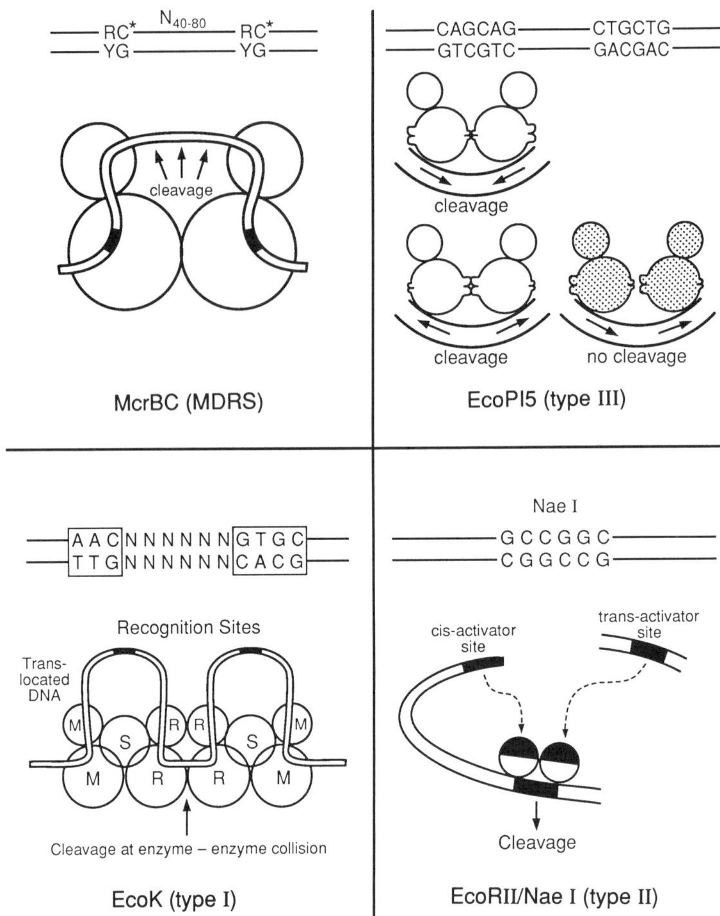

Figure 3. Restriction endonucleases requiring two sites for cleavage. At least one member of each type of RM system has been found or suggested to require two recognition sites for cleavage. Hemi-methylation of either strand at two purine-C sites suffices for McrBC cleavage. The type III *Eco*P15 requires two inverted (head-to-head or tail-to-tail) recognition sites. Type I enzymes may cleave at collisions between two enzyme-DNA complexes. Several type II enzymes require activator DNAs *in cis* or *trans* to promote efficient cleavage.

In contrast, at least the *Eco*RI RM system has no apparent *C* gene. What sets the four RM systems with *C* genes apart from at least *Eco*RI is that the *Eco*RI *R* and *M* genes are transcribed in the same direction. Perhaps this gene orientation serves an analogous regulatory role as the *C* genes, or the *Eco*RI RM system is unregulated or regulated in a different fashion. For example, a putative LexA binding site with a single base pair substitution from the canonical (CTtTataaaataACAG in *Eco*RI vs. the canonical CTGTatatatat-

ACAG) lies just upstream of the *Eco*RI endonuclease gene, suggesting the LexA repressor may transiently repress endonuclease expression when the *Eco*RI RM system enters a naive cell.

*Eco*RII and *Nae*I Allosteric Activation of Type II Enzymes

The type II restriction endonuclease *Eco*RII displays the unusual property that some DNA sequences containing *Eco*RII recognition sites are refractory to cleavage unless sensitive activator *Eco*RII sites are present in the reaction, either on the same DNA molecule *in cis* or on other DNA molecules *in trans* (84,85) (Figure 3). For example, an *Eco*RII-resistant site from phage T7 becomes sensitive when cloned into a plasmid with sensitive *Eco*RII sites (86). A similar phenomenon has been described for *Nae*I, *Nar*I, *Bsp*MI, *Hpa*II, and *Sac*II restriction endonucleases and, in these cases, either spermidine or cleavable activator DNA enhances cleavage of refractory or poorly cleaved sites (87,282). Activator DNA increased V_{max} for *Bsp*MI and *Nae*I, whereas for *Hpa*II, *Nar*I and *Sac*II, K_m decreased with no change in V_{max}. These findings suggest these enzymes have two DNA binding sites, an active site and an activator site, and that either spermidine or occupancy of the activator site can increase active site affinity or activity. For both *Eco*RII and *Nae*I, DNAs containing noncleavable methylated or inosine- or phosphorothioate-substituted recognition sites suffice to activate; thus DNA scission is not required for activated DNA cleavage (88,89). Electron microscopy of DNA-*Nae*I complexes reveals that *Nae*I loops DNA between *Nae*I sites (90), and one tantalizing possibility is that two different binding sites are responsible (283). Kinetic analysis supports the view that *Eco*RII has two binding sites (91). Moreover, *Eco*RII forms a complex with short activator DNAs that can be detected by band shift analysis (91). Remarkably, the *Eco*RII-DNA complex is stable in the presence of Mg^{2+} and cleavage does not occur, opening the door to physical analysis of an endonuclease-substrate-cofactor ternary complex if Mg^{2+} is indeed bound in the complex (91). Because enzymes involved in DNA rearrangements such as recombinases, transposases and invertases interact with and bring together two different DNA sequences or molecules, the subset of restriction enzymes that are activated by one DNA site to cleave another may be derived from or related to enzymes involved in transposition and recombination (Figurc 3).

Little is known about the active site structures or mechanism of catalysis for the majority of restriction endonucleases. The Mg^{2+} cofactor essential for DNA cleavage is thought to chelate the scissile phosphodiester bond and promote formation of the transition state trigonal bipyramidal configuration from the tetrahedral ground state. In the case of *Eco*RI, catalysis proceeds with inversion of stereochemical configuration about the cleaved bond, indicative of an *odd* number of reaction steps (92) and consistent with models in which a nucleophile on the protein abstracts a proton from water to generate a reactive hydroxyl ion that cleaves DNA. Because, unlike RNA, DNA is not inherently base labile, this model must further invoke additional residues that chelate cofactor and scissile bond to promote attack and escape of the leaving group.

X-ray Crystal Structures of Type II Restriction Endonucleases: Known and in Progress

The X-ray crystal structure of an *Eco*RI endonuclease-DNA substrate complex was our first detailed view of any DNA-protein co-crystal complex and the first known structure of any restriction endonuclease (32). This structure was solved in the absence of the essential Mg^{2+} cofactor to prevent DNA scission. Cleavage occurs when Mg^{2+} is diffused into the crystals and solution of the enzyme-product structure is in progress.

The *Eco*RI-DNA structure divulged that the DNA in the sequence specific complex is distorted from B-form DNA: the major groove widens to accommodate the protein's recognition machinery, the backbone kinks at the center of the *Eco*RI site, and the base pairs within and flanking the site have increased propeller twist. The original chain tracing of the endonuclease in the complex has been recently revised (30), revealing a symmetric homodimer with four α-helical bundles and two random chains that project into the major groove and interact with all of the substrate purines and pyrimidines (31).

The structures of free *Eco*RV endonuclease, a nonspecific *Eco*RV-DNA complex, and a specific *Eco*RV-DNA complex have all been recently solved (41) (Winkler, personal communication). Like the *Eco*RI-DNA complex, the DNA in the specific *Eco*RV-DNA complex is drastically distorted from B-DNA. In contrast, DNA within the nonspecific *Eco*RV-DNA complex is largely B-DNA. Thus both *Eco*RI and *Eco*RV either capture or drive conformational changes that distort the substrate, possibly as a prelude to scission.

The structure of the *Eco*RV-substrate complex and subsequent mutagenic studies (discussed above) have betrayed that the active sites of *Eco*RV and *Eco*RI may be similar. Although these two endonucleases share no overt homology in primary sequence, four residues are arrayed at similar locations about the scissile phosphodiester bond and lie at nearly identical positions within the two primary sequences. These residues are P90, D91, E111 and K113 in *Eco*RI and P73, D74, D90 and K92 in *Eco*RV.

Crystals of several other restriction endonucleases, either free or in complex with DNA, have been grown and solutions of their structures are in progress. These include *Hha*II (93), *Bam*HI (I. Schildkraut, personal communication), *Taq*I (A. Aggarwal, personal communication), *Pvu*II (J. Anderson, personal communication), and *Sma*I (A. Mikhailov, personal communication). High-resolution X-ray crystal structures are known for four nonspecific or DNA repair endonucleases: Staphylococcal nuclease (94), DNase I (95), T4 endonuclease V (96), and *E. coli* endonuclease III (97). While these structures have so far uncovered no similarity with restriction endonucleases, the notion that restriction enzymes arose from nonspecific or DNA repair endonucleases by dimerization and the addition of sequence specificity is sufficiently alluring to warrant detailed comparisons as additional structures emerge.

TYPE III RESTRICTION-MODIFICATION SYSTEMS

For most RM systems, both DNA strands of palindromic recognition sites are marked by methylation. Thus, following DNA replication, each new daughter molecule is hemi-methylated and this suffices to prevent endonuclease cleavage. In contrast, the unusual type III RM systems recognize non-palindromic DNA sequences and methylate only one DNA strand (5). In these cases one would have expected DNA replication to produce one hemi-methylated protected daughter molecule and one unmethylated sensitive daughter molecule. If this were the case, a type III RM system should destroy the host cell genome following replication. That this does not occur was a mystery only recently solved.

It has been known for some time that the related T phages, T3 and T7, have multiple recognition sites for the *Eco*P15 type III RM system but only T3 is sensitive to *Eco*P15 restriction. Meisel et al. (98) noticed that all *Eco*P15 sites in T7 phage face in one direction whereas in T3 *Eco*P15 sites point in both directions. By constructing M13 phage containing three *Eco*P15 sites in a variety of orientations, Meisel et al. (98) elegantly demonstrated that *Eco*P15 DNA cleavage and restriction, but not methylation, require two *Eco*P15 sites in opposite directions (Figure 3). Because DNA replication of *Eco*P15-modified DNA yields unmethylated *Eco*P15 sites that all face in the same direction, replication of the host chromosome does not produce *Eco*P15-sensitive DNA. Incoming foreign DNA containing two unmethylated sites in inverse orientation will, however, be restricted. Thus, like the type II *Nae*I RM systems, the type III *Eco*P15 RM system requires two sites for DNA scission, suggesting that RM systems may not simply serve to recognize and cleave lone sites but rather to bring together two different sites in the process of catalysis (Figure 3). This analysis again supports the view that RM systems share features with other DNA metabolizing enzymes such as recombinases.

TYPE I RESTRICTION-MODIFICATION SYSTEMS

Type I RM systems were among the first discovered (3,5). Subsequently, during the search for convenient molecular biological reagents, it was discovered that type I RM systems are biochemically complex. The active enzymes are comprised of multisubunit protein complexes encoded by three *hsd* genes: *R* (restriction), *M* (methyltransferase or modification), and *S* (specificity). Mutations in the *R* gene abolish restriction but permit methylation, mutations in the *M* gene abolish methylation and are inviable unless an *R* or *S* mutation is present, and mutations in *S* abolish both restriction and modification. The three encoded subunits associate into R_2M_2S complexes that have endonuclease, methyltransferase and ATPase activities or M_2S complexes with methyltransferase activity. Type I RM systems require Mg^{2+}, ATP, and SAM as cofactors. The recognition sites for type I RM systems are nonpalindromic, comprised of two domains of specific sequence separated by a spacer of defined length but nonspecific sequence. For example, the *Eco*K system recognizes 5′-AACN$_6$GTGC-3′. The preferred substrate for type I methyltransferases is hemi-methylated DNA, which is modified roughly 100-fold faster than unmethylated DNA. The endonuclease activity recognizes the unmethylated substrate and then translocates in an ATP-dependent fashion to cleavage sites, located up to several thousand base pairs away from the recognition site. The products are not cleaved at a defined position. Studier and Bandyopadhyay (99) proposed a model in which type I restriction enzymes cleave when two enzymes, translocating towards each other from different sites, collide (Figure 3).

Type I RM enzymes and their products have so far not proven useful as molecular biology reagents. However, studies on type I enzymes have provided insights into DNA-protein interactions. Type I RM systems fall into three families: *Eco*K (or IA-family), *Eco*A (or IB-family) and *Eco*R124/*Eco*DXXI (or IC-family) (7,9). The IA- and IC-families are encoded from the same chromosomal location and are thus, despite lack of homology, allelic. IC-family systems are plasmid encoded. Among but not between families, R, M and S subunits can be exchanged between members (for example *Eco*K and *Eco*B) and the corresponding genes cross-hybridize and share sequence homology. Comparison of the *S* genes for *Eco*K enzymes revealed two variable domains that correspond to two independent recognition elements for the bipartite recognition site (100). A similar analysis of two IA-family *S* genes with different substrate specificity again uncovered two non-conserved regions imbedded in otherwise conserved sequences, implicating these non-conserved regions in sequence-specific DNA-protein interactions (101). Some limited sequences are conserved between the *S* genes from the A, B, and C families (101–103).

Elegant pioneering work by Murray and collaborators (108,111) and more recently by Bickle and collaborators (103–106,109,110) has illuminated sequence-specific DNA recognition by type I enzymes.

Three members of the type IC family are known, *Eco*R124, *Eco*R124/3 and *Eco*DXXI. Unlike the chromosomal type IA and IB families, type IC RM systems are borne by large conjugative plasmids. *Eco*R124 is encoded by an IncFIV incompatibility group plasmid. Remarkably, the plasmid encoding *Eco*R124 can yield derivatives expressing the related *Eco*R124/3 specificity. Even more remarkable, these *Eco*R124/3 derivatives can switch to once again express *Eco*R124. Price et al. purified and determined the recognition specificities for *Eco*R124 and *Eco*R124/3 enzymes (103). Because type I endonucleases cleave nonspecifically, making it difficult to assign the specific recognition sites, Price et al. devised an ingenious strategy and instead determined the sites of methylation by incubating [methyl-^{3}H] S-adenosyl methionine, the purified enzyme, and DNA substrates of known sequence. Next, labelled restriction fragments were compared with the known sequence leading to the identification of 5-GAA(N_6)RTCG-3′ (R=A or G) and 5′-GAA(N_7)RTCG-3′ as the recognition sites for *Eco*R124 and *Eco*R124/3. Thus the two sections of the recognition sites, GAA and RTCG, are identical. The only difference is that *Eco*R124/3 recognizes a 7-nucleotide and *Eco*R124 a 6-nucleotide spacer. This analysis implied that the two protein-DNA recognition modules would be conserved between *Eco*R124 and *Eco*R124/3 and that some

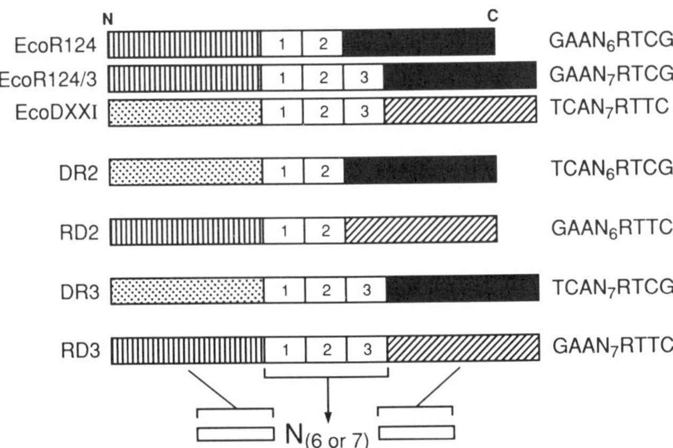

Figure 4. Hybrid type I *S* genes exhibit hybrid substrate specificity. From the type IC RM systems *Eco*R124 and *Eco*DXXI, Gubler et al. (105) constructed hybrid *S* genes containing the N- and C-terminal domains of different enzymes linked by either two or three central repeats. The resulting hybrid *S* genes yield predictable combinatorial substrate recognition specificities. R represents either adenine or guanine and N equals any DNA base. Adapted from Gubler et al. (105) with permission.

other feature would account for the different spacer length preference. By DNA sequence analysis, the two differ only in a repetitive element dividing the two putative DNA recognition elements: *Eco*R124 has two repeats (TAELTAEL) whereas *Eco*R124/3 has three (TAELTAELTAEL) (106) (Figure 4). This four amino acid insertion may force the DNA binding domains one nucleotide apart. Some sequence alterations in this repeat yield mutant enzymes with promiscuous methyltransferase activity that recognize both *Eco*R124 and *Eco*R124/3 recognition sites, possibly by increasing flexibility between the recognition domains (104).

Lastly, Gubler et al. (105) performed experiments demonstrating that hybrid *S* genes constructed from the type IC systems *Eco*R124 (GAA(N_6) RTCG) and *Eco*DXXI (TCA(N_7)RTTC) result in enzymes of hybrid DNA specificity. By electron microscopy of heteroduplex DNA, the *Eco*R124 and *Eco*DXXI *R* and *M* genes share extensive homology whereas the two *S* genes contain two bubbles of nonhomology. The length of the protein spacer between these two nonhomologous domains is associated with the length of the DNA spacer between the halves of type I RM recognition sites and both *Eco*R124/3 and *Eco*DXXI *S* genes contain three spacer repeats and both recognize a 7-bp spacer. Gubler et al. (105) constructed four hybrid *S* genes containing N- and C-terminal domains from different enzymes tethered by either two or three central repeats. Remarkably, the substrate specificity of these hybrid methyltransferases follows predictably from the parental specificities (Figure 4). Enzymes with two central repeats recognize 6-bp spacers, whereas enzymes with three repeats see 7-bp spacers. The N-terminal domain recognizes the same 5′ half-site as the parent (*Eco*R124 is GAA and *Eco*DXXI is TCA) and the C-terminal domain the 3′ half-site. Thus the four recognition domains of two enzymes and two different spacer lengths can be reassorted, either by *in vitro* (105) or *in vivo* recombination (presumed to occur in nature), to produce enzymes with 8 different specificities. Addition of several more domains or cassettes would rapidly generate significantly greater diversity.

The *Salmonella* type I RM systems *Sty*SPI and *Sty*SBI can similarly parent recombinants with a novel DNA recognition specificity called *Sty*SQI (107,108). As is the case with *Eco*R124/ *Eco* DXXI hybrid enzymes, the hybrid *Sty*SQI system recognizes a site comprised of this half-site from the *Sty*SPI and the 3′ half-site from the *Sty*SBI recognition site (109,110). This reciprocal recombinant enzyme *Sty*SJI recognizes the reciprocal hybrid site (111).

Because type I RM systems from three different families share homology in regions flanking the *HsdRMS* genes, undergo either natural (*Sty*SPI and *Sty*SBI → *Sty*SQI or *Eco*R124 → *Eco*R124/3 → *Eco*R124) or constructed (*Eco*R124 and *Eco*DXXI →hybrid specificity) recombination events that alter DNA sequence-specific recognition, and reside on both the chromosome and plasmids, the *hsd* genes may be capable of switching events, by either transposition or a cassette mechanism, that generate diversity in restriction specificities (8,9).

The genes encoding two additional *E. coli* restriction systems that target methylated DNA, Mrr and McrBC, lie adjacent to the *hsd* encoded type I RM systems in a cluster of six genes involved in restriction-modification. Because in all known cases type I RM systems methylate adenine in host DNA and at least Mrr recognizes adenine methylated DNA, these genes may have become linked to avoid situations in which independently segregating *mrr* and *hsd* genes would result in Mrr restriction of the host genome at sites modified by a type I RM system. If additional Mrr and McrBC homologs with different sequence specificities exist, this gene cluster could give rise to combinatorial rearrangements that uniquely define a multitude of different restriction enzyme patterns. Additional gene rearrangements to produce hybrid *S* genes or mutations that alter specificity would further diversify the population.

With these considerations, RM systems share several features with immune systems from multicellular eukaryotes in that both (i) protect the organism from intruders, (ii) exhibit target specificity, either based on protein antigens or DNA sequences, (iii) generate uniquely derived specificities by clonal events that are determined by genetic rearrangements driven by recombination between repeated gene elements, and (iv) distinguish self from non-self (tolerance in the immune system and avoidance of self-restriction in bacteria by modification and nonrandom gene assortment). One poignant distinction is that immunoglobulin and T-cell receptor gene rearrangements in lymphocytes occur *via* site-specific recombination whereas recombinational events that underlie diversification of restriction-modification specificities are thought to arise *via* homologous recombination. Perhaps site-specific recombinases also drive restriction system diversification.

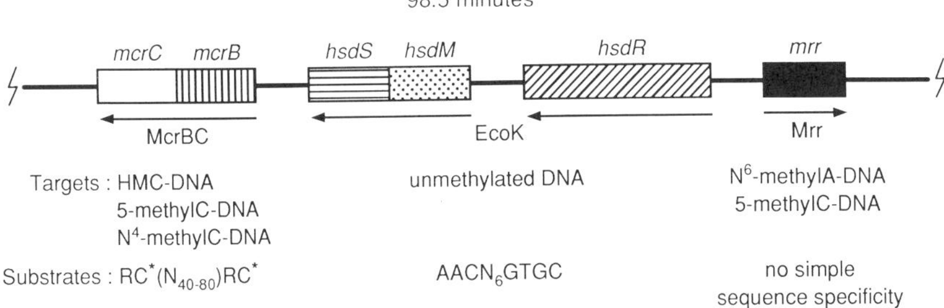

Figure 5. The Ellis Island of *E. coli*. The genes encoding three different restriction systems lie clustered in this region of the chromosome, possibly to permit transfer of the genes as a set. Promoters have been identified upstream of both *hsdR* and *hsdM* and a transcriptional terminator lies between *hsdR* and *M* (280). Promoters for *mcrBC* and *mrr* have not yet been identified *in vivo* . Adapted from Raleigh (8) with permission.

METHYLATION DEPENDENT RESTRICTION SYSTEMS (MDRS)

Three restriction systems that recognize methylated DNA (MDRS) have been discovered in *E. coli* : Mrr, Mcr A (RglA), and McrBC (RglB). The genes encoding McrBC and Mrr lie in a gene cluster (*mcrCB hsdSMR mrr*), called the Ellis Island of *E. coli* or the immigration control region (ICR) (8), which lies at 98.5 minutes on the *E. coli* chromosome and also includes the *hsdRMS* genes for the *E. coli* type I RM system (8,12,13,16,112–114) (Figure 5).

Mrr- and Mcr-like activities have been observed in a variety of other bacteria, suggesting that MDRS may be widespread in nature. MDRS may have evolved before the methylation-inhibited type I, II, and III restriction systems, especially if methylation arose early in evolution to serve functions other than restriction-modification. Alternatively, methylation-inhibited systems may have arrived first, and MDRS could then have evolved as a second round of defense targeted against DNA with foreign methylation patterns.

Mrr

Mrr was originally identified as a restriction system specific for DNA containing N^6-methyladenine (16). Subsequently, three groups showed that Mrr can also restrict C^5-methylcytosine in certain DNA sequences (115–117). The *mrr* gene has been cloned and overexpressed (116), and the purified Mrr protein found to have endonuclease activity against C^5-methylcytosine DNA from the *Xanthomonas* phage XP-12 (116). An extensive analysis of Mrr action *in vivo* (116) reveals that, in addition to *Hha* II ($G^{me}ANTC$) and *Pst*I ($CTGC^{me}AG$) methylated DNA, the system recognizes DNA modified by the adenine methyltransferases *Acc*I (GTMKAC) (M = A or C; K = T or G),

*Cvi*RI (TGC^{me}A), *Hinc*II (GTYR^{me}AC) (Y = T or C; R = A or G), *Hpa*I (GTTAAC), *Nla*II (C^{me}ATG), *Taq*I (TCG^{me}A), and the cytosine methyltransferases *Sss*I (C^{me}G) and *Hha*I (G^{me}CGC). This analysis reveals no simple Mrr consensus sequence. Further studies should be directed at understanding the remarkable ability of Mrr to recognize both adenine and cytosine methylated DNA.

McrBC

The McrBC restriction system was originally identified as an activity, named RglB, that restricts non-glucosylated T-even phages in which 5-hydroxymethylcytosine (5-HMC) replaces cytosine in DNA. This unusual substrate only occurs in T-phage mutants lacking glycosyltransferase activity and is presumed not to occur outside the laboratory. The finding that McrBC recognizes 5-methylcytosine and N^4-methylcytosine, both naturally-occurring DNA modifications, suggests these are the relevant substrates for McrBC in nature. An excellent review on the McrBC restriction system has recently appeared (8).

The McrBC restriction system is encoded by two adjacent genes that lie next to the *hsdRMS* gene cluster (Figure 5) (8,12,112–114,118–121). McrBC restricts DNA containing 5-methylcytosine in the sequence G^{me}C or A^{me}C, N^4-methylcytosine, or 5-hydroxymethylcytosine. *mcrB* encodes at least two polypeptides of 55 and 35 kD and possibly a third of 34 kD. The smaller protein(s) arises from an independent translation start site. *mcrC* encodes two polypeptides of 37 and 16 kD. *In vivo*, restriction of most but not all substrates requires both McrB and McrC. However. DNA methylated by *Msp*I (^{me}CCGG), *Sss*I (^{me}CG), or *in vivo* by the endogenous mouse methyltransferase (^{me}CG), are restricted by McrB alone. Mutations have been described which separate the Mcr and Rgl functions of McrBC. Piekarowicz et al. (122) found an *mcrB* mutation that confers a temperature-sensitive *mcrBC*$^-$ phenotype for *Hae*III methylated DNA but is *rglB*$^-$ (unable to restrict 5-HMC DNA *in vivo*) at all temperatures. This mutation may alter the region of McrB that interacts with methylated cytosine, or with the T4 antirestriction protein.

In early attempts to purify RglB, the enzyme rapidly lost activity during chromatography due to the loss of a heat-labile, nondialyzable factor (123). McrBC cleavage activity has recently been reconstituted *in vitro* by Sutherland et al. (124). These efforts required the use of a *recD*$^-$ host strain missing exonuclease V activity. Sutherland et al. (124) found that McrBC cleavage activity was rapidly lost during any chromatographic step. A major advance was the finding that activity could be restored by adding a heat-stable dialyzable crude extract component from an *mcrB*$^-$*C*$^-$ strain. Addition of a variety of small molecules revealed that GTP supported the reaction. Several other GTP analogs (rGTP, dGTP, and ddGTP) also restored McrBC activity whereas rGDP, rGMP, or the nonhydrolyzable analogs GTPγS, GMP-PNP, or GMP-PCP did not. Interestingly, by DNA sequence analysis, both the 55 and the 35 kD McrB proteins contain a putative GTP binding site.

With the discovery of this GTP requirement (124), Sutherland et al. established an *in vitro* complementation assay by mixing *mcrB*$^+$*C*$^-$ and *mcrB*$^-$

C^+ extracts and then purified McrBL (the 55 kD McrB protein) and McrC by supplementation. Notably, McrB and McrC do not copurify. With the purified McrBC enzyme, Sutherland et al. proceeded to define the substrate specificity, first with a pBR322 derived fragment and subsequently with synthetic substrates. Two methylated sites, spaced 40 to 80 bp apart (an 11, 19 or 21 bp gap was not recognized), are required for cleavage. A *single* methyl group at each site suffices, and these methyl groups can be either on the same (*cis*) or opposite strands (*trans*). This is in keeping with the finding that hemimethylated DNA delivered by F-factor mating is McrBC restricted *in vivo* (8). Scission then occurs at a number of sites between the recognition sites. The McrBC substrate can be defined as $RC(N_{40-80})RC$, and the reaction requires $McrB_L$, McrC, Mg^{2+} and GTP. McrBC is the first GTP-dependent restriction enzyme and may share features with other GTP-binding proteins, such as translation elongation factors or ras.

McrA

Like McrBC/RglB, the McrA restriction system was originally identified as the RglA restriction system, which restricts non-glucosylated T-even phages which sport DNA containing 5-hydroxymethylcytosine (121). It was later discovered that McrA also restricts DNA containing 5-methylcytosine, and the name McrA was proposed which, like McrBC, stands for modified cytosine restriction (12,112). The *mcrA* gene is located at the right end on the e14-excisable prophage-like element. e14 can be cured by DNA damage, and thus many *E. coli* laboratory strains are spontaneously $e14^-$ and hence *mcrA*$^-$. So far, McrA is known to restrict DNA modified by *Hpa*I at $C^{me}CGG$ or *Sss*I at ^{me}CG. In some strain backgrounds, McrA activity is temperature sensitive (113). Thus, when the McrA status of an unknown strain is being assayed, it is advisable to test for restriction at both 30° C and 37° C.

MDRS in other Species

The first known methylation-dependent restriction system (MDRS) was a type II RM system expressed by *Diplococcus pneumoniae, Dpn*I, which cleaves the sequence GATC only when N^6-methylated at the central adenines (11). So far, *Dpn*I and isoschizomers are the only known methylation- dependent type II RM systems. MDRS identified in other bacteria are more akin to McrA, McrBC and Mrr of *E. coli.* For example, Sladek et al. (125) identified a restriction system in the mycoplasma *Acholeplasma laidlawii* (strain JA1) which restricts DNA containing 5-methylcytosine produced by *Alu*I($AG^{me}CT$), *Hpa*II ($C^{me}CGG$), *Msp*I ($^{me}CCGG$), and *Hha*I ($G^{me}CGC$), suggesting that at least $G^{me}C$ and $C^{me}C$ are substrates for this MDRS. Similarly, a variety of *Streptomyces* restricts both adenine and cytosine methylated DNA (126). Targets include *dam* ($G^{me}ATC$) or *Taq*I ($TGC^{me}A$) adenine methylated DNA, and *dcm* ($GGA/T^{me}CC$), *Alu*I ($AG^{me}CT$), *Hha*I ($G^{me}CGC$), or *Hph*I ($T^{me}CACC$) cytosine methylated DNA. Of nine different strains only one, *Streptomyces avermitilis,* restricted both adenine and cytosine methylated

DNA, three restricted only adenine methylated, three restricted only cytosine methylated DNA, and one restricted neither.

Several other bacteria restrict *dam* or *dcm* methylated DNA, which are not restricted in *E. coli.* For example, *Bacillus thuringiensis* restricts DNA modified by Dam or Dcm *E. coli* methyltransferases (127). Restriction of *dcm* methylated DNA may result from an Mcr restriction system with altered sequence specificity. Restriction of *dam* methylated DNA could be similar to the type II MDRS *Dpn*I or to the phenomenon described by Russell and Zinder (128) whereby *dam* methylation inhibits transformation because hemi-methylated DNA, produced by one round of replication, blocks further replication. A similar *dam*-dependent phenomenon in *Halofax volcanii* was attributed to a restriction system specific for DNA modified by the Dam (G^{me}ATC) adenine methyltransferase of *E. coli* (129). As in *Bacillus thuringiensis,* restriction in *H. volcanii* could involve a *Dpn*I isoschizomer or inhibition of DNA replication by hemi-methylated *dam* sites in plasmid replication origins (128).

MDRS are probably underrepresented amongst known RM systems because: (i) MDRS were only discovered recently, (ii) many screens for endonuclease activity contained only *dam, dcm,* and *Eco*K methylated DNAs, which might not be relevant substrates, (iii) other MDRS as enzymatically complex as McrBC might have escaped detection by *in vitro* cleavage assays, and (iv) nucleotide-dependent systems would not be detected because nucleotides (ATP.GTP) are not routinely included.

TWO SITES FOR CLEAVAGE

From studies on type I, type II, type III, and MDRS RM systems, a common emerging theme is that two recognition sites are in several examples required for cleavage (Figure 3).

In the collision-cleavage model for type I systems, two independent substrate-enzyme complexes translocate from sites of recognition to cleave at random sites located at points where two enzymes collide. This model suggests interactions between not only two different sites but also two distinct enzyme complexes (99). Because site-specific recombinases also seek out two different DNA sequences, on the same or different DNA molecules, type I enzymes may share similar properties involving translocation and enzyme-enzyme interactions.

Several type II endonucleases, most notably *Nae*I and *Eco*RII, poorly cleave unless additional recognition sites are available as activator DNA (85,87) (Figure 3). Activator sites can be present *in cis* or *trans* and activation does not require cleavage of activator DNA (88,89). This suggests that some type II enzymes have two DNA binding domains that communicate, possibly by allosteric activation, to trigger scission. Whether these represent two identical binding sites or distinct active and activator sites is not yet known. With at least *Nae*I, looping occurs between *Nae*I sites (90) but it is not known if these loops are tethered *via* single endonuclease dimers or dimer–dimer interactions forming tetramers. These findings again support the view that at least some restriction endonucleases require two recognition sites for action, underscoring that their mechanism of action is likely more complex than simple recognition and catalysis.

The P15 type III RM system requires two recognition sites for cleavage but not for methylation (Figure 3). In this case, the nonpalindromic sites must be in inverse orientation (either head-to-head or tail-to-tail) and this serves to protect newly-replicated DNA from cleavage (98). Possibly two inverted sites allow enzyme–enzyme interactions required for scission but dispensable for methylation.

Lastly, the McrBC MDRS recognizes two methylated cytosines, each preceded by a purine, that are separated by 40 to 80 nucleotides (124) (Figure 3). Shorter spacers are not recognized, suggesting that either translocation or higher-order DNA structures, such as loops, may be required for scission.

Many restriction endonucleases, especially type II enzymes, effectively cleave at a single recognition site. Moreover, simple recognition and cleavage within a specific DNA sequence should be sufficient for restriction-modification system function. This raises the question as to why some enzymes appear to be more complex. For type III enzymes, a likely explanation is that this is their solution to the dilemma of nonpalindromic sites at which a single methyltransferase would be expected only to hemi-methylate and thus not protect newly replicated DNA. It seems unlikely that these endonucleases would have been paired with two different methylases, so perhaps only those endonucleases that could acquire another level of regulation involving protein-protein interactions at inverted sites survived.

Interactions between two different sites are common in other instances. For example, in DNA mismatch repair in *E. coli* (130), the MutS and MutH proteins collaborate to recognize both mispaired DNA bases and the methylation state at a *dam* site to check which is the template strand. The MutH endonuclease then nicks the unmethylated strand at a hemi-methylated GATC *dam* site and an excision tract removes the mismatched base and flanking nucleotides followed by resynthesis. Thus DNA repair, recombination, or transposase enzymes that normally conduct DNA transactions requiring two DNA sites may be relevant in considering the origins and function of restriction endonucleases requiring two sites for cleavage.

METHYLTRANSFERASES

We have concentrated our attentions in this chapter on endonucleases, but no discussion of RM would be complete without considering their partners, the DNA methyltransferases (131–133,284). This family of interesting enzymes was initially somewhat neglected, possibly because DNA cleavage is simpler to study compared to DNA methylation. However, an explosion of information promises to further our understanding of the mechanisms of sequence-specific DNA recognition and catalysis by DNA methyltransferases. DNA methyltransferases can be distinguished by the modification introduced: N^6-methyladenine, 5-methylcytosine, or N^4-methylcytosine. In general prokaryotic and some simple eukaryotic organisms contain both adenine and cytosine methylated DNA;

higher multicellular eukaryotes such as mice and man are only known to contain cytosine methylation, and some organisms, such as *S. cerevisiae* and *D. melanogaster,* lack detectable DNA methylation. *E. coli* mutants missing the three known chromosomally encoded DNA methyltransferases (Hsd,Dcm,Dam) are viable and lack detectable methylation (132,133). Thus DNA methylation is not essential for viability. In the case of cells expressing a sequence-specific endonuclease, DNA methylation is required to protect the host chromosome from destruction and, under these conditions, a methyltransferase is essential for viability.

Sequence-specific methyltransferases fall into several distinct families. Interestingly, the adenine and N^4-cytosine modifying enzymes fall into one loosely related family, possibly because both types methylate amino groups (7). The most highly conserved region contains the tetrapeptide DPPY (or F); a second less well conserved region is also present (134). The DPPY tetrapeptide has been implicated in sequence-specific DNA recognition by studies on a P22 phage Mnt repressor mutant in which a His to Pro substitution (DPHF in wild-type to DPPF in mutant) confers the ability to recognize an AT instead of GC base pair. Notably, this AT base pair occurs in a GATC *dam* site, and *dam* adenine methylation is required for binding by the mutant repressor (135). 5-methylcytosine methyltransferases comprise two families: one composed of members that modify multiple different recognition sites (136), and a second group that modifies unique recognition sites (137). In general, the 5-methylcytosine methyltransferases share greater identity than the adenine and N^4-cytosine methyltransferase family.

Sequence Specific DNA Recognition

DNA methyltransferases are typically monomers and must therefore recognize substrate in a different way than the partner endonucleases, which are usually dimers. Methyltransferases are probably monomers because (i) the typical substrate is DNA that is hemi-methylated (and therefore asymmetric) following DNA replication and (ii) only a single methyl transfer is required to return to the fully methylated state.

The multispecific cytosine methyltransferases of *B. subtilis* phage SPR and ø3T are unusual and modify more than one specific DNA recognition sequence. Analysis of mutant enzymes that have lost ability to methylate one substrate but still modify others disclosed that these enzymes contain a variable region constituted of multiple ~ 50 amino acid segments, termed target recognition domains (TRD), each conferring specificity for a different substrate (138,285). In this view, the more typical monospecific DNA methyltransferases may contain only one TRD in the variable region.

In an analysis of enzymes recognizing the tetranucleotide CCGG, Walter et al. (139) found that *Bsu*FI and *Msp*I methyltransferases share 56% identity, including the TRD, and both enzymes modify the outer cytosines (^{me}CCGG). *Bsu*FI shared less identity (46%) with *Hpa*II, which modifies the inner cytosines, and in this case the TRDs differ. Lastly, the TRD of the multi-specific SPR enzyme was sufficiently different to imply that mono- and multi-specific enzymes may require or have acquired different TRDs that recognize the same DNA sequence. Precisely how target recognition domains, substrate, and enzyme active site cooperate awaits solution of a DNA methyltransferase X-ray crystal structure.

5-methylcytosine DNA methyltransferases share ten conserved regions with a variable domain of 80 to 120 amino acids between conserved regions VIII and IX. Klimasauskas et al. (140) constructed hybrid enzymes containing portions of the *Hha*I ($G^{me}CGC$) and *Hpa*II ($C^{me}CGG$) methyltransferases. Several hybrid enzymes were active and in each case retained the substrate specificity of the parent contributing the variable domain, clearly demonstrating that the variable domain conducts sequence-specific DNA recognition in the monospecific methyltransferases.

Like *E. coli*, the T-even phages T2 and T4 express a Dam methyltransferase that methylates adenines at GATC sites (131). A set of remarkable T-phage mutants have been isolated that escape restriction by the P1 RM system. Such mutant phage express Dam mutants, known as Dam^h, which exhibit relaxed substrate specificity and modify both GATC and GACC sites to confer protection against P1 restriction at AGACC sites (131). The gene encoding a T2 dam^h mutation has been cloned and sequenced, revealing a Pro126 to Ser substitution (141). Interestingly, a nearby mutation altering Phe127 to Val yields a mutant enzyme that recognizes DNA containing cytosine but not 5-hydroxymethylcytosine (141). Similar screens and selections could be devised to isolate altered or reduced specificity mutants of other methyltransferases.

Sequence-specific binding by DNA methyltransferases is also amenable to non-denaturing DNA agarose gel shift analysis. For example, *Eco*RI methyltransferase forms a sequence-specific ternary complex in the presence of substrate and sinefungin (142). Dubey and Roberts (143) found that, in the absence of cofactor, *Msp*I methyltransferase bound substrate-containing DNA with only five-fold greater affinity compared to DNA lacking substrate. In contrast, cofactor (SAM) or cofactor analogs (SAH, sinefungin) allowed the enzyme to bind substrate containing DNAs 100-fold more tightly than non-substrate DNAs. Thus *Msp*I methyltransferase may be similar to the type II *Eco*RV endonuclease family which also requires cofactor for high affinity substrate binding.

Mechanics of Catalysis

Based on a mechanistic catalytic model for thymidylate synthase, Santi et al. (144) proposed a prescient analogous model for 5-methylcytosine DNA methyltransferases. Santi envisioned methyltransferases covalently attacking the C6 position of the pyrimidine ring to generate a net negative charge at and activate the C5 carbon to react with an electrophile, the methyl group of SAM. Following electrophilic methyl addition, the C5 proton is abstracted and β-elimination ejects the enzyme, restores the C5–C6 double bond, and leaves C5 now decorated with a methyl group. This model nicely accounted for the toxicity and inhibition of methylation and methyltransferase function by the cytosine analogs 5-aza-C and 5-fluoro-C, which were proposed to act as transition state analogs that prevent release of the DNA-methyltransferase covalent complex (144).

This model has now been largely confirmed by genetic and biochemical means. The methylation inhibitor 5-aza-C is indeed only toxic to cells expressing a 5-methylcytosine transferase (145) and therefore subject to formation of lethal DNA-protein adducts. The *Hha*I methyltransferase ($G^{me}CGC$) is inactivated by the cysteine specific reagent *N*-ethylmaleimide, and DNA protects the enzyme from NEM inactivation (146). Based on these findings, Wu and Santi proposed that a cysteine is the active site nucleophile and, in fact, a Pro-Cys dipeptide is conserved in all 5-methylcytosine transferases. The same conserved cysteine in *Eco*RII methyltransferase ($C^{me}C^{A}$/TGG), Cys186, can be UV photo-crosslinked to SAM (147). Moreover, the enzyme forms adducts with substrates containing 5-fluoro-C, from which peptide-DNA complexes have been purified and sequenced. His177 and Cys186 were missing from the sequenced peptide, implicating Cys186 as the site of covalent attachment to DNA (148). Similar covalent adducts between 5-fluoro-C DNA and enzyme have been observed with human CG methyltransferase (149). In the case of the *Hae*III methyltransferase ($GG^{me}CC$), the site of covalent link is again the conserved cysteine, in this case Cys71 (150). Lastly, Wyszynski et al. (151) have performed a genetic analysis of *Eco*RII methyltransferase mutants substituted at Cys186. Substitution by glycine, valine, tryptophan and serine destroyed methyltransferase activity. Interestingly, the Cys186Ser mutant enzyme renders *E. coli* sensitive to 5-aza-C, suggesting that this mutant enzyme may still form toxic covalent DNA-protein adducts (see also (145)). In addition, the Cys186Gly mutant enzyme is toxic to *E. coli* (151). One possibility is the Cys186Gly mutant enzyme interacts with normal cytosine-containing DNA to form adducts that are toxic to the cell. These genetic findings clearly imply that Cys186 is critical for *Eco*RII enzyme action. However, if Cys186 is the active site nucleophile as proposed, based on homology and biochemical studies, the toxicities of the Cys186Ser and Cys186Gly mutants are difficult to reconcile with nucleophilic attack because one would not expect the serine or glycine substitutions to be effective nucleophiles. Perhaps Cys186 is not the only residue involved in complex formation or the toxicity of the serine and glycine substituted mutants results from tight binding rather than covalent addition. Further studies on these extremely interesting *Eco*RII mutant methyltransferases are clearly needed.

In comparison, much less is known about the adenine and C^4-methyltransferases. Following photo-crosslinking of the *Eco*RI N^6-adenine methyltransferase to 8-azido-SAM, two labelled subtilisin cleavage products were identified and sequenced (152). Surprisingly, the resulting peptides, residues 206–212 and 213–221, do not correspond to either conserved region of adenine methyltransferases, including the DPPY (F) region. Previous genetic analysis of an Mnt mutant repressor (135) strongly implicates a DPPF peptide in sequence-specific recognition of an N^6-methyladenine containing GATC site. DPPY tetrapeptides in methyltransferases could possibly be required for substrate recognition and not SAM binding, but how a conserved region would allow interaction with many different DNA sequences is unclear. Alternatively, the DPPY region could be involved in SAM binding but fail to crosslink to *Eco*RI methyltransferase.

Beyond Bacteria

Multicellular eukaryotic DNA is methylated by a 5-methylcytosine methyltransferase specific for CG dinucleotides, whose role may be to regulate gene expression, X-chromosome lyonization, and imprinting. The eukaryotic CG methyltransferase has been cloned and sequenced from mice (153) and humans (154), and further studies on this interesting enzyme are in progress. Unlike type II RM methyltransferases, the mammalian CpG methyltransferase is a memory methyltransferase that preferentially modifies hemi-methylated sites. Type I RM methyltransferases also prefer hemi-methylated substrates and are thus prokaryotic memory methyltransferases. Kelleher et al. (155) have isolated mutations that destroy the ability of the *Eco*K-type I RM system to distinguish between unmethylated and hemi-methylated sites and may shed light on this interesting enzymatic property.

The yeast *S. cerevisiae* lacks detectable DNA methylation (156). In pioneering studies, Fehér et al. (157) and Brooks et al. (158) expressed the *Bsp*RI cytosine or Dam adenine methyltransferases in *S. cerevisiae* and observed incomplete methylation of genomic recognition sites. Partial methylation was attributed to inaccessibility of some regions of chromatin. Subsequently, Hoekstra and Malone (159,160) re-examined Dam methyltransferase in *S. cerevisiae* , and (i) reproduced the observation that only some GATC sites are modified and even then only incompletely, (ii) noted a weak increase in the frequency of mitotic but not meiotic recombination, and (iii) made the intriguing observation that Dam more fully modifies GATC sites when expressed in yeast *rad1* or *rad3* mutants which are defective in the UV excision DNA repair pathway (160). Similar findings were observed with either Dam methyltransferase or the *Bacillus subtilis* SPR phage multispecific cytosine methyltransferase expressed in *RAD* versus *rad2* or *rad3* mutant strains (161). Thus N^6-methyladenine and 5-methylcytosine bases may be recognized as DNA lesions in *S. cerevisiae* and repaired by DNA excision repair. If so, what recognizes the modified base to generate a nick for the excision repair tract? Candidates include the yeast equivalent of the UvrABC excinuclease that nicks adjacent to pyrimidine-pyrimidine dimers or yeast homologs of the bacterial Mrr and Mcr enzymes. *RAD1, RAD2* and *RAD3* are required for removal of methylated bases and all three have been cloned (162). *RAD3* encodes a DNA helicase whereas the functions of the *RAD1* and *RAD2* products are not yet known. An alternative possibility is that methylated bases are not recognized and actively excised but rather gratuitously removed from the DNA by excision repair tracts initiated at spontaneous DNA lesions.

Although the extent of methylation increases in excision repair mutant yeast strains expressing foreign DNA methyltransferases, other factors contribute to the accessibility of yeast chromatin to methylation. *S. cerevisiae* chromosomes end in telomeres, specialized nucleoprotein complexes that protect and replicate the linear DNA ends. A *dam* site present in telomeric DNA repeats is not accessible to *E. coli* Dam methyltransferase (163). Similarly, genes placed near telomeres are poorly expressed in *S. cerevisiae*, and this position effect has been attributed to telomere DNA-protein complexes that block transcription (164). A similar phenomenon, known as silencing, occurs at the silent mating type cassettes HML and HMR and prevents expression of the MAT cassettes at these loci. *sir* mutations release silencing, permit expression of telomerase adjacent genes, and allow Dam methyltransferase access to *dam* sites at telomeres (165), presumably by altering chromatin structure. Singh and Klar (166) expressed Dam and SPR methyltransferases in *rad3-2* mutant yeast strains and found that DNA methylation significantly increases in regions of active transcription, consistent with the view that transcription relaxes chromatin structure.

The *E. coli* Dam methyltransferase has also been expressed in *Bacillus subtilis* (167). As in *S. cerevisiae, dam* methylation in *B. subtilis* is incomplete unless DNA repair pathways are disabled by mutation. In this case, increased methylation occurs in strains defective in either the UV excision repair pathway (*uvrB*) or in the adaptive response pathway (*ada1*) that is induced by and repairs alkyl-DNA damage (167). The resulting increase in *dam* methylation in *B. subtilis* DNA repair mutants impairs viability and induces some features of the SOS response.

We can now consider the effects of heterologous DNA methylation in three different organisms. In *E. coli*, methyltransferases subject to Mrr, McrA, or McrBC restriction severely impair growth and induce the SOS response in wild-type strains but have little or no consequences in restriction-defective mutants. Thus the methylated bases lead to the production of DNA damage (DNA DSBs) but are not themselves lesions. In *S. cerevisiae*, methylation is removed by DNA excision repair but the physiological effects of methylation in either wild-type or mutant strains are minor, such as a two- to four-fold increase in mitotic recombination. Again, the methylated bases do not seem to be severe DNA lesions. Exogenous methylation has recently proven to be an insightful probe of chromatin structure in *S. cerevisiae* (163,165,166), and more studies along these lines are clearly warranted. Lastly, in *B. subtilis* methylation is removed by either excision or alkylation DNA repair pathways. In the absence of repair, methylation is detrimental to *B. subtilis,* suggesting that N^6-methyladenine in GATC sites is a DNA lesion in *B. subtilis,* possibly by interfering with an essential DNA-protein interaction(s) unique to *B. subtilis.*

ANTI-RESTRICTION SYSTEMS

Bacterial Anti-Restriction Functions

Host bacteria typically protect their genome from restriction enzymes by coexpressing DNA methyltransferases that modify the recognition sites (Figure 1). In some situations, this may not be sufficient. For example, repair of damaged DNA may produce unmethylated DNA sensitive to cleavage. In the bacterium *E. coli*, DNA damage evokes the SOS response, a coordinated induction of about 20 genes whose products are involved in DNA repair and alter the physiology of the cell to promote repair (168,169). One of the consequences of SOS induction is a phenomenon, known as restriction alleviation (170), during which restriction of foreign DNA by the host *Eco*K-type I RM system is significantly diminished. For example a 60 J/m^2 UV dose reduces restriction of λ phage roughly 5,000-fold (1 out of 2 phage survive restriction versus 1 out of 10,000 without UV irradiation). Other SOS-inducing conditions (*dam*$^-$ mutations, nucleotide analogs) also alleviate restriction (171,172). SOS induction alleviates restriction by type I RM systems (such as *Eco*K), but not by the type II *Eco*RI system or the type III *Eco*P1 RM system. This suggests that an SOS-induced protein(s) acts specifically on the *Eco*K-type I RM system (or a unique property of its substrate or product such as DNA sequence) rather than by a more general mechanism, such as repair of DNA double-strand breaks (DSB).

Hiom and Sedgwick (173) have recently made the provocative finding that the UmuDC proteins are required not only for SOS-induced DNA repair but also for restriction alleviation (173). Both the *umuC122::Tn5* and the *umuD44* mutations reduced restriction alleviation roughly 100-fold such that one in five phage survive *Eco*K restriction. Plasmids expressing the UmuDC proteins failed to complement the *umuC122* mutation to restore restriction alleviation but did complement the induced mutagenesis defect of the *umuC* mutation. In contrast, overexpression of the cloned *umuD* gene partially complemented *umuC122* and restored restriction alleviation, but did not complement for induced mutagenesis since UmuC was still missing and both UmuD and C are required for induced mutagenesis. RecA cleavage of UmuD is essential for restriction alleviation, as noncleavable UmuD mutants failed to complement. Provision of UmuD in its cleaved form was not sufficient to restore restriction alleviation in *umuC122* strains, and UV irradiation was still required. Interestingly, plasmids expressing UmuD and C-terminally truncated UmuC mutant proteins spontaneously induced 100-fold alleviation of λ restriction by *Eco*K. In summary, these observations suggest that the UmuDC proteins, previously identified as essential for induced mutagenesis and substrates (UmuD) for RecA cleavage, play a central and complex part in the SOS-induced phenomenon of restriction alleviation.

While the precise role of UmuDC is not yet known, Hiom and Sedgwick (173) rule out several possible models for restriction alleviation showing, for example, that *Eco*K modification is not altered or increased during restriction alleviation. They go on to suggest that the induced mutagenic pathway, or partial reactions thereof, may interfere with restriction by type I RM systems. Their findings suggest UmuD plays a central and UmuC an ancillary role in this process. The type I RM systems, which extensively translocate DNA between recognition and cleavage sites, may be uniquely sensitive to inhibition by other DNA metabolic processes. The finding that UmuDC participates in restriction alleviation suggests a link between SOS-induced mutagenic repair and restriction alleviation and further study may uncover functions of type I RM systems beyond defense against foreign DNA.

Plasmid-borne Anti-Restriction Functions

Like host bacteria, foreign DNA molecules (phage, plasmids) have found it advantageous to circumvent restriction upon entering a new host cell. Several self-transmissible plasmids are known to express anti-restriction systems. In general, these plasmid-borne anti-restriction systems do not induce SOS-sponsored restriction alleviation in the recipient cell (174,175). Plasmid pKM101, a member of the IncN incompatibility group, expresses ArdA (for alleviation of restriction of DNA) which inhibits restriction by type I RM systems but not by type III RM systems or the MDRS McrA or McrBC (174). Plasmid ColIb-P9 of the IncI1 incompatibility group expresses a related protein, Ard, with 60% identity to ArdA (174,175). The *ard* gene is transferred early and overcomes the restriction barrier during conjugation. Like ArdA, Ard blocks type I RM systems, including *Eco*K. When present in multiple copies, Ard also has an effect on *Eco*K methylation and may therefore target the *Eco*K RM enzyme complex (175). Curiously, ArdA has been reported to confer a 100-fold alleviation of restriction by the type II *Eco*RI enzyme (174). The IncI1 plasmid ColIb-P9 also alleviates *Eco*RI restriction in transconjugation but not in transformation. In this case, the effect is independent of Ard and has been attributed to prolonged contact between donor and recipient cell and multiple conjugative attempts (175). The finding that, despite 60% identity, ArdA inhibits *Eco*RI but Ard does not, suggest a unique feature of ArdA may target some structure or function common to *Eco*K and *Eco*RI or their products.

Phage-borne Anti-Restriction Functions

Like conjugative plasmids, phage have mechanisms to evade RM systems during invasion of a new host cell. For example, some phage DNA is decorated with unusual chemical modifications that, like methylation, serve to protect DNA from restriction. In contrast to sequence-specific methyltransferases associated with RM systems, phage DNA modifications have limited or no sequence specificity and therefore protect phage DNA from a variety of RM systems of different specificities. For example, Mu phage modifies virtually every adenine to acetamido adenine (176). T-even phages (T2,T4,T6) incorporate 5-hydroxymethylcytosine (HMC) in place of cytosine. HMC-DNA is resistant to most RM systems but, unlike cytosine-containing DNA, is sensitive to MDRS such as McrBC and McrA (121). To avoid MDRS. T-even phages modify HMC-DNA by glycosylation at the hydroxyl groups, and this bulky modification prevents cleavage by McrBC or McrA as well as by most, if not all, other RM systems. A variety of other unusual phage DNA modifications is known and presumed to block RM system actions (177).

In addition to unusual DNA modifications, phage evade RM systems by expressing anti-restriction functions. The best characterized of these anti-RM systems are the T3 SAMase, which destroys the required S-adenosyl-methionine cofactor (178), and the T7 gene 0.3 protein, which is not a SAMase (179–181). Gene 0.3 protein has been purified to homogeneity (179). The purified homodimeric protein binds stoichiometrically to the partially purified *Eco*B-type I restriction enzyme and inhibits both nuclease and ATPase activity. Purified gene 0.3 protein did not inhibit several type II endonucleases. T7 phage also appears to escape restriction in part by entering a unique compartment of the cell where phage DNA can be transcribed but not restricted (181). Bacteriophage N4 is resistant to many RM systems. In this case, the phage neither methylates its DNA nor expresses a SAMase and at present the mechanism(s) of N4 anti-restriction is unknown (S. Hattman, personal communication). Other examples of anti-restriction modification systems have been extensively and recently reviewed (9,17).

IN VIVO ACTIONS OF RESTRICTION ENZYMES AND ENDONUCLEASES

It is clear that at least some RM systems can function *in vivo* as cellular defense systems and it is likely that some indeed perform this function in bacterial natural habitats. Where does one begin in considering other possible functions and the origins of sequence-specific endonucleases? Because restriction enzymes produce DNA double-strand breaks, possible vantage points are to examine (i) pathways for repair of DNA DSBs, (ii) the roles of DNA lesions in initiation of DNA recombination, and (iii) the functions of sequence-specific endonucleases that are not partners in RM systems.

DNA DSB Repair Pathways

In *E. coli*, recombination and multiple genome copies are required to repair DNA double-strand breaks (DSBs) delivered by X-rays or γ-rays or by the enzyme λ terminase, and prior induction of the SOS response improves survival (182–185). Similarly, in *S. cerevisiae* γ-rays and the HO endonuclease produce DNA breaks that require recombination for repair (186,187), linear plasmid DNA is recombinogenic (188), and hot spots for meiotic recombination suffer double-strand scission during meiosis (189–191). These findings have been elaborated into a model, the double-strand break repair model (27,186) which suggests that recombination is initiated by DNA double-strand breaks and that DNA double-strand breaks are repaired by recombination (18,19,21–24,27,186,192,193). However, two points are important to bear in mind. First, although DSBs can initiate recombination, this does not prove they do so in all situations. Second, DSBs could promote but not be repaired by recombination if, for example, DNA breaks serve as entry sites for recombination machinery and are then repaired by ligation (21–24,29,194).

Normally linear plasmid DNA transforms *E. coli* poorly because of rapid exonucleolytic destruction. However, plasmid-borne staggered or blunt DSBs can be repaired following transformation in certain mutant *E. coli* strains (195–198). Strains which lack the RecBCD exonuclease and carry suppressor mutations restoring recombination proficiency (*recBCD sbcA* or *sbcBC* mutant strains, for example) are efficiently transformed by linear DNA when the plasmid shares homology with the chromosome and can stably integrate (196). Although these mutant strains do not support efficient plasmid replication, under selective pressure plasmids which lack chromosomal homology are maintained as episomes (199). These observations suggest that linear DNA integrates more frequently than it recircularizes, implying that plasmid-borne double-strand breaks are usually repaired by recombination rather than ligation. Plasmids bearing a DNA double-strand gap are repaired by gene conversion with the RecE homologous recombination pathway, which is active in *recBC sbcA* mutant strains (199). In *recBCD*$^+$ hosts, linear plasmid DNA circularizes and yields transformants at 1% the frequency observed with circular DNA. Transformation is largely dependent on recombination functions and deleted plasmids often occur (197,198). Infecting phage that suffer *Eco*RI cleavage are degraded by the RecBCD nuclease (200) and do not form plaques. At a low frequency restricted phage can be repaired by recombination with an endogenous intact prophage (19). These findings are in contrast to our observations that *Eco*RI breaks in the chromosome require ligation but not RecA-sponsored recombination for repair (29). This difference is further underscored by the observations that transformation by linear DNA (195) and phage restriction by *Eco*RI (Heitman, unpublished data) are not affected by DNA ligase mutations. Thus plasmid- or phage-borne DNA DSBs may be treated differently compared to chromosomal breaks. Small DNA molecules may lack the higher order structure of the chromosome. After DNA scission, plasmid- and phage-borne DNA ends may be more exposed to nuclease degradation, which prevents ligation and necessitates repair by other means.

We have studied the repair of DNA double-strand breaks produced *in vivo* in *E. coli* with *Eco*RI endonuclease mutants. We found that *Eco*RI DSBs induce the SOS DNA repair system but, in contrast to γ-rays or λ terminase lesions, strains defective in SOS induction (*lexA3*) and DNA recombination (*recA,recB, recN*) were no more sensitive to *Eco*RI DSBs than isogenic wild-type strains (29). Our model is that different types of DNA double-strand breaks are repaired by different pathways. For example, γ-rays break DNA by shattering sugar residues, releasing the attached base to leave one base pair gaps that have 5′-phosphate and in some cases 3′-phosphoglycolate ends that cannot be repaired by ligation and require exonuclease trimming at the 3′ end and recombination for repair (201). Similary, λ terminase (and probably also HO endonuclease) remains bound to the DNA strand break (184); such lesions exclude DNA ligase and require recombinational repair. In contrast, the staggered breaks produced by the *Eco*RI endonuclease are substrates for DNA ligase. How does DNA ligase know which ends to ligate together? One possible model is that the higher order structure of the *E. coli* chromosome holds the severed ends in proximity and promotes their ligation.

Repair of *Eco*RI-induced DNA double-strand breaks has been previously studied in yeast (202,203). Barnes and Rine expressed the wild-type *Eco*RI endonuclease under control of the tightly regulated *GAL1* promoter on a low copy number CEN plasmid. Repression was sufficient in this system to permit uninduced cells to survive in the absence of *Eco*RI methyltransferase. Based on preliminary observations, they concluded that *rad52* mutants were much more sensitive to *Eco*RI DSBs (202). Because *rad52* mutants have a defect in repairing γ-ray or HO-inflicted DSBs, these findings suggested *Eco*RI DSBs also require *RAD52* mediated repair. However, when Rine and Barnes subsequently measured the fraction of cells surviving increasing periods of *Eco*RI expression, *rad52* mutants were in fact no more than 1.5-fold more sensitive to *Eco*RI action compared with the isogenic *RAD52* parent (203). Although not explicitly discussed, this clearly raises the question of whether

*Eco*RI breaks are repaired by *RAD52*-dependent recombination in yeast. Where could they have gone wrong? Barnes and Rine found that a plasmid expressing *Eco*RI was more toxic in *rad52* mutants, but only in the case where the plasmid has an *Eco*RI site near the *GAL1* promoter (202). *RAD52* and *rad52* mutant strains were roughly equally sensitive to *Eco*RI endonuclease expressed from a plasmid with no *Eco*RI site (203). An alternative interpretation of their findings is that *RAD52* is not required to repair *Eco*RI DSBs in the chromosome, but rather to repair the *Eco*RI break in the plasmid upstream of the *Eco*RI endonuclease gene by recombination between flanking repetitive elements. Following induction of *Eco*RI, Barnes and Rine (202) did in fact observe deleted plasmids which lack the *Eco*RI site and *GAL1* promoter and no longer express *Eco*RI, lending credence to this alternative interpretation.

In related studies, *Eco*RI was found to induce expression of the *RAD54* gene which, like *RAD52*, is required for high efficiency meiosis and to repair γ-ray and HO-inflicted DNA DSBs in yeast (204). In contrast to *rad54* null mutations, which increase γ-ray sensitivity by three orders of magnitude, mutations that block induction of *RAD54* by DNA damage but do not alter the constitutive level of expression have no affect on sensitivity to HO or γ-ray breaks (205). Thus, *RAD54* is induced by DSBs but the induced level is not required to repair these breaks, which mirrors our finding that *Eco*RI breaks induce SOS but do not require SOS function for repair (29). It is not yet known if *rad54* mutants have increased sensitivity to *Eco*RI DSBs.

Prokaryotic restriction endonucleases have also been introduced into tissue culture cell lines. Morgan et al. (206) isolated stably-transfected CHO lines which constitutively express the *Eco*RI endonuclease gene from the mouse metallothionein promoter. The enzyme is nuclear based on immunolocalization, demonstrating that *Eco*RI has fortuitous nuclear localization signals and CHO cells can tolerate some level of *Eco*RI expression and the resulting DNA DSBs. Induction of the promoter with heavy metals led to a cell cycle delay and a profound increase in chromosome aberrations but not in sister chromatid exchanges (SCE). Aberrant chromosomes may arise through ligation of *Eco*RI breaks or recombination initiated at *Eco*RI lesions. DSBs have been implicated in the formation of SCE; however these results suggest this view is either incorrect or *Eco*RI DSBs differ from those generating SCEs.

The Roles of DNA Lesions in Initiation of DNA Recombination

DNA recombination allows organisms to reassort genes contributed to progeny. Recombination also permits the cell to repair DNA lesions, such as breaks and gaps, which destroy the genetic information borne by a chromosome (207). Two models have been proposed for the events that initiate recombination. One suggests that DNA nicks promote the entry of helicases or exonucleases that generate recombinogenic single-strand DNA gaps or tails (208,209), while the second envisions DNA double-strand breaks that promote recombination (27). A variety of experiments suggests that both nicks and breaks can stimulate recombination in *E. coli* and in yeast (21–24,27,186–189,192–194,210–213). For example, enzymes that nick DNA or mutations in DNA ligase enhance recombination in both *E. coli* and in *S. cerevisiae*

(211–214), suggesting that DNA nicks can initiate recombination. The sequence-specific gene II replicator protein of bacteria phage f1 can stimulate recombination at recognition sites in the yeast genome (214). However, in this example the possibility that nicks are processed to recombinogenic breaks cannot be excluded.

On the other hand, DNA double-strand breaks or ends stimulate recombination in *E. coli* and in yeast (21–24,27). In *E. coli* DNA breaks (including *Eco*RI breaks) activate the λRed pathway, chi-mediated RecBCD recombination, and the RecE and RecF pathways (19,21–24,194). DNA that is transferred by transduction or conjugation is in a linear double- or single-stranded DNA form prior to recombination. Similarly, in yeast and certain mutant strains of *E. coli* , linear DNA readily transforms cells by recombining into homologous regions of the chromosome (188,196). DNA scission has been proposed to initiate yeast meiotic recombination in general (27), as it does in the specific case of mating type switching in which HO endonuclease cleavage at the *MAT* locus provokes recombination (210). DNA scission by the Flp recombinase behaves similarly (215). In addition, a DNA double-strand break occurs in the yeast *ARG4* gene and other recombination hotspots concomitant with commitment to meiotic recombination (189–191). The responsible endonuclease has not yet been identified but may be encoded by the *NUC1* gene (216).

Furthermore, a mutation (*rad52*) has been described which increases sensitivity to γ-irradiation (186). *rad52* strains are sensitive to a single γ-ray induced DNA break (217–219), do not switch mating type and are in fact killed by the single break that occurs at *MAT*(187,220), and are not transformed by linear DNA (188). Since *RAD52* is required for mitotic and meiotic recombination (reviewed in (221)), these findings implicate recombination in the repair of some types of DNA breaks.

It has been suggested that DNA breaks may promote exchange after exonuclease processing to a strand-invasive substrate which increases recombination at the break, or by allowing recombinases entry to stimulate recombination away from the break (21–24). Our findings (29) suggest that *Eco*RI DNA breaks act as transient entry points for the RecBCD complex and are then resealed by DNA ligase. Because RecBCD requires DNA ends for its recombination activity, there may be an *Eco*RI cellular analog to provide RecBCD entry points. Mahan and Roth (222) have suggested that the recombination process itself may generate DNA double-strand breaks that act as RecBCD entry points and RecA protein can itself promote DNA cleavage under some conditions (223). In *S. cerevisiae* Nickoloff et al. (224) observed that HO DSBs in duplicated DNA were repaired with different kinetics and yielded different recombination products compared to HO DSBs in unique regions of the genome. They proposed that severed DNA ends are held together at HO breaks which lack homology elsewhere in the genome, and that the DNA ends may not directly promote recombination but rather serve as entry points for recombination enzymes. Fishman-Lobell and Haber (225) have discovered that repair of HO breaks in unique regions of the genome requires RAD1. Although *RAD1* is essential for DNA excision repair and also participates in recombination, *rad1* mutants are not X-ray sensitive and can switch mating type. RAD1 apparently plays a unique role in the repair of DNA DSBs lacking homology, possibly by trimming the ends to reach homology that can sponsor recombinational repair.

Functions of Sequence-specific Endonucleases not Partners in RM Systems

Homothallic *S. cerevisiae* strains express the HO sequence-specific endonuclease in the absence of any DNA methyltransferase. HO creates DSBs that provoke recombinational mating type switching in *S. cerevisiae* . Ectopic HO breaks are also repaired by recombination and, in some cases, repair is by deletions between flanking homologous regions (226,227). In the absence of repair, HO breaks are long lived (228). HO sponsored recombination requires protein synthesis (cycloheximide inhibited) following HO action, and $5' \rightarrow 3'$ exonuclease action converts the initial HO cut to $3'$ single-stranded invasive tails (229).

A number of sequence-specific endonucleases have been described which, similar to the yeast HO enzyme, are not paired with methyltranferases in RM systems but rather promote recombination. In many cases, these endonucleases are expressed by retrotransposable elements or group I introns in phage and yeast (reviewed in (230–232)). Prokaryotes were thought to be devoid of introns until the revolutionary discovery that bacteriophage T4 contains group I (self-splicing) introns in the *td* , *sunY* and *nrdB* genes (230,231,233,234). Later, a similar intron was found in the *B. subtilus* SPOI phage DNA polymerase gene. Remarkably, the introns can home to allelic intronless genes or possibly transpose to new genes. These phage introns contain ORFs encoding proteins not required for splicing. In the case of the *td* and *sunY* introns, which have been demonstrated to be mobile, these ORFs encode sequence-specific endonucleases that cleave at or near the site for intron homing or transposition. The *td* intron expresses I-TevI and the *sunY* intron I-TevII (Figure 6), endonucleases that recognize large asymmetric sequences, cleave to yield $3'$-single-stranded tails with $3'$-hydroxyl and $5'$ phosphate ends, and have the unusual property of recognizing and cleaving T-phage DNA containing either cytosines or the bulky glucosylated hydroxymethylcytosine bases. Intron homing also leads to gene conversion of adjacent polymorphic exon sequences, implicating exonucleolytic degradation during recombination (230).

The proposed model for intron homing, derived from the DSB repair model (27), envisions the intron-encoded endonucleases recognizing the intronless allele, inflicting DSBs that are processed to $3'$ ends that strand invade the intact intron-containing homolog, priming DNA replication from the free $3'$-OH end, and gene converting the intronless allele to an intron-containing allele (230,232). Experimental support for this model reveals that mobilization frequency is increased 10- to 1000-fold by endonuclease, requires exon homology and RecA function, is RNA-independent, and results in nonreciprocal exchanges (230). The observation that the phage intron-encoded endonucleases cleave at secondary sites (230) suggests that introns may mobilize

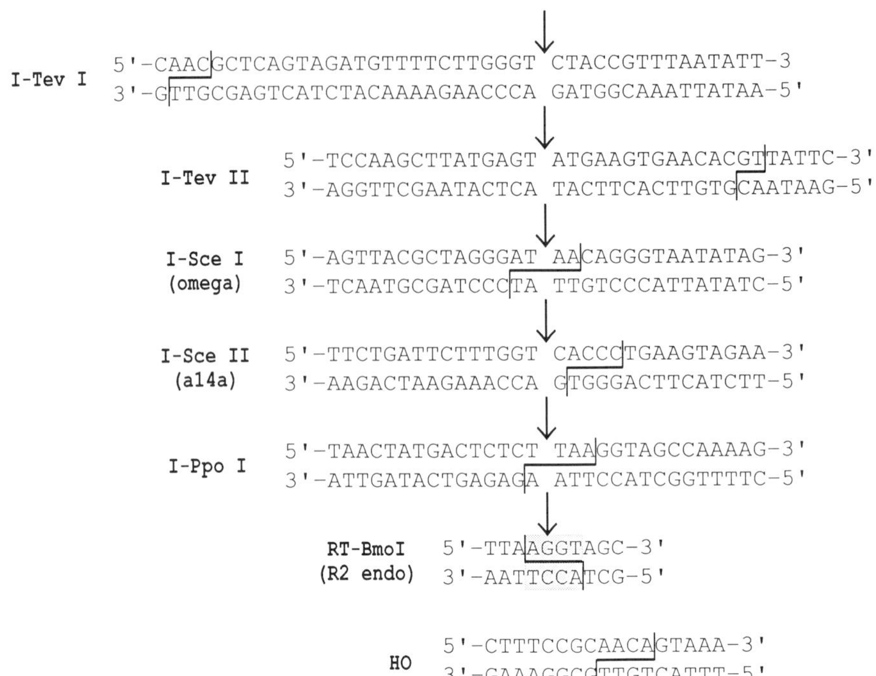

Figure 6. Recognition sites for recombination-promoting endonucleases. Several sequence-specific endonucleases are not paired in RM systems, but rather generate DSBs that promote recombination and mobilization of introns, retrotransposons, or mating type cassettes. These enzymes recognize large asymmetric recognition sites to generate 3′ or 5′ single-stranded tails with 3′-hydroxyls that are thought to promote recombination by DNA double-strand break repair. Arrows indicate sites of homing or transposition, staggered lines the cleavage positions and the shaded box bases lost during recombination. Adapted from Belfort (230) with permission.

to new locations by transposition, providing sufficient homology to the intron donor site flanking these secondary sites of cleavage.

Clyman and Belfort (26) have defined enzymes which function *in trans* during T4 intron mobilization. In the absence of T4 infection, intron mobilization is decreased 500-fold in *recA* mutants, but not at all in *lexA3* mutants in which RecA can no longer induce SOS. Thus, the RecA recombinase is itself required. Mutations in other host homologous (*recJ*, *recN*, *ruvB*) or site-specific (*fis*, *himA*, *hupA*, *hupB*) recombination functions had no effect. The recombinase requirement could also be satisfied by the T4 UvsX enzyme. In the absence of T4 infection, $recA^+$ *E. coli* strains require λ functions to mobilize T4 introns, suggesting the λRed recombination system (*redα* -encoded 5′ → 3′ exonuclease, *redβ* -encoded annealing activity), known to utilize DNA DSBs (21,22), is involved. Indeed, *redα* or *redβ* mutant phage poorly supported intron movement, arguing that phage λ can supply a required $5' \rightarrow 3'$ exonuclease to process the DSB to invasive tails. Clyman and Belfort's findings (26) also implicate involvement of a $3' \rightarrow 5'$ exonuclease, in that *recBC* mutants reduce mobility two-fold, exonuclease III deficiency (*xthA* mutation) two- to eight-fold, and double *recBC xthA* mutants lacking both exonuclease V and III do not support intron mobility. Lastly, $3'$-overhanging, $5'$-overhanging, or blunt DSBs delivered by restriction endonucleases *in vitro,* sufficed to target movement. These genetic results clearly support a DSB repair model for intron movement and reveal a recombinase and both $5' \rightarrow 3'$ and $3' \rightarrow 5'$ exonucleases are required.

A similar group of intron-encoded endonucleases are found in mitochondria or chloroplasts of eukaryotes such as *S. cerevisiae* (235–241). Two examples are known in *S. cerevisiae* : the r1 intron in the 21S mitochondrial RNA gene expresses the omega (ω) or I-*Sce*I endonuclease (235), and the a14 intron in the gene encoding cytochrome oxidase subunit I encodes the a14α or I-*Sce*II endonuclease (236) which also has latent RNA maturase activity (240). The original descriptions of the unusual ω allele in yeast long predated the discovery of splicing (see (242)). During crosses, yeast mitochondrial DNA recombines readily and only markers separated by 1 kb or less are linked. In contrast, the polar region of the mitochondrial genome, encompassing the ω locus and flanked by antibiotic resistance markers, exhibits a bizarre polar behavior. Crosses between ω^+ strains or between ω^- strains behave normally. In contrast, when ω^+ and ω^- strains are crossed, reciprocal recombinants between flanking antibiotic resistance markers occur rarely and the majority of the progeny inherit ω^+. We now know ω is an intron in the mitochondrial 21S rRNA gene, and the intron contains an ORF encoding the I-*Sce*I endonuclease. Conversion is promoted when the I-*Sce*I endonuclease creates a DSB at the ω^- allele, which is then repaired by gene conversion from the ω^+ allele (237).

In the silkworm *Bombyx mori*, the 28S ribosomal genes are interrupted by two retrotransposable elements, R1 and R2, which express a sequence-specific endonuclease that targets integration (243). Expression of this endonuclease in *E. coli* from a plasmid also carrying the 28S gene results in DSBs at or near the site of retrotransposon insertion. The specific recognition site, deduced from sites of retrotransposon insertion and primer extension across *in vivo* cleavage sites, occurs at the asymmetric site TTAcutAGGTAGC embedded within a larger 25 to 30 nucleotide sequence. Cleavage generates a 3 or 4- bp $5'$-overhang, in contrast to the $3'$-overhang generated by the intron-encoded endonucleases. Similarly, the rRNA gene of *Physarum polycephalum* contains a group I intron that encodes I-*Ppo*I, an endonuclease involved in intron homing (238,244).

Mobile elements expressing sequence-specific endonucleases are not restricted to RNA introns. Astonishingly, two composite genes encoding a 69 kD vacuolar H^+-ATPase subunit and the DNA polymerase of *Thermococcus litoralis* (245), have been discovered that contain intervening sequences that, in contrast to other known introns, contain ORFs in frame with both flanking exons (246,247). Even more astoundingly, these intervening sequences are removed from the initial protein product by protein splicing (247), and the excised protein segments are sequence-specific endonucleases that mobilize homing and possibly transposition of their encoding gene (248). Thus both types of sequence-specific mobilizing endonucleases have found safe harbors in genes which, through RNA or protein splicing, can tolerate a guest endonuclease gene.

In contrast to the majority of restriction endonucleases, which bear little or no resemblance to each other, these intron-encoded and several additional endonucleases share sequence identity and define two distinct families (232). One family shares the signature dodecapeptide element, LAGLI-DADG, reiterated twice flanking a region comprised of about 115 residues. This family includes the *S. cerevisiae* endonucleases I-*Sce*I, I-*Sce*II, HO, and VDE (vacuolar derived endonuclease) (249), and the *T. litoralis* I-*Tli*I endonuclease. The second family (232) shares the conserved motif GIY-$N_{10\text{-}11}$-YIG and includes a number of group I intron encoded endonucleases, such as T4 I-*Tev*I, and also the non-intron encoded T4 *Seg*A endonuclease (250).

Based on these examples of endonuclease-provoked recombination and transposition, Eddy and Gold (25) tested if the type II restriction enzyme *Eco*RI could promote similar events. Indeed, when imbedded in the *lacZ* gene the *Eco*RI RM system could direct its own mobilization to a cleaved *Eco*RI recognition site present in another copy of *lacZ* in appropriate mutant host strains (for example, *recBC sbcA*) (25). In this case, *Eco*RI cleavage reduced viability of the recipient molecule 10- to 20-fold whereas, in contrast, I-*Tev*I or I-*Tev*II cleavage does not impair T4 phage viability. Thus while a type II restriction system can foster recombination events, the intron-expressed endonucleases may have evolved to prevent degradation and promote recombination, possibly by binding to the cleaved product.

Other DNA metabolic enzymes that conduct DNA repair (MutH), DNA replication (fl gene II protein), and DNA recombination (RecBCD, RAG1, RAG2, SCID), or scavenge host DNA for nucleotides (T4 endonuclease II) are known or proposed to be sequence-specific endonucleases. Base pair mismatches that arise during DNA replication are corrected by DNA mismatch repair systems. In *E. coli* , the MutH protein participates in mismatch repair as a Mg^{2+}-dependent endonuclease that discriminates the methylation state at GATC sequences, sites of *dam* methylation, to discern the parental versus newly-replicated strand (130). Hemi-methylated GATC sites are nicked 5′ to the guanine on the unmethylated newly-replicated DNA strand likely to contain the error, yielding 3′-hydroxyl termini that can act as primers for replication/exonucleolytic excision from the GATC *dam* site to and across the mismatched base pair, recognized by the MutS protein. Purified MutH protein is a sequence-specific Mg^{2+}-dependent endonuclease that: (i) nicks unmethylated GATC sites on either strand, (ii) nicks hemi-methylated GATC sites on the unmodified strand, and (iii) does not nick fully methylated GATC sites (251). Thus, the MutH endonuclease/Dam methyltransferase pair shares some features with RM systems except that in this case a DNA nick targets single-stranded excision and repair rather than a DSB eliciting destruction and restriction.

The fl phage replicator protein, gene II protein, is a sequence-specific endonuclease that, as a tetrameric complex, recognizes the repeated DNA sequence elements of the phage origin of replication and specifically nicks one

DNA strand in a Mg^{2+}-dependent fashion to provide a 3′-hydroxyl for replication (252). Gene II is further required during unwinding and replication termination and presumably stays bound during an entire round of replication. *In vitro* , substitution of the Mg^{2+} cofactor by Mn^{2+} enables gene II protein to cleave rather than simply nick the replication origin. Unlike nicking, this Mn^{2+} cleavage reaction is independent of supercoiling and proceeds as two kinetically-independent nicking events that yield an unusual product in which one side of the break is sealed to a covalent hairpin (252). With the exception of this unusual hairpin product, the gene II protein Mn^{2+} cleavage activity is similar to the relaxation of type II restriction enzyme substrate specificity under some conditions, known as star activity. While gene II protein initiates replication of f1, ectopic expression stimulates recombination at gene II recognition sites in yeast, reflecting a role for DNA nicks or breaks in recombination initiation (214).

RecBCD is a multifunctional enzyme involved in recombination in *E. coli. recB* or *recC* mutations reduce recombination following conjugation by 100- to 1000-fold, whereas *recD* mutants are hyperrecombinogenic. RecBCD also has an unusual property, first described in λ phage crosses, of promoting homologous recombination around specific DNA sequences, GCTGGTGG, known as chi sites (253,254). The RecABCD recombination pathway is the only one known to utilize chi sites, suggesting some specific interaction between RecBCD and chi. The RecBCD enzyme has been purified and found to have helicase, exonuclease and ATP stimulated single-strand endonuclease activities. Most interestingly, RecBCD is also a sequence-specific endonuclease that nicks double-stranded DNA 4 to 6 nucleotides to the 3′ side of chi sites (255).

T4 phage expresses a lone endonuclease, Endo II, that selectively cleaves the host genome at CCGNNTTGGC sites, enabling exonucleases to scavenge the host DNA as nucleotides for phage DNA synthesis (256).

In the immune systems of multicellular eukaryotes, genomic recombinational rearrangements mediate the expression of a vast array of different proteins, antibodies in B-cells and antigen receptors in T-cells, that enable the organism to recognize and defend against foreign intruders. Three genes have been identified, *RAG1, RAG2* and *scid* , whose products participate in these sequence-specific recombination events. Expression of *RAG1* and *RAG2* enables fibroblasts to undergo V(D)J recombination (257–259) and mice lacking either *RAG1* or *RAG2* have no mature T- or B-cells because V(D)J recombination is blocked (260,261). Thus *RAG1* and *RAG2* mutations confer novel forms of severe combined immunodeficiency (scid) syndrome. These findings support models in which *RAG1* and *RAG2* are either subunits or regulators of the V(D)J recombinase (see (262)).

In addition to RAG1 and RAG2, a third protein has been implicated in V(D)J recombination. This protein is the product of the *scid* gene, originally identified in a mutant mouse strain with severe combined immunodeficiency (scid) syndrome (263,264). Lymphocytes bearing the *scid* mutation initiate V(D)J recombination but fail to reseal the broken ends (265,266). In addition, the *scid* mutation confers a more general DNA repair defect, impairing DNA DSB repair and rendering cells γ-ray sensitive (267–269). Roth et al. have recently demonstrated that following V(D)J initiation in *scid* cells, the coding joint ends are covalently closed hairpins (266,270), possibly similar to those observed with the f1 gene II protein Mn^{2+} reaction (252). The proposed model envisions that DSBs formed during V(D)J recombination are first sealed to hairpins, which in a second step are nicked to provide single-stranded ends that are trimmed and ligated. The more general DNA repair defect conferred by *scid* suggests that an early step in DSB repair might be sealing to hairpins to prevent exonuclease destruction. The SCID protein may encode the hairpin nickase and its absence would lead to a defect in completing V(D)J recombination and repairing other DSBs. These recent studies on V(D)J recombination again underscore the links between endonucleases and DNA recombination.

Schiestl and Petes (20) have recently made the provocative observation that restriction enzymes can promote recombination in *S. cerevisiae* and Kuspa and Loomis (271) have made very similar findings in *Dictyostelium* . When plasmid DNA was cleaved and transformed in the presence of a restriction endonuclease, it undergoes nonhomologous or illegitimate recombination at corresponding restriction sites in the yeast genome. One possible model is that proteins can enter the cell by transformation and, in the case of restriction enzymes, enter the nucleus and cause DSBs subsequently repaired, in some cases, by ligation to linear plasmid DNA. It will be interesting to see if restriction endonuclease expression in the host cell will also promote illegitimate recombination of incoming linear DNA or sensitive plasmid DNA. If so, restriction enzymes could clearly promote genome rearrangements *in vivo* .

Collectively, these observations support the views that (i) some types of DNA breaks, such as γ-ray breaks, are repaired by DNA recombination, (ii) nicks and breaks may initiate recombination and (iii) some natural recombination initiator sites suffer DNA scission prior to recombination. One possibility is that sequence-specific endonucleases are paired with methyltransferases in RM systems and with exonucleases that process DSBs to invasive tails in recombination systems.

Endonucleases in Eukaryotes

With the exception of RM systems encoded by viruses, which infect the algae *Chlorella*, RM systems have not been found in eukaryotic organisms, presumably because at least multicellular eukaryotes have evolved other defenses such as the immune system. Why unicellular eukaryotes lack RM systems and in some cases, such as *S. cerevisiae*, lack detectable DNA methylation altogether, is less clear. Although recent findings suggest genetic exchange can occur between pro- and eukaryotes (272,273), this may be less frequent in nature. Introduced RM systems may have lacked appropriate nuclear targeting signals and therefore might not have survived. DNA methylation is removed by excision repair enzymes in yeast (160,161) and might therefore not protect the yeast genome from restriction. An additional barrier to gene transfer between species is the recipient DNA mismatch repair system, which recognizes and repairs mismatches in heteroduplex DNA thereby impairing genetic exchange (274,275).

Gingeras and co-workers have artificially expressed the *Pae*R7 type II RM system in mouse L cells (276). Transfectants expressing both *Pae*R7 methyltransferase and endonuclease were, however, unable to restrict growth of DNA viruses, including murine adenovirus, herpes simplex virus and vaccinia. Possible explanations offered for the lack of restriction were: (i) too little endonuclease was expressed because extracts contained about 1000-fold less activity than bacterial strains expressing cloned *Pae*R7, (ii) too much methyltransferase expression led to preferential modification rather than restriction, (iii) viral DNA and RM enzymes were in different intracellular compartments, or (iv), the transfected genes were unstable. Thus transfer of RM systems from prokaryotic to eukaryotic cells may require more than gene transfer and adequate expression to explain the absence of RM systems in eukaryotes.

Despite the absence of RM systems, some eukaryotic organisms do express endonucleases and methyltransferases. As discussed in the previous section, *S. cerevisiae* expresses HO endonuclease to promote genetic rearrangements that switch mating type. *In vitro*, purified HO endonuclease recognizes a 15 to 16 bp sequence and cleaves in a Mg^{2+}-dependent fashion (277). Thus HO mirrors features of type II restriction endonucleases. One difference may be that HO binds tightly to the cleaved substrate and may recruit accessory factors, such as exonucleases, through protein-protein interactions to coordinate generation of the recombination initiator. However it is not difficult to imagine HO endonuclease could become paired with a DNA methyltransferase and then suffer mutations that relax specificity and increase turnover rate.

CONCLUSIONS AND SPECULATIONS ON RESTRICTION ENZYMES

At least four broad classes of RM systems are known (Types I, II, III, and MDRS), some of which can be further categorized into subtypes (Type IA, IB, IC and Type IIS, IIi, IId), and over 400 different systems have been identified. From where did this vast array arise?

As their name implies, many restriction-modification systems have been demonstrated to restrict phage growth. However, for most type II restriction systems this has not been tested due to the difficulties of working with unusual bacteria which may or may not have characterized phages. A large number of cloned RM systems do not restrict phage growth when expressed in *E. coli* even though in several cases the enzyme is clearly well expressed. These enzymes may fulfill a purpose other than restriction. One alluring suggestion has been that DNA scission by restriction enzymes may stimulate recombination *in vivo*, perhaps similar to the ways we use them to stimulate recombination *in vitro* (4,18). Evidence supporting this putative role is mounting. Stahl et al. (19) showed that *Eco*RI restriction stimulates recombination of an infecting λ phage with an intact prophage. However this effect is not dramatic (five-fold at best), perhaps because phage DNA is rapidly destroyed by exonuclease degradation after suffering scission (200).

More recently, Thaler (21–24) has shown that restriction enzyme initiated DSBs sponsor recombination by the λ Red pathway. Restriction endonucleases promote recombination in *S. cerevisiae* and *Dictyostelium* (20,271), and Eddy and Gold (25) demonstrated that *Eco*RI can stimulate homologous recombination by the *E. coli* RecE pathway. Even more thought-provoking are a number of bacteriophage and fungal intron-associated site-specific endonucleases that promote intron homing and transposition. These intron-associated endonucleases, like yeast HO endonuclease, are not paired with a DNA methyltransferase and clearly promote genetic recombination (230).

Sequence comparisons of a large collection of different cloned RM systems reveal surprisingly limited sequence homologies. With the exception of type I RM systems, which share considerable homology within but not between families, there has been scant homology identified between endonucleases of different sources, whereas methyltransferases share limited degrees of homology that may be more related to enzymatic constraints than a reflection of common origin. Even in cases of isoschizomeric type II RM systems, which recognize and cleave identical DNA sequences, the vast majority share no homology between endonucleases. One poignant exception is *Eco*RI and *Rsr*I, which recognize GAATTC and are expressed by *E. coli* and *Rhodobacter spheroides*, where the endonucleases share 50% identity and are thought to have arisen by horizontal gene transfer and divergence. Curiously, the *Eco*RI and *Rsr*I methyltransferases do not share more sequence homology than with other type II methyltransferases. The *Taq*I and *Tth*HB8I RM systems are another example of isoschizomeric enzymes with homology and in this case endonucleases and methyltransferases both share a high degree of identity (77 and 79%). Similarly, the intron-encoded endonucleases define two families of related enzymes (232).

The emerging view is that sequence-specific endonucleases and DNA methyltransferases arose independently in evolution to serve functions other than restriction-modification. Dam methyltransferase, which recognizes ubiquitous GATC sites and was first identified in *E. coli* and later found in a large variety of other bacteria, is not associated with a restriction endonuclease and is not part of an RM system. Instead, Dam regulates a disparate variety of intracellular processes including transcription, transposition, strand-directed DNA mismatch repair, and the initiation of DNA replication (132). One can imagine that other DNA methyltransferases play or once played similar roles and originally arose, like the *E. coli* Dam enzyme, to regulate by altering DNA-protein interactions at GATC versus N^6-adenine methylated GATC sites.

An independent evolutionary origin for restriction endonucleases was once considered heretical because it was widely assumed, based on kinetics of γ-ray killing, that a single DNA double-strand break would be lethal to any cell. Thus how could an endonuclease arise independent of a partner DNA methyltransferase? We now know that: (i) endonuclease-inflicted breaks have different intracellular consequences and are less toxic than γ-ray breaks, (ii) DSBs participate as intermediates in recombination, (iii) a panoply of anti-restriction functions is expressed by a variety of plasmids, phage and bacteria, and could protect cells expressing an endonuclease prior to acquisition of a protective methyltransferase and (iv) like other DNA lesions, DNA DSBs are subject to DNA repair.

Studies on endonucleases involved in promoting recombination make it poignantly clear that, like the Dam methyltransferase of *E. coli*, endonucleases exist which are not partnered with a methyltransferase and satisfy needs in the cell other than RM. Collectively, these considerations support the view that sequence-specific endonucleases and methyltransferases arose independently in evolution and subsequent events, including gene transfer and mutagenesis, paired and then tailored endonuclease and methyltransferase specificities. Our own work on the *Eco*RI RM system provides one vantage point. Promiscuous *Eco*RI endonuclease mutants were isolated that exhibit increased star activity and cleave both canonical and star sites. Despite the presence of *Eco*RI methyltransferase, these mutant endonucleases damage the DNA of the host cell (as evidenced by SOS induction) but the vast majority of cells survive (40). Perhaps similar situations arose during initial pairings of RM enzymes of incompletely-matched sequence specificities. Mutagenesis and selection could then give rise to better matches. Because DNA nicks and breaks induce the SOS response, SOS induction is mutagenic, and SOS DNA repair systems are widespread among bacteria, mismatched RM enzymes could themselves have promoted mutagenesis. This model envisioning pairings of independent enzymes is in some ways similar to the story that God makes lovers at the same time and then places them at different points on the earth to wander until they find each other.

From where did sequence-specific endonucleases arise? Since DNA is the information currency of the cell and a vast array of enzymes play intimate DNA metabolic roles, RM enzymes perhaps emanate from these. For example, Ung uracil N-glycosylase is a DNA repair enzyme that, with an AP endonuclease, can restrict DNA containing uracil (278). In this view, RM enzymes are not so much toxic weapons, but rather altered DNA metabolizing enzymes. As an example, DNA topoisomerases are largely viewed as beneficient and essential for life, whereas the presence of certain inhibitors turns them into potent toxins. Endonucleases may also have evolved from sequence-specific DNA binding proteins by incorporating catalytic potential. Experimentally at least one observes events in the opposite direction: *Eco*RI endonuclease mutants that bind DNA with high affinity but are catalytically defective (63) block transcriptional elongation by RNA polymerase *in vitro* (279).

The findings that members from each of the four families of RM systems require two recognition sites for cleavage (Figure 4) further suggests that restriction endonucleases may originally have arisen from other DNA metabolizing enzymes, such as recombinases or integrases, which require two sites for action. Alternatively, restriction enzymes may have evolved from sequence-specific endonucleases that target recombination. Until now, DNA sequences have not revealed homologies between these different classes of enzymes. This may reflect an early divergence (because endonucleases participate in recombination and may therefore be quite ancient), rapid divergence under strong selective pressure for sequence specificity and restriction function or, at the worst, an incorrect model. We suggest that further sequence, structural and enzymatic analysis as well as studies on DNA repair, DNA recombination, and *in vivo* consequences of endonuclease and restriction enzyme action, may reveal further shared properties between endonucleases and other DNA metabolic enzymes and shed light on the origins of this interesting and valuable collection of genetic engineering tools and windows on DNA-protein interactions.

Acknowledgments: We thank Mario Noyer-Weidner for drawing to our attention two additional examples of related isoschizomeric restriction endonucleases: *Bsu*FI and *Msp*I endonucleases, which both recognize CCGG and share 45% overall sequence similarity and three smaller regions with 60% identity (286); and *Bsu*BI and *Pst*I endonucleases, which recognize CTGCAG and share 46% amino acid identity (287). These examples bring the number of related isoschizomeric enzyme pairs to four (*Eco*RI/*Rsr*I, *Taq*I/*Tth*HB8I, *Msp*I/*Bsu*FI, and *Pst*I/*Bsu*BI), suggesting that at least in some cases isoschizomers may be derived from a common ancestor. These related enzyme pairs should also provide valuable insights into conserved regions required for sequence-specific recognition and catalysis.

I thank Kan Agarwal, Malcolm Casadaban, Norton Zinder and Peter Model for fostering and supporting my interests, David Waugh, Francis Barany, Tom Bickle, John Zebala, Rich Roberts, Linda Jen-Jacobson, Xavier Soberón, Lise Raleigh, Fritz Winkler, Detlev Krüger, Michael McClelland, Dick Gumport, Geoff Wilson, Stephen Halford, John Taylor, Stan Hattman, Paul Modrich, Tom Gingevas, Gerry Smith, Sam Gabbara, David Thaler, Michael Topal, Arvydas Janulaitis, Mario Noyer-Weidner and Marlene Belfort for reprints, preprints, discussions, and comments on the manuscript. My work, some of which is summarized here, was supported by an NSF grant to Norton Zinder and Peter Model, and an MSTP fellowship from the NIH. Lastly, I thank Bonnie Kissell for her patience and diligence in preparing this chapter.

REFERENCES

1 Roberts, R.J. (1976) CRC Critical Review in Biochemistry 4, 123–164.
2 Luria, S.E. and Human, M.L. (1952) J. Bacteriol. 64, 557–569.
3 Bertani, G. and Weigle, J.J. (1953) J. Bacteriol. 65, 113–121.
4 Arber W. (1979) Science 205, 361–365.
5 Bickle, T. (1987) in *Escherichia coli* and *Salmonella typhimurium* (Neidhardt, F.C., Ingraham, J.L., Low, K.B, Magasanik, B., Schaechter, M. and Umbarger, H.E., eds.), pp. 692–696, American Society for Microbiology, Washington, DC.
6 Wilson, G.G. (1988) Gene 74, 281–289.
7 Wilson, G.G. and Murray, N.E. (1991) Annu. Rev. Genet. 25, 585–627.
8 Raleigh, E.A. (1992) Mol. Microbiol. 6, 1079–1086.
9 Bickle, T.A. and Krüger, D.H. (1993) Microbiol. Rev. (in press).
10 Wilson, G.G. (1991) Nucl. Acids Res. 19, 2539–2566.
11 Lacks, S. and Greenberg, B. (1977) J. Mol. Biol. 114, 153–168.
12 Raleigh, E.A. and Wilson, G. (1986) Proc. Nat. Acad. Sci. U.S.A. 83, 9070–9074.
13 Noyer-Weidner, M., Diaz, R. and Reiners, L. (1986) Mol. Gen. Genet. 205, 469–475.

14 Blumenthal, R.M., Gregory, S.A. and Cooperider, J.S. (1985) J. Bacteriol. 164, 501–509.
15 Kiss, A., Pósfai, G., Keller, C.C., Venetianer, P. and Roberts, R.J. (1985) Nucl. Acids Res. 13, 6403–6421.
16 Heitman, J. and Model, P. (1987) J. Bacteriol. 169, 3243–3250.
17 Krüger, D.H. and Bickle, T.A. (1983) Microbiology Rev. 47, 345–360.
18 Chang, S. and Cohen, S.N. (1977) Proc. Nat. Acad. Sci. U.S.A. 74, 4811–4815.
19 Stahl, M.M., Kobayashi, I., Stahl, F.W. and Huntington, S.K. (1983) Proc. Nat. Acad. Sci. U.S.A. 80, 2310–2313.
20 Schiestl, R.H. and Petes, T.D. (1991) Proc. Nat. Acad. Sci. U.S.A. 88, 7585–7589.
21 Thaler, D.S., Stahl, M.M. and Stahl, F.W. (1987) Genetics 116, 501–511.
22 Thaler, D.S., Stahl, M.M. and Stahl, F.W. (1987) J. Mol. Biol. 195, 75–87.
23 Thaler, D.S., Stahl, M.M. and Stahl, F.W. (1987) EMBO J. 6, 3171–3176.
24 Thaler, D.S. and Stahl, F.W. (1988) Annu. Rev. Genet. 22, 169–197.
25 Eddy, S.R. and Gold, L. (1992) Proc. Nat. Acad. Sci. U.S.A. 89, 1544–1547.
26 Clyman, J. and Belfort, M. (1992) Genes Dev. 6, 1269–1279.
27 Szostak, J.W., Orr-Weaver, T.L., Rothstein, R.J. and Stahl, F.W. (1983) Cell 33, 25–35.
28 Heitman, J. (1992) BioEssay 14, 445–454.
29 Heitman, J., Zinder, N.D. and Model, P. (1989) Proc. Nat. Acad. Sci. U.S.A. 86, 2281–2285.
30 Kim, Y., Grable, J.C., Love, R., Greene, P.J. and Rosenberg, J.M. (1990) Science 240, 1307–1309.
31 Rosenberg, J.M. (1991) Curr. Opin. Struc. Biol. 1, 104–113.
32 McClarin, J.A., Frederick, C.A., Wang, B.-C., Greene, P.J., Boyer, H.W., Grable, J. and Rosenberg, J.M. (1986) Science 234, 1526–1541.
33 Heitman, J. and Model, P. (1990) Proteins: Structure, Function and Genetics 7, 185–197.
34 Oelgeschläger, T., Geiger, R., Rüter, T., Alves, J., Fliess, A. and Pingoud, A. (1990) Gene 89, 19–27.
35 Needels, M.C., Fried, S.R., Love, R., Rosenberg, J.M., Boyer, H.W. and Greene, P.J. (1989) Proc. Nat. Acad. Sci. U.S.A. 86, 3579–3583.
36 Osuna, J., Flores, H. and Soberón, X. (1991) Gene 106,7–12.
37 Brennan, C.A., Van Cleve, M.D. and Gumport, R.I. (1986) J. Biol. Chem. 261, 7270–7278.
38 Becker, M.M., Lesser, D., Kurpiewski, M., Baranger, A. and Jen-Jacobson, L. (1988) Proc. Nat. Acad. Sci. U.S.A. 85, 6247–6251.
39 Lesser, D.R., Kurpiewski, M.R. and Jen-Jacobson, L. (1990) Science 250, 776–786.
40 Heitman, J. and Model, P. (1990) EMBO J. 9, 3369–3378.
41 Winkler, F.K. (1992) Curr. Opin. Struc. Biol. 2, 93–99.
42 Aiken, C., Milarski-Brown, K. and Gumport, R.I. (1986) Fed. Proc. 45, 1914.
43 Aiken, C. and Gumport, R.I. (1988) Nucl. Acids Res. 16, 7901–7916.

44 Greene, P.J., Ballard, B.T., Stephenson, F., Kor, W.J., Rodriguez, H., Rosenberg, J.M. and Boyer, H.W. (1988) Gene 68, 43–52.
45 Stephenson, F.H., Ballard, B.T., Boyer, H.W., Rosenberg, J.M. and Greene, P.J. (1989) Gene 85, 1–13.
46 Kaszubska, W., Aiken, C., O'Connor, C.D. and Gumport, R.I. (1989) Nucl. Acids Res. 17, 10403–10425.
47 Rubin, R.A. and Modrich, P. (1977) J. Biol. Chem. 252, 7265–7272.
48 Aiken, C.R., Fisher, E.W. and Gumport, R.I. (1991) J. Biol. Chem. 266, 19063–19069.
49 Aiken, C.R., McLaughlin, L.W. and Gumport, R.I. (1991) J. Biol. Chem. 266, 19070–19078.
50 Taylor, J.D. and Halford, S.E. (1989) Biochemistry 28, 6198–6207.
51 Halford, S.E., Taylor, J.D., Vermote, C.L.M. and Vipond, I.B. (1993) in Nucleic Acids and Molecular Biology (Eckstein, F. and Lilley D.M.J., eds.), Springer-Verlag, Berlin (in press).
52 Taylor, J.D., Goodall, A.J., Vermote, C.L. and Halford, S.E. (1990) Biochemistry 29, 10727–10733.
53 Taylor, J.D., Badcoe, I.G., Clarke, A.R. and Halford, S.E. (1991) Biochemistry 30, 8743–8753.
54 Taylor, J.D. and Halford, S.E. (1992) Biochemistry 31, 90–97.
55 Vermote, C.L.M. and Halford, S.E. (1992) Biochemistry 31, 6082–6089.
56 Vermote, C.L.M., Vipond, I.B. and Halford, S.E. (1992) Biochemistry 31, 6089–6097.
57 Thielking, V., Selent, U., Köhler, E., Wolfes, H., Pieper, U., Geiger, R., Urbanke, C., Winkler, F.K. and Pingoud, A. (1991) Biochemistry 30, 6146–6422.
58 Selent, U., Rüter, T., Köhler, E., Liedtke, M., Thielking, V., Alves, J., Oelgeschläger, T., Wolfes, H., Peters, F. and Pingoud, A. (1992) Biochemistry 31, 4808–4815.
59 Thielking, V., Selent, U., Köhler, E., Landgraf, A., Wolfes, H., Alves, J. and Pingoud, A. (1992) Biochemistry 31, 3727–3732.
60 Bougueleret, L., Schwartzstein, M., Tsugita, A. and Zabeau, M. (1984) Nucl. Acids Res. 12, 3659–3676.
61 Bougueleret, L., Tenchini, M.L., Botterman, J. and Zabeau, M. (1985) Nucl. Acids Res. 13, 3823–3839.
62 Winkler, F.K., D'Arcy, A., Blocker, H., Frank, R. and van Boom, J.H. (1991) J. Mol. Biol. 217, 235–238.
63 Wright, D.J., King, K. and Modrich, P. (1989) J. Biol. Chem. 264, 11816–11821.
64 King, K., Benkovic, S.J. and Modrich, P. (1989) J. Biol. Chem. 264, 11807–11815.
65 Brooks, J.E., Benner, J.S., Heiter, D.F., Silber, K.R., Sznyter, L.A., Jager-Quinton, T., Moran, L.S., Slatko, B.E., Wilson, G.G. and Nwanko, D.O. (1989) Nucl. Acids Res. 17, 979–997.
66 Jack, W.E., Greenough, L., Dorner, L.F., Xu, S.-y., Strzelecka, T., Aggarwal, A.K. and Schildkraut, I. (1991) Nucl. Acids Res. 19, 1825–1829.
67 Xu, S.-y. and Schildkraut, I. (1991) J. Bacteriol. 173, 5030–5035.

68 Xu, S.-y. and Schildkraut, I. (1991) J. Biol. Chem. 266, 4425–4429.
69 Sato, S., Hutchinson, C.A., III and Harris, J.I. (1977) Proc. Nat. Acad. Sci. U.S.A. 74, 542–546.
70 Slatko, B.E., Benner, J.S., Jager-Quinton, T., Moran, L.S., Simcox, T.G., Van Cott, E.M. and Wilson, G.G. (1987) Nucl. Acids Res. 15, 9781–9796.
71 Barany, F. (1988) Gene 63, 167–177.
72 Barany, F., Slatko, B., Danzitz, M., Cowburn, D., Schildkraut, I. and Wilson, G.G. (1992) Gene 112, 91–95.
73 Barany, F. and Zebala, J. (1992) Gene 112, 13–20.
74 Barany, F. (1987) Gene 56, 13–27.
75 Zebala, J.A., Choi, J. and Barany, F. (1992) J. Biol. Chem. 267, 8097–8105.
76 Ghosh, S.S., Obermiller, P.S., Kwoh, T.J. and Gingeras, T.R. (1990) Nucl. Acids Res. 18, 5063–5068.
77 Zebala, J.A., Choi, J., Trainor, G.L. and Barany, F. (1992) J. Biol. Chem. 267, 8106–8116.
78 Barany, F., Danzitz, M., Zebala, J. and Mayer, A. (1992) Gene 112, 3–12.
79 Li, L., Wu, L.P. and Chandrasegaran, S. (1992) Proc. Nat. Acad. Sci. U.S.A. 89, 4275–4279.
80 Gingeras, T.R., Greenough, L., Schildkraut, I. and Roberts, R.J. (1981) Nucl. Acids Res. 9, 4525–4536.
81 Butkus, V., Klimasauskas, S., Petrauskiene, L., Maneliene, Z., Lebionka, A. and Janulaitis, A. (1987) Biochim. Biophys. Acta 909, 201–207.
82 Tao, T., Walter, J., Brennan, K.J., Cotterman, M.M. and Blumenthal, R.M. (1989) Nucl. Acids Res. 17, 4161–4175.
83 Tao, T., Bourne, J.C. and Blumenthal, R.M. (1991) J. Bacteriol. 173, 1367–1375.
84 Hattman, S., Gribbin, C. and Hutchinson, C.A., III (1979) J. Virol. 32, 845–851.
85 Krüger, D.H., Barcak, G.J., Reuter, M. and Smith, H.O. (1988) Nucl. Acids Res. 16, 3997–4008.
86 Krüger, D.H., Prosch, S., Reuter, M. and Goebel, W. (1990) J. Basic Microbiol. 30, 679–683.
87 Oller, A.R., Vanden Broek, W., Conrad, M. and Topal, M.D. (1991) Biochemistry 30, 2543–2549.
88 Pein, C.-D., Reuter, M., Meisel, A., Cech, D. and Krüger, D.II. (1991) Nucl. Acids Res. 19, 5139–5142.
89 Conrad, M. and Topal, M.D. (1992) Nucl. Acids Res. 19, 5127–5130.
90 Topal, M.D., Thresher, R.J., Conrad, M. and Griffith, J. (1991) Biochemistry 30, 2006–2010.
91 Gabbara, S. and Bhagwat, A.S. (1992) J. Biol. Chem. 267, 18623–18630.
92 Connolly, B.A., Eckstein, F. and Pingoud, A. (1984) J. Biol. Chem. 259, 10760–10763.
93 Chandrasegaran, S., Smith, H.O., Amzel, M.L. and Ysern, X. (1986) Proteins: Structure, Function and Genetics 1, 263–266.
94 Cotton, F.A., Hazan, E.E., Jr. and Legg, M.J. (1979) Proc. Nat. Acad. Sci. U.S.A. 76, 2551–2555.
95 Suck, D., Lahm, A. and Oefner, C. (1988) Nature 332, 464–468.

96 Morikawa, K., Matsumoto, O., Tsujimoto, M., Katayanagi, K., Ariyoshi, M., Doi, T., Ikehara, M., Inaoka, T. and Ohtsuka, E. (1992) Science 256, 523–526.
97 Kuo, C.-F., McRee, D.E., Fisher, C.L., O'Handley, S.F., Cunningham, R.P. and Tainer, J.A. (1992) Science 258, 434–440.
98 Meisel, A., Bickle, T.A., Krüger, D.H. and Schroeder, C. (1992) Nature 355, 467–469.
99 Studier, F.W. and Bandyopadhyay, P.K. (1988) Proc. Nat. Acad. Sci. U.S.A. 85, 4677–4681.
100 Fuller-Pace, F.V. and Murray, N.E. (1986) Proc. Nat. Acad. Sci. U.S.A. 83, 9368–9372.
101 Kannan, P., Cowan, G.M., Daniel, A.S., Gann, A.A.F. and Murray, N.E. (1989) J. Mol. Biol. 209, 335–344.
102 Argos, P. (1985) EMBO J. 4, 1351–1355.
103 Price, C., Shepherd, J.C.W. and Bickle, T.A. (1987) EMBO J. 6, 1493–1497.
104 Gubler, M. and Bickle, T.A. (1991) EMBO J. 10, 951–957.
105 Gubler, M., Braguglia, D., Meyer, J., Piekarowicz, A. and Bickle, T.A. (1992) EMBO J. 11, 233–240.
106 Price, C., Lingner, J., Bickle, T.A., Firman, K. and Glover, S.W. (1989) J. Mol. Biol. 205, 115–125.
107 Bullas, L.R., Colson, C. and van Pel, A. (1976) J. Gen. Microbiol. 95, 166–172.
108 Fuller-Pace, F.V., Bullas, L.R., Delius, H. and Murray, N.E. (1984) Proc. Nat. Acad. Sci. U.S.A. 81, 6095–6099.
109 Nagaraja, V., Shepherd, J.C., Pripfl, T. and Bickle, T.A. (1985) J. Mol. Biol. 182, 579–587.
110 Nagaraja, V., Shepherd, J.C.W. and Bickle, T.A. (1985) Nature 316, 371–372.
111 Gann, A.A.F., Campbell, A.J.B., Collins, J.F., Coulson, A.F.W. and Murray, N.E. (1987) Mol. Microbiol. 1, 13–22.
112 Raleigh, E.A., Trimarchi, R. and Revel, H. (1989) Genetics 122, 279–296.
113 Krüger, T., Grund, C., Wild, C. and Noyer-Weidner, M. (1992) Gene 114, 1–12.
114 Dila, D., Sutherland, E., Moran, L., Slatko, B. and Raleigh, E.A. (1990) J. Bacteriol. 172, 4888–4900.
115 Kelleher, J.E. and Raleigh, E.A. (1991) J. Bacteriol. 173, 5220–5223.
116 Waite-Rees, P.A., Keating, C.J., Moran, L.S., Slatko, B.E., Hornstra, L.J. and Benner, J.S. (1991) J. Bacteriol. 173, 5207–5219.
117 Kretz, P.L., Kohler, S.W. and Short, J.M. (1991) J. Bacteriol. 173, 4707–4716.
118 Ross, T.K., Achberger, E.C. and Braymer, H.D. (1987) Gene 61, 277–289.
119 Ross, T.K., Achberger, E.C. and Braymer, H.D. (1989) Mol. Gen. Genet. 216, 402–407.
120 Ross, T.K., Achberger, E.C. and Braymer, H.D. (1989) J. Bacteriol. 171, 1974–1981.

121 Carlson, K., Raleigh, E.A. and Hattman, S. (1992) in The Molecular Biology of Bacteriophage T4 (Karam, J., ed.), American Society for Microbiology, Washington, DC (in press).
122 Piekarowicz, A., Yuan, R. and Stein, D.C. (1991) J. Bacteriol. 173, 150–155.
123 Fleishman, R.A., Campbell, J.L. and Richardson, C.C. (1976) J. Biol. Chem. 251, 1561–1570.
124 Sutherland, E., Coe, L. and Raleigh, E.A. (1992) J. Mol. Biol. 225, 327–348.
125 Sladek, T.L., Nowak, J.A. and Maniloff, J. (1986) J. Bacteriol. 165, 219–225.
126 MacNeil, D.J. (1988) J. Bacteriol. 170, 5607–5612.
127 Macaluso, A. and Mettus, A.-M. (1991) J. Bacteriol. 173, 1353–1356.
128 Russell, D.W. and Zinder, N.D. (1987) Cell 50, 1071–1079.
129 Holmes, M.L., Nuttal, S.D. and Dyall-Smith, M.L. (1991) J. Bacteriol. 173, 3807–3813.
130 Modrich, P. (1989) J. Biol. Chem. 264, 6597–6600.
131 Hattman, S. (1983) in Bacteriophage T4 (Mathews, C.K., Kutter, E.M., Mosig, G. and Berget, P.B., eds.), pp. 152–155, American Society for Microbiology, Washington, DC.
132 Marinus, M.G. (1987) Annu. Rev. Genet. 231, 113–131.
133 Marinus, M.G. (1987) in *Escherichia coli* and *Salmonella typhimurium* (Neidhardt, F.C., Ingraham, J.L., Low, K.B., Magasanik, B., Schaechter, M. and Umbarger, H.E., eds.), pp. 697–702, American Society of Microbiology, Washington, DC.
134 Lauster, R., Kriebardis, A. and Guschlbauer, W. (1987) FEBS Lett. 220, 167–176.
135 Vershon, A.K., Youderian, P., Weiss, M.A., Susskind, M.M. and Sauer, R.T. (1985) (Calendar, R. and Gold, L., eds.), pp. 209–218, Alan R. Liss, New York, NY.
136 Lauster, R., Trautner, T.A. and Noyer-Weidner, M. (1989) J. Mol. Biol. 206, 305–312.
137 Pósfai, J., Bhagwat, A.S., Pósfai, G. and Roberts, R.J. (1989) Nucl. Acids Res. 17, 2421–2435.
138 Wilke, K., Rauhur, E., Noyer-Weidner, M., Lauster, R., Pawlek, B., Behrens, B. and Trautner, T.A. (1988) EMBO J. 7, 2601–2609.
139 Walter, J., Noyer-Weidner, M. and Trautner, T.A. (1990) EMBO J. 9, 1007–1013.
140 Klimasauskas, S., Nelson, J.L. and Roberts, R.J. (1991) Nucl. Acids Res. 19, 6183–6190.
141 Miner, Z., Schlagman, S. and Hattman, S. (1988) Gene 74, 275–276.
142 Reich, N.O. and Mashhoon, N. (1990) J. Biol. Chem. 265, 8966–8970.
143 Dubey, A.K. and Roberts, R.J. (1992) Nucl. Acids Res. 20, 3167–3173.
144 Santi, D.V., Garrett, C.E. and Barr, P.J. (1983) Cell 33, 9–10.
145 Bhagwat, A.S. and Roberts, R.J. (1987) J. Bacteriol. 169, 1537–1546.
146 Wu, J.C. and Santi, D.V. (1987) J. Biol. Chem. 262, 4778–4786.
147 Som, S. and Friedman, S. (1991) J. Biol. Chem. 266, 2937–2945.
148 Friedman, S. and Ansari, N. (1992) Nucl. Acids Res. 20, 3241–3248.

149 Smith, S.S., Kaplan, B.E., Sowers, L.C. and Newman, E.M. (1992) Proc. Nat. Acad. Sci. U.S.A. 89, 4744–4748.
150 Chen, L., MacMillan, A.M., Chang, W., Ezaz-Nikpay, L., Lane, W.S. and Verdine, G.L. (1991) Biochemistry 30, 11018–11025.
151 Wyszynski, M.W., Gabbara, S. and Bhagwat, A.S. (1992) Nucl. Acids Res. 20, 319–326.
152 Reich, N.O. and Everett, E.A. (1990) J. Biol. Chem. 265, 8929–8934.
153 Bestor, T., Laudano, A., Mattaliano, R. and Ingram, V. (1988) J. Mol. Biol. 203, 971–983.
154 Yen, R.-W.C., Vertino, P.M., Nelkin, B.D., Yu, J.J., El-Deiry, W., Cumaraswamy, A., Lennon, G.G., Trask, B.J., Celano, P. and Baylin, S.B. (1992) Nucl. Acids Res. 20, 2287–2291.
155 Kelleher, J.E., Daniel, A.S. and Murray, N.E. (1991) J. Mol. Biol. 221, 431–440.
156 Proffitt, J.H., Davie, J.R., Swinton, D. and Hattman, S. (1984) Mol. Cell. Biol. 4, 985–988.
157 Fehér, Z., Kiss, A. and Venetianer, P. (1983) Nature 302, 266–268.
158 Brooks, J.E., Blumenthal, R.M. and Gingeras, T.R. (1983) Nucl. Acids Res. 11, 837–851.
159 Hoekstra, M.F. and Malone, R.E. (1985) Mol. Cell. Biol. 5, 610–618.
160 Hoekstra, M.F. and Malone, R.E. (1986) Mol. Cell. Biol. 6, 3555–3558.
161 Fehér, Z., Schlagman, S.L., Miner, Z. and Hattman, S. (1989) Current Genetics 16, 461–464.
162 Friedberg, E.C. (1988) Microbiol. Rev. 52, 70–102.
163 Wright, J.H., Gottschling, D.E. and Zakian, V.A. (1992) Genes Dev. 6, 197–210.
164 Gottschling, D.E., Aparicio, O.M., Billington, B.L. and Zakian, V.A. (1990) Cell 63, 751–762.
165 Gottschling, D.E. (1992) Proc. Nat. Acad. Sci. U.S.A. 89, 4062–4065.
166 Singh, J. and Klar, A.J.S. (1992) Genes Dev. 6, 186–196.
167 Guha, S. and Guschlbauer, W. (1992) Nucl. Acids Res. 20, 3607–3615.
168 Walker, G.C. (1984) Microbiol. Rev. 48, 60–93.
169 Walker, G.C. (1985) Annu. Rev. Biochem. 54, 425–457.
170 Day, R.S., III (1977) J. Virol. 21, 1249–1251.
171 Efimova, E.P., Delver, E.P. and Belogurov, A.A. (1988) Mol. Gen. Genet. 214, 313–316.
172 Efimova, E.P., Delver, E.P. and Belogurov, A.A. (1988) Mol. Gen. Genet. 214,317–320.
173 Hiom, K.J. and Sedgwick, S.G. (1992) Mol. Gen. Genet. 231, 265–275.
174 Belogurov, A.A., Delver, E.P. and Rodzevich, O.V. (1992) J. Bacteriol. 174, 5079–5085.
175 Read, T.D., Thomas, A.T. and Wilkins, B.M. (1992) Mol. Microbiol. 6, 1933–1941.
176 Swinton, D., Hattman, S., Crain, P.F., Cheng, C.S., Smith, D.L. and McCloskey, J.A. (1983) Proc. Nat. Acad. Sci. U.S.A. 80, 7400–7404.
177 Warren, R.A.J. (1980) Annu. Rev. Microbiol. 34, 137–158.
178 Studier, F.W. and Movva, N.R. (1976) J. Virol. 19, 136–145.
179 Mark, K.K. and Studier, F.W. (1981) J. Biol. Chem. 256, 2573–2578.

180 Bandyopadhyay, P.K. and Studier, F.W. (1985) J. Mol. Biol. 182, 567–578.
181 Moffatt, B.A. and Studier, F.W. (1988) J. Bacteriol. 170, 2095–2105.
182 Krasin, F. and Hutchinson, F. (1977) J. Mol. Biol. 116, 81–98.
183 Krasin, F. and Hutchinson, F. (1981) Proc. Nat. Acad. Sci. U.S.A. 78, 3450–3453.
184 Murialdo, H. (1988) Mol. Gen. Genet. 213, 42–49.
185 Picksley, S.M., Attfield, P.V. and Lloyd, R.G. (1984) Mol. Gen. Genet. 195, 267–274.
186 Resnick, M.A. (1969) Genetics 62, 519–531.
187 Malone, R.E. and Esposito, R.E. (1980) Proc. Nat. Acad. Sci. U.S.A. 77, 503–507.
188 Orr-Weaver, T.L., Szostak, J.W. and Rothstein, R.J. (1981) Proc. Nat. Acad. Sci. U.S.A. 78, 6354–6358.
189 Sun, H., Treco, D., Schultes, N.P. and Szostak, J.W. (1989) Nature 338, 87–90.
190 Cao, L., Alani, E. and Kleckner, N. (1990) Cell 61, 1089–1101.
191 Sun, H., Treco, D. and Szostak, J.W. (1991) Cell 64, 1155–1161.
192 Roeder, G.S. and Stewart, S.E. (1988) Trends Genet. 4, 263–267.
193 Smith, G.R. (1991) Cell 64, 19–27.
194 Stahl, F.W. (1986) in Progress in Nucleic Acid Research and Molecular Biology (Cohn, W. E. and Moldave, K.S., eds.), pp. 169–194, Academic Press, New York, NY.
195 Conley, E.C. and Saunders, J.R. (1984) Mol. Gen. Genet. 194, 211–218.
196 Winans, S.C., Elledge, S.J., Krueger, J.H. and Walker, G.C. (1985) J. Bacteriol. 161, 1219–1221.
197 Conley, E.C., Saunders, V.A. and Saunders, J.R. (1986) Nucl. Acids Res. 14, 8905–8917.
198 Conley, E.C., Saunders, V.A., Jackson, V. and Saunders, J.R. (1986) Nucl. Acids Res. 14, 8919–8932.
199 Kobayashi, I. and Takahashi, N. (1988) Genetics 119, 751–757.
200 Simmon, V.F. and Lederberg, S. (1972) J. Bacteriol. 112, 161–169.
201 Henner, W.D., Rodriguez, L.O., Hecht, S.M. and Haseltine, W.A. (1983) J. Biol. Chem. 258, 711–713.
202 Barnes, G. and Rine, J. (1985) Proc. Nat. Acad. Sci. U.S.A. 82, 1354–1358.
203 Rine, J. and Barnes, G. (1986) in Yeast Cell Biology (Hicks, J., ed.), pp. 395–413, Alan R. Liss, New York, NY.
204 Cole, G.M., Schild, D., Lovett, S.T. and Mortimer, R.K. (1987) Mol. Cell. Biol. 7, 1078–1084.
205 Cole, G.M. and Mortimer, R.K. (1989) Mol. Cell. Biol. 9, 3314–3322.
206 Morgan, W.F., Fero, M.L., Land, M.C. and Winegar, R.A. (1988) Mol. Cell. Biol. 8, 4204–4211.
207 Cox, M.M. (1991) Mol. Microbiol. 5, 1295–1299.
208 Meselson, M.S. and Radding, C.M. (1975) Proc. Nat. Acad. Sci. U.S.A. 72, 358–361.

209 Radding, C.M., Flory, J. Wu, A., Kahn, R., Dasgupta, C., Gonda, D., Bianchi, M. and Tsang, S.S. (1982) Cold Spring Harbor Symp. Quant. Biol. 47, 821–828.
210 Strathern, J.N., Klar, A.J.S., Hicks, J.B., Abraham, J.A., Ivy, J.M., Nasmyth, K.A. and McGill, C. (1982) Cell 31, 183–192.
211 Konrad, E.B. (1977) J. Bacteriol. 130, 167–172.
212 Seifert, H.S. and Porter, R.D. (1984) Proc. Nat. Acad. Sci. U.S.A. 81, 7500–7504.
213 Game, J.C., Johnston, L.H. and von Borstel, R.C. (1979) Proc. Nat. Acad. Sci. U.S.A. 76, 4589–4592.
214 Strathern, J.N., Weinstock, K.G., Higgins, D.R. and McGill, C.B. (1991) Genetics 127, 61–73.
215 Jayaram, J. (1986) in Mechanisms of Yeast Recombination (Klar, A. and Strathern, J., eds.), pp. 19–24, Cold Spring Harbor Laboratory, Cold Spring Harbor, NY.
216 Burbee, D., Campbell, J.L. and Heffron, F. (1988) in Mechanisms and Consequences of DNA Damage Processing (Friedberg, E.C. and Hanawalt, P.C., eds.), pp. 223–230, Alan R. Liss, New York, NY.
217 Ho, K.S.Y. and Mortimer, R.K. (1973) Mutat. Res. 30, 327–334.
218 Ho, K.S.Y. (1975) Mutat. Res. 30, 327–334.
219 Resnick, M.A. (1976) J. Theor. Biol. 59, 97–106.
220 Weiffenbach and Haber, J. (1981) Mol. Cell. Biol. 1, 522–534.
221 Haynes, R.H. and Kunz, B.A. (1981) in The Molecular Biology of the Yeast *Saccharomyces* (Strathern, J. and Broach, J.R., eds.), pp. 371–414, Cold Spring Harbor Laboratory, Cold Spring Harbor, NY.
222 Mahan, M.J. and Roth, J.R. (1989) Genetics 121, 433–443.
223 Bedale, W.A., Inman, R.B. and Cox, M.M. (1991) J. Biol. Chem. 266, 6499–6510.
224 Nickoloff, J.A., Singer, J.D., Hoekstra, M.F. and Heffron, F. (1989) J. Mol. Biol. 207, 527–541.
225 Fishman-Lobell, J. and Haber, J.E. (1992) Science 258, 480–484.
226 Rudin, N. and Haber, J.E. (1988) Mol. Cell. Biol. 8, 3918–3939.
227 Sugawara, N. and Haber, J.E. (1992) Mol. Cell. Biol. 12, 563–575.
228 Raveh, D., Hughes, S.H., Shafer, B.K. and Strathern, J.N. (1989) Mol. Gen. Genet. 220, 33–42.
229 White, C.I. and Haber, J.E. (1990) EMBO J. 9, 663–673.
230 Belfort, M. (1990) Annu. Rev. Genet. 24, 363–385.
231 Belfort, M. (1991) Cell 64, 9–11.
232 Lambowitz, A.M. and Belfort, M. (1993) Annu. Rev. Biochem. (in press).
233 Chu, F.K., Maley, G.F., Maley, F. and Belfort, M. (1984) Proc. Nat. Acad. Sci. U.S.A. 81, 3049–3053.
234 Chu, F.K., Maley, F.G., West, D.K., Belfort, M. and Maley, F. (1986) Cell 45, 157–166.
235 Colleaux, L., d'Auriol, L., Betermier, M., Cottarel, G., Jacquier, A., Galibert, F. and Dujon, B. (1986) Cell 44, 521–533.
236 Delahodde, A., Goguel, V., Becam, A.M., Creusot, F., Perea, J., Banroques, J. and Jacq, C. (1989) Cell, 56, 431–441.
237 Dujon, B. (1989) Gene 82, 91–114.

238 Muscarella, D.E., Ellison, E.L., Ruoff, B.M. and Vogt, V.M. (1990) Mol. Cell. Biol. 10, 3386–3396.
239 Perlman, P.S. and Butow, R.A. (1989) Science 246, 1106–1109.
240 Wenzlau, J.M., Saldanha, R.J., Butow, R.A. and Perlman, P.S. (1989) Cell 56, 421–430.
241 Michel, F. and Dujon, B. (1986) Cell 46, 323.
242 Dujon, B. (1981) in The Molecular Biology of the Yeast *Saccharomyces* (Strathern, J. and Broach, J.R., eds.), pp. 505–635, Cold Spring Harbor Laboratories, Cold Spring Harbor, NY.
243 Xiong, Y. and Eickbush, T.H. (1988) Cell 55, 235–246.
244 Muscarella, D.E. and Vogt, V.M. (1989) Cell 56, 443–454.
245 Perler, F.B., Comb, D.G., Jack, W.E., Moran, L.S., Qiang, B., Kucera, R.B., Benner, J., Slatko, B.E., Nwankwo, D.O., Hempstead, S.K., Carlow, C.K.S. and Jannasch, H. (1992) Proc. Nat. Acad. Sci. U.S.A. 89, 5577–5581.
246 Hirata, R., Ohsumi, Y., Nakano, A., Kawasaki, H., Suzuki, K. and Anraku, Y. (1990) J. Biol. Chem. 265, 6726–6733.
247 Kane, P.M., Yamashiro, C.T., Wolczyk, D.F., Neff, N., Goebl, M. and Stevens, T.H. (1990) Science 250, 651–657.
248 Gimble, F.S. and Thorner, J. (1992) Nature 357, 301–306.
249 Nakagawa, K., Morishima, N. and Shibata, T. (1991) J. Biol. Chem. 265, 1977–1984.
250 Sharma, M., Ellis, R.L. and Hinton, D.M. (1992) Proc. Nat. Acad. Sci. U.S.A. 89, 6658–6662.
251 Welsh, K.M., Lu, A.-L., Clark, S. and Modrich, P. (1987) J. Biol. Chem. 262, 15624–15629.
252 Greenstein, D. and Horiuchi, K. (1989) J. Biol. Chem. 264, 12627–12632.
253 Smith, G.R. (1988) Microbiol. Rev. 52, 1–28.
254 West, S.C. (1992) Annu. Rev. Biochem. 61, 603–640.
255 Ponticelli, A.S., Schultz, D.W., Taylor, A.F. and Smith, G.R. (1985) Cell 41, 145–151.
256 Krabbe, M. and Carlson, K. (1991) J. Biol. Chem. 266, 23407–23415.
257 Schatz, D.G. and Baltimore, D. (1988) Cell 53, 107–115.
258 Schatz, D.G., Oettinger, M.A. and Baltimore, D. (1989) Cell 59, 1035–1048.
259 Oettinger, M.A., Schatz, D.G., Gorka, C. and Baltimore, D. (1990) Science 248, 1517–1523.
260 Mombaerts, P., Iacomini, J., Johnson, R.S., Herrup, K., Tonegawa, S. and Papaioannou, V.E. (1992) Cell 68, 869–877.
261 Shinkai, Y., Rathbun, G., Lam, K.-P., Oltz, E.M., Stewart, V., Mendelsohn, M., Charron, J., Datta, M., Young, F., Stall, A.M. and Alt, F.W. (1992) Cell 68, 855–867.
262 Schatz, D.G., Oettinger, M.A. and Schlissel, M.S. (1992) Annu. Rev. Immunol. 10, 359–383.
263 Bosma, G.C., Custer, R.P. and Bosma, M.J. (1983) Nature 301, 527–530.
264 Bosma, M.J. and Carroll, A.M. (1991) Annu. Rev. Immunol. 9, 323–344.
265 Carroll, A.M. and Bosma, M.J. (1991) Genes Dev. 5, 1357–1366.

266 Roth, D.B., Nakajima, P.B., Menetski, J.P., Bosma, M.J. and Gellert, M. (1992) Cell 69, 41–53.
267 Fulop, G.M. and Phillips, R.A. (1990) Nature 347, 479–482.
268 Biedermann, K.A., Sun, J., Giaccia, A.J., Tosto, L.M. and Brown, J.M. (1991) Proc. Nat. Acad. Sci. U.S.A. 88, 1394–1397.
269 Hendrickson, E.A., Qin, X.-Q., Bump, E.A., Schatz, D.G., Oettinger, M. and Weaver, D.T. (1991) Proc. Nat. Acad. Sci. U.S.A. 88, 4061–4065.
270 Lieber, M.R. (1992) Cell 70, 873–876.
271 Kuspa, A. and Loomis, W.F. (1992) Proc. Nat. Acad. Sci. U.S.A. 89, 8803–8807.
272 Davies, J. (1990) Trends in Biotech. 8, 198–203.
273 Amabile-Cuevas, C.F. and Chicurel, M.E. (1992) Cell 70, 189–199.
274 Rayssiguier, C., Thaler, D.S. and Radman, M. (1989) Nature 342, 396–401.
275 Riele, H.T:, Maandag, E.R. and Berns, A. (1992) Proc. Nat. Acad. Sci. U.S.A. 89, 5128–5132.
276 Kwoh, T.J., Obermiller, P.S., McCue, A.W., Kwoh, D.Y., Sullivan, S.A. and Gingeras, T.R. (1988) Nucl. Acids Res. 16, 11489–11506.
277 Kostriken, R., Strathern, J.N., Klar, A.J.S., Hicks, J.B. and Heffron, F. (1983) Cell 35, 167–174.
278 Duncan, B.K. (1985) J. Bacteriol. 164, 689–695.
279 Pavco, P.A. and Steege, D.A. (1990) J. Biol. Chem. 265, 9960–9969.
280 Loenen, W.A.M., Daniel, A.S., Braymer, H.D. and Murray, N.E. (1987) J. Mol. Biol. 198, 159–170.
281 Jen-Jacobson, L., Lesser, D.R. and Kurpiewski, M.R. (1991) in Nucleic Acids and Molecular Biology, Vol. 5 (Eckstein, F. and Lilley, D.M.J., eds.), pp. 141–170, Springer-Verlag, Berlin, Heidelberg.
282 Conrad, M. and Topal, M.D. (1989) Proc. Nat. Acad. Sci. U.S.A. 86, 9707–9711.
283 Yang, C.C. and Topal, M.D. (1992) Biochemistry 31, 9657–9664.
284 Noyer-Weidner, M. and Trautner, T.A. (1992) in DNA Methylation: Molecular Biology and Biological Significance (Jost, J.P. and Saluz, H.P., eds.), pp. 39–108, Birkhauser Verlag, Basel, Switzerland.
285 Walter, J., Trautner, T.A. and Noyer-Weidner, M. (1992) Embo J. 11, 4445–4450.
286 Kapfer, W., Walter, J. and Trautner, T.A. (1991) Nucl. Acids. Res. 19, 6457–6463.
287 Xu, G.-L., Kapfer, W., Walter, J. and Trautner, T.A. (1993) Nucl. Acids Res. (in press).

MANIPULATION OF AMINO ACID BALANCE IN MAIZE SEEDS

Takashi Ueda and Joachim Messing

Waksman Institute, Rutgers University
P.O. Box 759
Piscataway, NJ 08855

INTRODUCTION

One of the major determinants that influence the nutritive values of plant seeds is the amino acid composition of seed proteins. A large fraction of seed proteins is constituted by the protein reserves referred to as seed storage proteins. They are synthesized in abundance in the embryo or the endosperm through a complex plant developmental program. These proteins serve as an essential nutrient source to support early plant seedling growth upon germination. Because of their abundance in plant seeds and their favorable amino acid composition, seed storage proteins of many agricultural crops have become the major source of nutrition for monogastric animals, including humans.

The major storage protein in maize (*Zea mays* L.), the highest-yielding cereal in the world, is the prolamin protein fraction referred to as zein. As for other cereal storage proteins, zeins are differentially synthesized at a high level in a specialized triploid endosperm tissue during seed development. In normal maize genotypes, zeins constitute 50 to 60% of the total endosperm protein at seed maturity. Since the endosperm tissue constitutes about 90% of mature kernel weight, zeins serve as the primary determinant of the amino acid composition of maize seed protein. However, zeins by themselves do not provide a well-balanced source of essential amino acids for the monogastric animal nutrition. Like most other cereal storage proteins, zeins are extremely deficient in lysine and tryptophan, and to a lesser extent in methionine (1–3). Predominant synthesis of zein in developing maize kernels amplifies its

Genetic Engineering, Vol. 15, Edited by J.K. Setlow
Plenum Press, New York, 1993

unbalanced amino acid composition, and consequently, maize seed protein is deficient in these three essential amino acids. To overcome this limitation of maize seed protein, earlier efforts were focused on the identification of "high-lysine" maize mutants with reduced zein synthesis (4,5). Although these mutants improved overall amino acid balance in maize seed protein, their unfavorable agronomic traits prevented their extensive commercialization (6).

Since the cloning of genes encoding various zein proteins, many speculations and strategies have been suggested on improving the nutritive limitation of maize seed protein through genetic engineering of zein genes. However, successful genetic manipulation of zein genes would require a comprehensive understanding of the molecular mechanisms involved in zein gene regulation as well as in zein protein synthesis and assembly. In recent years, the combination of genetic and molecular approaches has made a significant advancement in the elucidation of various mechanisms involved in zein synthesis. In this review article, we have summarized current understanding of these regulatory mechanisms. We hope that this review will serve as a useful guideline for designing novel strategies to improve the nutritive quality of maize seed protein in the future.

STRUCTURE AND AMINO ACID BALANCE OF ZEIN PROTEINS

Zein Protein Families

Zein, the alcohol-soluble fraction of endosperm proteins (classified as prolamin by Osborne (7)), is the most abundant storage protein in maize which constitutes 50 to 60% of the total endosperm protein in mature seeds (8,9). Zein consists of a group of heterologous, generally hydrophobic proteins which are divided into several classes and subclasses based on their structural similarities (10,11), their differential alcohol-solubilities (12), and their molecular weights as determined by SDS-PAGE. The α-class zeins, which include the 22- and 19-kD zeins, are the most abundant of the zeins, constituting the majority of total zeins (13). In addition to these major α-class zeins, four additional zein proteins with molecular weights of 27-, 16-, 15- and 10-kD have been identified by SDS-PAGE. To distinguish these proteins from the α-class zeins, the 15-, 16- and 27-kD, and 10-kD zeins are now classified as β-, γ-, and δ-class zeins, respectively, based on their distinctive protein structures (10,11). Whereas the α-class zeins can be extracted in alcohol (70% ethanol or 55% isopropanol), the β-, γ-, and δ-class zeins require an additional reducing reagent such as β-mercaptoethanol for their extraction because of their high cysteine contents. The β-, γ-, and δ-class zeins are thought to serve as the major source of sulfur storage for young seedlings after germination. In addition, α-class zeins exhibit extensive charge heterogeneity when fractionated by isoelectric focusing or by two-dimensional gel electrophoresis (14–16), which is thought to be due to the differences in the structure of multicopy genes encoding these proteins (17). Since there is no enzymatic function associated with these zein proteins, they serve as ideal candidates for amino acid modifications through genetic engineering.

Zein Protein Structure

As deduced from the cloned genomic and cDNA sequences of various zein genes, all different classes of zein proteins share amino acid sequence homology in their signal peptides located at the aminotermini (18). These signal peptides, usually consisting of 19 to 21 amino acids, are present in the zein precursor polypeptides, and they mediate the targeting of these zein polypeptides into the lumen of the endoplasmic reticulum (ER). Besides the signal peptide sequence, very little homology exists among different classes of zein proteins. The two α-class zeins, the 19- and 22-kD zeins, share very similar primary structures and are thought to be under a similar processing mechanism involved in protein translocation and assembly (19). Both zeins contain a conserved 20 amino acid sequence motif which is tandemly repeated in the central part of the polypeptides (20–23). The differences in the number of repeats in these two zein polypeptides account, in part, for the difference in their apparent molecular sizes. The repeated peptide sequences are predicted to form α-helical secondary structures and they are thought to play functional roles in protein assembly in the protein body (24). The β-class 15-kD zein does not contain any repetitive structure. Circular dichroism analysis has revealed that this protein is primarily composed of β-sheet and turn structures (25). The two γ-class zeins, the 27-kD and 16-kD zeins, are rich in proline and share very similar protein structures. They are sometimes classified as alcohol-soluble glutelins (ASG) because of their unique solubility in water at a high pH (26). The two zeins differ mainly in the number of proline-rich hexapeptide tandem repeats and in the 21 amino acid long peptide stretch that is only present in the 27-kD zein (26–28). These two zeins are encoded by the genes believed to be derived from a common ancestral gene (26). Among different zein classes, the γ-class zeins are the only ones which exhibit a structural similarity to other cereal storage proteins (11,27). Finally, no unique structure is found in the δ-class 10-kD zein. However, secondary structure predictions have indicated that both methionine-rich 10-kD and 15-kD zein polypeptides share a similar pattern of turns and α-helical regions following the signal peptide sequence (11), which is also found in a methionine-rich storage protein in Brazil nut (29).

Amino Acid Balance in Zein Proteins

As for other cereal prolamin storage proteins, zein proteins are rich in proline and amino acids with amide nitrogens (particularly glutamine) (Table 1). On the other hand, they all are deficient in two essential amino acids, lysine and tryptophan. These two essential amino acids are the most limiting amino acids in maize seed protein for animal nutrition as demonstrated by feeding trials with laboratory and farm animals (3). In addition, the level of another essential amino acids, methionine, is low in most zein classes except in the β-class and the δ-class zeins (Table 1). As compared with the α-class zeins, β-, γ- and δ-class zeins contain a higher level (by 3- to 7-fold) of cysteine residues, thus requiring an additional reducing reagent for their extraction (11). Since α-class zeins constitute the major fraction (70 to 80%) of total zeins, their amino

Table 1
Amino Acid Composition of Mature Zein Polypeptides

Zein Class	α-Class		β-Class	γ-Class		δ-Class
Apparent MW	19-kD[1]	22-kD[2]	15-kD[3]	16-kD[4]	27-kD[4]	10-kD[5]
Amino Acid						
Leu	44	42	16	14	19	15
Gln	39	31	26	31	30	15
Ala	31	34	22	13	10	7
Pro	21	22	14	25	51	20
Ser	15	18	8	9	8	8
Phe	13	8	0	7	2	5
Asn	9	13	3	1	0	3
Ile	9	11	1	1	4	3
Tyr	8	6	14	9	4	1
Val	6	17	3	8	15	5
Gly	4	2	14	15	12	4
Thr	4	7	4	5	9	5
Arg	3	4	5	3	5	0
His	3	3	0	4	16	3
Cys	2	1	7	12	15	5
Glu	1	1	3	3	2	0
Met	1	5	18	3	1	29
Asp	1	0	1	0	0	1
Lys	0	0	0	0	0	0
Trp	0	0	0	1	0	0
Total	214	225	159	164	203	129

References: 1 (22), 2 (21), 3 (25), 4 (26) and 5 (32).

acid compositions strongly influence the final amino acid balance in maize seed protein.

The limitation in the essential amino acid balance in maize seed protein makes maize seeds nutritionally very poor, particularly for the monogastric animals which cannot synthesize essential amino acids (3). Thus, corn meals used in animal feeds (particularly for monogastric livestock such as poultry) are usually supplemented with legume (mainly soybean) meals to increase the level of lysine. However, this corn-legume mixture is still deficient in methionine (30), and synthetic methionine is supplemented with this mixture. Similarly, it is likely that the methionine level is limited in the cereal-legume mixtures in human diets in many Third World communities (31). Therefore, for the nutritive quality of zein proteins, the 10-kD and the 15-kD zeins are essential since they contain high levels of methionine residues. The methionine contents in the 10-kD and 15-kD zein proteins are 22.5% and 11%, respectively, which are significantly higher than those in α- and γ-class zein proteins (less than 2%) (25,32). The methionine residues in the 15-kD zein protein are clustered in three regions separated by 30 amino acid residues but no apparent organization of the methionine residues is found with respect to other amino acid residues (25). The 10-kD zein contains 29 methionine residues, 17 of which are present in the central region of the protein as methionine doublets separated by two or three amino acid residues (11). The relative levels of synthesis and accumulation of these two proteins in maize seeds would influence the final methionine content of seed protein. In summary, for the improvement of the nutritive quality of maize seed protein, it is most desirable to increase the levels of the three essential amino acids: lysine, tryptophan and methionine. This may be achieved through the manipulation of zein protein structures and their gene expression.

ZEIN MULTIGENE FAMILIES

Zeins are encoded by a complex large multigene family consisting of about 100 genes present per haploid genome (33–36). Many of these zein genes are clustered at several loci in the genome as revealed by *in situ* hybridization (37) and pulse field gel electrophoresis (38). Whereas the most abundant α-class zeins including the 22- and 19-kD zeins are encoded by a large gene family of 25 to 50 members (12,33,35,38), the β-class (15-kD), γ-class (16- and 27-kD) and δ-class (10-kD) zeins are encoded by genes present in only a few copies (32,39–42). Although a large number of genes comprise the α-class zein gene family, more than half of the gene members are thought to be pseudogenes containing one or more in-frame stop codons in their coding sequences (12,43–45). No intervening sequences have been found in any of the genomic zein gene clones sequenced so far (11,22,25,39,46).

REGULATORY STEPS CONTROLLING THE ZEIN ACCUMULATION IN MAIZE SEEDS

Synthesis and Assembly of Zein Proteins

During maize kernel development, zein synthesis takes place in the endosperm between 10 and 50 days after pollination (DAP) (2). Zein proteins are synthesized initially as precursor polypeptides containing signal peptides by membrane-bound polyribosomes (47,48). These precursor polypeptides are co-translationally transported into the lumen of the ER, during which their signal peptides are processed (proteolytic cleavage) (19,49,50). Subsequently, the mature zein polypeptides aggregate in protein bodies, vesicles formed in the ER (51). Sizes of protein bodies in developing endosperm cells increase progressively with the distance from the aleurone cell layer to the central part of endosperm. This reflects the level of zein accumulation as well as a process of cell maturation. The diameter of protein bodies in the immature endosperm cells located at the periphery (subaleurone cell layer or outer cell layer) of the endosperm ranges from 0.1 to 0.2 μm, whereas in the mature cells located at the center of the endosperm (starchy endosperm), it ranges from 1 to 2 μm (52).

Mechanisms involved in the assembly and the packaging of zein proteins in protein bodies are not clear at present. Immunochemical localization studies have indicated that different classes of zein polypeptides are distributed heterogeneously within protein bodies in developing endosperm cells, which occurs in an ordered manner (52). In immature endosperm cells, protein bodies are primarily composed of β- and γ-class zeins. In mature endosperm cells, a high level of α-class zein accumulation occurs in protein bodies (52–54). These α-class zeins generally assemble into central aggregates within the matrix of β- and γ-class zeins, displacing the latter towards the surface of protein bodies (52,53). Recently δ-class zein has been shown to be localized in the central core region of protein bodies along with α-class zeins (54a). Based on these cytological observations and the analysis of primary structures of zein proteins (24,47), it has been postulated that the assembly of zeins into protein bodies is mediated by hydrophobic interactions between zein proteins belonging to different classes. This idea has been further strengthened by the observations of the zein assembly in the protein body-like vesicles in *Xenopus* oocytes (55–57). When the oocyte was injected with α-class zein mRNA alone, the protein bodies formed were slightly less dense than the native endosperm protein bodies (56). On the other hand, coinjection of α-class zein mRNA with β- and γ-class zein mRNAs resulted in the formation of much denser protein bodies (57).

To examine the physical properties of zein proteins mediating their assembly into protein bodies, the molecular conformation of the 19- and 22-kD zein polypeptides has been analyzed by circular dichroism (24). From the analysis, these zein polypeptides are likely to exhibit α-helical secondary structures through their characteristic 20 amino acid long peptide repeats described earlier. According to the structural model proposed, nine helices of the repeated peptides are placed in an antiparallel arrangement. Through the intra- and inter-molecular hydrogen bonds and van der Waals interactions, these nine helices would aggregate into roughly cylindrical, rod-shaped molecules. As their deposition in the protein body continues, these rod-shaped zein molecules can stack on top of each other by the hydrogen bonding between the glutamine residues located at the helical turns and the cylindrical cap. Whether or not the 19- and 22-kD zein proteins actually conform to the proposed structure in protein bodies and how these α-class zeins interact with other zeins in protein body assembly remain to be investigated.

The observations described above suggest the self-assembly of zein proteins into the protein bodies through intermolecular interactions between zein proteins belonging to the different classes. However, the involvement of an ER-residing protein in zein secretion and assembly has also been proposed (58,59). The b-70 protein, an ER protein with an apparent molecular weight of about 70,000, has been shown to be associated with the membranes of protein bodies and the rough ER (59,60). b-70 protein has been found to be overexpressed in the endosperm of zein regulatory mutants with reduced zein synthesis, including *floury-2* (*fl2*), *defective endosperm-b30* (*de*-B30)*, and *mucronate* (*mc*) (6,59,61). In the endosperm cells of these regulatory mutants, abnormal zein aggregations occur in protein bodies. In *fl2* plants, an increase in the b-70 protein level occurs during seed maturation and it is specific to the endosperm tissue. b-70 protein shares amino acid sequence homology with mammalian immunoglobulin binding protein (BiP) (62) which functions as a "molecular chaperone" in facilitating protein transport from the ER by mediating the proper protein folding and assembly (58,63). Based on the structural similarity of b-70 protein to BiP and the correlation between the elevated b-70 expression and the altered zein aggregation pattern in mutant endosperm cells, b-70 has been suggested to play a role as a molecular chaperone in zein secretion and assembly.

In summary, the accumulation and assembly of zein proteins into protein bodies take place in an ordered manner and involve a precise arrangement of protein structure and various protein-protein interactions. The efficiency of zein accumulation and assembly would significantly influence the final zein content in maize seeds. Thus, the structural requirements of zein proteins for their proper assembly into protein bodies must be taken into consideration when designing modifications in zein proteins.

Transcriptional Regulation of Zein Genes

The differential synthesis of zeins in developing maize kernels is mediated, at the initial step, by the endosperm-specific expression of zein genes (64,65). All zein genes are coordinately expressed during endosperm development, which starts around 10 to 12 days after pollination (DAP) (66). The coordinated expression of zein genes with respect to the tissue- and developmental stage-specificities may be attained through a common regulatory mechanism. However, recent studies on a regulatory mutant, *opaque-2*, have clearly demonstrated the presence of an additional regulation, controlling the gene member-specific expression of zein genes (67,68). Thus, complex multiple regulatory mechanisms are likely to be involved in the transcriptional regulation of zein genes. Furthermore, the quantitative level of zein gene expression also seems to be mediated by the developmentally regulated cellular process of genome amplification, which will be discussed later.

Common Regulatory Mechanism. In search of a potential regulatory mechanism controlling the coordinated expression of all zein genes with respect to the tissue- and developmental stage-specificities, 5′ flanking sequences of various zein genes have been examined for the presence of conserved *cis*-acting elements. A highly conserved *cis*-acting element identified in most of the cloned zein genes is the 7-bp sequence motif, 5′-TGTAAAG-3′ generally located about 330-bp upstream from the first ATG initiator codon (39). This sequence motif, commonly referred to as the "-300 element" (also called "endosperm box" or "prolamin box"), shares homology to the SV40 core enhancer sequence motif (69). Among all α-class zein genes, the conservation of the sequence surrounding the -300 element further extends to a 15-bp stretch, 5′-CACATGTGTAAAGGT-3′ (70). However, in the 10-kD zein gene, only the first six nucleotides of the -300 motif are conserved, and its location is shifted further upstream with respect to the initiator codon of the gene (11). Interestingly, the -300 element motif is also conserved in the 5′ flanking regions of many other cereal storage protein genes (71–76), as well as of other endosperm specifically expressed maize genes. Thus, the -300 element has been proposed to play a role in the endosperm-specific regulation of zein genes. This notion has been strengthened by the specific binding activity of nuclear proteins to the - 300 element region of a 19-kD zein gene (77). Recently, a 43-bp sequence from a 19-kD zein promoter containing the -300 element has been shown to exhibit an enhancer-like activity in transiently transformed maize endosperm protoplasts (78).

In addition to the -300 element, specific interactions of multiple nuclear proteins with AT-rich sequences, including the TATA and CAAT elements, have been demonstrated for α- and β-class zein promoters by *in vitro* binding and DNase I foot-printing assays (79–82). Some of these DNA-binding proteins are specific to the endosperm tissue while others are present in both endosperm and seedling tissues. Among DNA-binding proteins, high mobility group (HMG) proteins from maize endosperm cells have been shown to form stable complexes with these AT-rich promoter sequences. Binding of multiple protein factors to zein promoters further indicates the complexity of mechanisms involved in zein gene transcription.

Functional analysis of zein promoters. Due to the difficulties in transforming maize plants, functional analyses of zein promoters have been conducted with the use of stable transformation and transient expression systems of heterologous dicot plant species. For a 22-kD zein (*Z4*) promoter, *cis*-acting elements responsible for the endosperm-specific and temporal regulation have been shown to reside within the -174 region (relative to the transcription start site) in transgenic tobacco seeds (83). On the other hand, when an intact 19-kD zein gene (*Z19ab* 1) was introduced into transgenic petunia plants, its expression was low but not endosperm tissue-specific (84). The subsequent functional analysis of this 19-kD zein promoter in the carrot transient expression system has revealed that three discrete regions located within -483 of the promoter are essential for the transcriptional activity (85). Interestingly, the -300 element present in these α-class zein promoters was found to be dispensable for the promoter function in the two analyses described above (83,85). It is not clear whether or not this result is due to the utilization of the heterologous dicot plant system. The functional role played by the -300 element as well as the *cis*-acting mechanisms involved in the tissue-specific and temporal regulation of zein genes still remains to be elucidated in the maize system.

In some α-class genes, including the 19-kD *PML1* and *PML2* genes and the 22-kD *PMS1* and *PMS2* genes, a complex duplicated promoter structure has been identified (86,87). These duplicated promoters, designated as P1 and P2, are localized around 900 to 1000 bp and 50 bp upstream of the translation start sites of the genes, respectively. Both promoters contain the general transcription *cis*-acting element such as TATA and CAAT boxes as well as the conserved -300 element (70,88). Analyses of zein mRNAs by northern blot, R-loop, S1-mapping and primer extension methods have indicated that two discrete transcripts with molecular sizes of 800 to 900 bases and 1600 to 1800 bases are generated from these two promoters (86,87,89–91). However, the larger transcripts seem to constitute a minor fraction of zein mRNA in the developing endosperm (64,65,92). Eighteen short open reading frames (ORFs) are present between the distal P1 promoter and the AUG initiation codon of the structural gene. Recently, a functional 3′ splice site has been identified around 40 nucleotides upstream from the initiation codon (93). Splicing at this site would remove all short ORFs from the mRNA. As for the proximal P2 promoter (discussed earlier), the distal P1 promoter region seems to be also functional when analyzed in a unicellular algae, *Acetabularia* (70,91) and maize endosperm protoplasts (94). Furthermore, in transgenic petunia plants, both P1 and P2 promoters independently activated endosperm-specific expression of a bacterial reporter gene (92). The functional significance of the duplicated promoter structure in zein gene regulation is not clear. It seems likely that these duplicated promoter structures have been generated as a result of gross rearrangement of multicopy α-class zein genes through gene duplication by unequal crossover, insertion and deletion (12,41,44). In a 19-kD zein gene, short GC-rich perfect palindromic sequences clustered in a 133-bp stretch have been found around 2 kb upstream from the ATG initiator codon (95).

Multigene member-specific regulation. Combination of genetic and molecular analyses of a maize regulatory mutant, *opaque-2* (*o2*) has provided a great deal of understanding on the molecular mechanisms involved in the multigene member-specific expression of zein genes. The mutation in the *Opaque-2* (*O2*) locus causes a significant reduction (50 to 70%) in zein content in mature seeds (4,96). In plants homozygous for *o2*, the synthesis of the α-class zeins, particularly of the 22-kD zein, is primarily reduced (97,98). This reduction in 22-kD zein synthesis has been shown to be, at least in part, due to a decreased transcription rate of their genes (65). The *O2* gene has been isolated by transposon tagging (99,100), and the *O2* cDNA has been cloned and sequenced (101,102). The amino acid sequence deduced from the *O2* cDNA suggests that the *O2* gene encodes a DNA binding protein belonging to the leucine zipper (bZIP) family. The expression of the *O2* gene in normal maize plants is specific to the endosperm tissue (68,100) and Opaque-2 (O2) proteins have been localized in the nuclei of endosperm cells (103). The O2 protein has been shown to bind to a specific *cis*-acting element, 5′-TCCACGT-AGA-3′, present in the 22-kD zein promoter (67,102) and to transactivate the promoter function (68). A single nucleotide substitution in the O2 binding site abolishes the high affinity binding of O2 *in vitro* as well as its response to transactivation by O2 *in vivo.* This O2 binding site in the 22-kD zein promoter is located about 20 bp downstream from the conserved -300 element (67), which suggests possible interaction between O2 proteins and the -300 element binding factor(s). Similar but not identical O2 target sequence motifs were also found in the promoters of zein genes belonging to other classes which are not presumably under the regulation of *O2.* However, nucleotide substitutions in the O2 binding motifs in their promoters seem to prevent the binding of O2 protein and consequently their response to the transactivation by O2. Introduction of an intact O2 binding site into these heterologous zein promoters has been shown to render them responsive to the transactivation by O2 protein (68). Thus, the multigene member-specific regulation of the 22-kD zein genes by O2 protein is mediated by its sequence-specific interaction with the 22-kD zein promoter. A similar molecular *trans*-acting mechanism of O2 protein has been shown to be involved in the regulation of *b-32* gene as well as in the auto-regulation of the *O2* gene (104).

Post-transcriptional regulation. Although zein synthesis in the developing endosperm seems to be regulated primarily at the transcriptional level, there has been some evidence indicating that the quantitative level of zein synthesis is also regulated at the post-transcriptional and translational levels. This has become evident when the transcription rates of various zein genes, the levels of the corresponding polysomal or steady-state zein mRNAs, and the levels of the corresponding zein proteins were compared. Generally, there is a good correlation between the transcription rate and the corresponding polysomal mRNA level for most zein genes in a normal maize genotype. However, this correlation is particularly poor for 22-kD zein genes (64,65). Whereas the transcription rate of 22-kD zein genes is higher than that of 19-kD zein genes, the level of the corresponding polysomal mRNA is significantly lower for the 22-kD zein genes than for the 19-kD zein genes. This suggests that the α-class zein genes are under a post-transcriptional regulation. Mechanisms involved in the post-transcriptional regulation of these two α-class zein genes are not clear. Differences in gene structures do not seem to contribute to this regulation since these two α-class zein genes share similar structures. In addition, translational efficiency of α-class zeins may be further controlled by their mRNA secondary structures. Some members of the α-class zein multigene family contain short imperfect inverted repeats in the 5′ and 3′ untranslated regions (23). These inverted repeats may allow the mRNA transcripts to fold back through hybrid formation between the two inverted repeats. Although the functional significance of these inverted repeats is not clear, the base pairing between the 5′ and 3′ ends of these α-class zein mRNAs seems to reduce their translational efficiency when examined in a cell-free translation system (105).

Post-transcriptional regulation has also been demonstrated for the δ-class 10-kD zein gene in a particular maize inbred. Among various maize inbreds, BSSS-53 has been found to overexpress the 10-kD zein protein (106). Quantitative analyses of the transcription rate of the 10-kD zein gene and the steady-state level of the corresponding mRNA has indicated that the overexpression of the 10-kD zein protein in BSSS-53 is regulated, in part, by a post-transcriptional mechanism mediating a higher accumulation of 10-kD zein mRNA (32,107,130). Whether the overexpression of 10-kD zein in BSSS-53 is mediated by an increased stability of mRNA or by other regulatory mechanisms controlling later steps in gene expression is not clear. However, a single Mendelian locus, *Zpr10/(22*), mapped to chromosome 4, has been shown to be involved in these regulatory mechanisms (107,130).

Genome amplification during endosperm development. A remarkably high level of zein protein synthesis attained during endosperm development is also correlated with the cellular DNA amplification process (108). A tremendous increase in the nuclear DNA content occurs in developing endosperm cells as a result of their cessation of cell divisions followed by their continuous active DNA replication (109). The cessation of cell divisions initially occurs in the cells present in the central part of the endosperm, while cells present at the periphery remain meristematic (110, 111). An average DNA level of 90 C is normally observed in the nuclei of centrally located endosperm cells around 16 DAP. It reaches an average peak level of 123 C at 18 to 19 DAP (109,112). In some nuclei, DNA level can reach as high as 200 to 690 C. With respect to the entire endosperm, the average DNA content per nucleus at 15 DAP is approximately 12.8 C (113). It appears that this DNA amplification is extensive to the entire nuclear genome (endoreduplication), rather than selective for the endosperm specifically expressed genes like zein genes, and that both maternal and paternal genomes undergo endoreduplication (113). According to the recent immuno-localization study, predominant syntheses of α-class zeins take place in the centrally located mature endosperm cells (52). It seems likely that this high level of α-class zein synthesis in the mature endosperm cells is attained not only by the expression of a large number of genes encoding these proteins but also by the spatially regulated cellular DNA amplification process. Thus, the general observations of weak promoter activities exhibited by the 5′ flanking sequences of many zein genes (114,115) may not be surprising if one considers that overexpression of zein genes in developing endosperm tissue relies, in part, on the cellular genome amplification process.

MANIPULATION OF AMINO ACID BALANCE IN MAIZE SEED PROTEIN

Genetic Mutations and Variations with Altered Zein Synthesis

High-lysine regulatory mutants. To date a number of genetic mutations have been identified which affect the zein synthesis, leading to an altered profile of zein accumulation in maize seeds (116). These mutations, referred to as regulatory mutations, include recessive mutations such as *opaque-2* (*o2*) *opaque-5* (*o5*), *opaque-6* (*o6*) and *opaque-7* (*o7*) as well as semi-dominant mutations such as *floury-2* (*fl2*), *mucronate* (*mc*) and *defective endosperm b-30* (*de*-B30*). These mutations generally affect the synthesis and accumulation of α-class zeins. For example, the synthesis of 22-kD zein is primarily reduced in the *o2* mutation whereas the syntheses of both 19- and 22-kD zeins are reduced in the *fl2* mutation. Molecular mechanisms mediated by these regulatory loci in zein synthesis are not well understood except for the *o2* mutation as described earlier.

After the discovery of the elevated lysine and tryptophan contents in the *o2* and *fl2* mutant seeds in the 1960s (4,5), attention was focused on these mutations as a potential means to improve the nutritive limitation of maize seed protein. Although these mutations improved the overall amino acid balance in maize seed protein, because of unfavorable agronomic traits associated with the mutations including reduced seed yield, reduced protein content, soft floury endosperm, and decreased disease resistance (117), their extensive commercialization was not as successful as expected. However, in certain genetic backgrounds, the soft and floury phenotype of the *o2* mutant seeds was found to be modified to a vitreous phenotype without affecting the elevated level of lysine content (117,118). This observation has led the breeders at the International Maize and Wheat Improvement Center (CIMMYT) to generate improved maize varieties, now referred to as Quality Protein Maize (QPM) (119–121).

Since the identification of the modified *opaque-2* variants, efforts have been made to elucidate the molecular mechanisms involved in this modification. The genes responsible for these phenotypic changes are termed *opaque-2* modifier genes. The conversion of soft opaque endosperm morphology to a vitreous type in the modified *o2* variants is also associated with an elevated (2- to 3-fold) synthesis of the γ-class 27-kD zein (117,122–125). The increase in the 27-kD zein synthesis seems to result from an increase (2- to 3-fold) in the gene expression (presumably regulated at the transcriptional or post-transcriptional level) rather than the differential amplification of the gene copies (122). Consequently, the increased synthesis of the 27-kD zein leads to its altered distribution in the endosperm. Whereas in the normal endosperm, a high level of 27-kD zein synthesis takes place predominantly in the first few subaleurone cells, it also occurs in the central part of the endosperm in modified *o2* seeds (122). The elevated 27-kD zein synthesis and its altered distribution in modified *o2* seeds are highly correlated with the activities of *o2* modifier genes and exhibit a dosage dependency on the *o2* modifier genes (122,126). The *o2* modifier genes seem to act in a semi-dominant manner and their action is independent of the *o2* mutation (122). Molecular mechanisms involved in the *o2* modifier gene regulation leading to an increase in lysine and the 27-kD zein contents in modified *o2* seeds are not clear at present. However, the increased 27-kD zein expression does not account for the high lysine content in the modified *o2* seeds since the lysine residues are not present in the primary structure of the 27-kD zein proteins (27,28).

Differential expression of methionine-rich δ-class zein in maize inbred BSSS-53. In addition to the genetic mutations described above, many standard maize inbreds display variabilities in the level of zein synthesis. Most of these variabilities are caused by the *cis*-acting mutations in individual zein genes. An exception to this is the quantitative variability in the expression of the δ-class 10-kD zein proteins among different standard maize inbreds. In a maize inbred, BSSS-53, the synthesis of the 10-kD zein has been found to be significantly (2-fold) elevated as compared with other inbreds such as W64A, W22 and W23 (106). Since the 10-kD zein, together with the 15-kD zein, is rich in methionine (Table 1), an essential amino acid highly deficient in the α- and γ-class zeins, the quantitative levels of the 10-kD zein synthesis and accumulation greatly influence the overall methionine balance of maize seed protein. Consequently, the overexpression of 10-kD zein in BSSS-53 results in a 30% increase in the overall methionine content in its seeds (106,127). Because of this elevated methionine content, hybrid BSSS-53 seeds have provided an improved nutritional source in poultry feeds (128).

The elevated synthesis of 10-kD zein in BSSS-53 appears to be regulated at the post-transcriptional level (107,129) by at least one *trans*-acting locus, *Zpr10/(22)* (130). When compared with other inbreds, an increase in the steady-state level of the 10-kD zein mRNA has been found in developing endosperm of BSSS-53, which coincides with the elevated level of 10-kD zein synthesis (32). Quantitative analyses of the steady-state level of 10-kD zein mRNA and the transcription rate of the 10-kD zein gene among different maize inbreds have indicated that the increase in the steady-state level of the 10-kD zein mRNA in BSSS-53 endosperm is caused by the increased stability of the mRNA rather than the increased transcription rate of the gene (107). This post-transcriptional regulation of the 10-kD zein mRNA appears to be controlled by a *trans*-acting mechanism which is specific to the BSSS-53 inbred (107). Although more than one segregating locus in BSSS-53 may contribute to the overexpression of the 10-kD zein protein, *Zpr10/(22)* appears to be the major regulatory locus controlling this phenotype (107,129). The molecular mechanism involved in the overexpression of 10-kD zein protein by *Zpr10/(22)* is not clear. *Zpr10/(22)* may indirectly regulate the methionine-rich 10-kD zein synthesis by controlling the availability of free methionine which limits the aminoacetylation of methionyl transfer RNA, as shown for developing soybean cotyledons (131). However, it seems unlikely, since the synthesis of another methionine-rich 15-kD zein is apparently unaffected by *Zpr10/(22)* in BSSS-53 (106).

The desirable methionine content in BSSS-53 seeds, attained through the action of *Zpr10/(22)* on the 10-kD zein synthesis, however, is of limited value when this *trans*-acting mechanism is introduced into hybrid backgrounds (128). When BSSS-53 is used as a paternal parent in crosses with other inbreds such as W23 that express low levels of the 10-kD zein, a drastic reduction in the 10-kD zein accumulation occurs in hybrid seeds. However, when it is used as a maternal parent, the 10-kD zein is often expressed at higher levels. This parental influence on the phenotype may be due to the dosage effect in triploid endosperm genome or genomic imprinting (132). Since hybrid seeds are commonly used in agriculture, the desirable effect of the *Zpr10/(22)* *trans*-acting mechanism on the seed methionine content cannot be optimized by conventional breeding. This is a good example in which genetic engineering of the *trans*-acting mechanism can be utilized to overcome problems often encountered in conventional breeding.

In addition, contrary to the overexpression of the 10-kD zein in BSSS-53, the repression of the 10-kD zein has also been found in Mo17 inbred (129). This repression in Mo17 does not appear to be derived from a *cis*-acting mutation of the 10-kD zein gene, a reduced transcription rate nor a reduced translatability of the 10-kD mRNA. Another *trans*-acting mechanism present in Mo17 seems to control the steady-state level of the 10-kD zein mRNA by narrowing the developmental time frame of mRNA accumulation.

Genetic engineering of zein genes. The recent breakthrough in the stable transformation of maize plants (133,134) has set a stage for the manipulation of the amino acid balance in maize seeds through genetic engineering. Because of the predominant synthesis of zein proteins in developing maize kernels, it is desirable to modify their amino acid compositions through genetic manipulations. Appropriate modifications must be incorporated into the structures of zein genes in such a way that these modifications will optimize not only the amino acid composition of zein proteins but also the quantitative levels of gene expression and subsequent protein synthesis and accumulation. For instance, even when a desirable level of synthesis is attained for engineered zein proteins through the manipulation of their gene structures, these proteins still have to be efficiently packaged into protein bodies to cause a significant impact on the final amino acid composition of seed protein. Because of a large number of genes encoding zeins in the maize genome, a simple manipulation of a few genes is not likely to result in a drastic change in the amino acid composition of zein protein fractions. This is particularly true for the α-class zein genes since 25 to 50 copies of genes already exist in the genome. However, an advantage of manipulating α-class zeins may be the large storage capacity for these proteins in protein bodies. On the other hand, manipulation of β-, γ-, and δ-class zein genes seems to be easier and more effective because only one or two copies of genes are present in the genome. Another consideration one has to take into account is how to increase the accumulation of genetically engineered proteins. These engineered proteins must compete with the abundant endogenous zein proteins in the packaging process. One could take advantage of the existing regulatory mutants such as *opaque-2* and *floury-2*, with a reduced synthesis of α-class zein, which may provide a less competitive environment for the accumulation of engineered zein proteins. There have been a few reports dealing with the manipulation of zein genes and protein structures, which are summarized below.

One approach to improve the amino acid composition of zein proteins is to modify the coding sequence of zein genes by introducing a large number of codons for lysine, tryptophan and methionine. Naturally, these novel codons should be introduced into a part of the coding sequence which encodes mature zein polypeptides, but not into the region encoding the signal peptides. The most critical test is whether or not these modified zein proteins will be properly processed and assembled into protein bodies. Such an analysis has been made for a 19-kD zein protein (56). To design novel zein proteins with elevated lysine and tryptophan contents, the amino acid sequence of a 19-kD zein protein was modified by lysine substitutions, insertions of lysine- and tryptophan-rich oligopeptides, or an insertion of a hydrophobic fragment (about 17 kD) from the simian virus 40 (SV40) coat protein through manipulations of a cDNA clone. In the mature 19-kD zein polypeptide domain, the lysine substitutions and the oligopeptide insertions were made in the long aminoterminal "turn" region, in the glutamine-rich turns between the α helices, or within the α helices. The SV40 coat protein fragment was inserted into the aminoterminal region. With the use of the *Xenopus* oocyte system, the efficiency of the synthesis and the assembly of these modified zein proteins was examined. When the *in vitro* synthesized zein mRNAs were injected into the oocytes, all three types of modified zein proteins were efficiently synthesized and properly processed. Subsequently, these modified zein proteins, except for the one with the insertion of a large SV40 coat protein fragment, self-aggregated in the ER into dense structures within membrane-bound vesicles resembling the protein bodies in maize endosperm cells. Thus, these observations have demonstrated that zein proteins can tolerate minor structural modifications without impeding their translation, their transport into the ER and their assembly into protein bodies. However, only two lysine and tryptophan residues were included in the minor modifications. Sufficient improvement of amino acid balance in zein proteins would require a higher content of these essential amino acids. Therefore, it would be very critical to determine the extent of amino acid modifications these zein proteins can tolerate without hindering their processing and assembly into protein bodies.

Although the *Xenopus* oocyte system offers an efficient system to study the synthesis and assembly of zein proteins, it cannot provide information concerning the stability and the distribution of zein proteins synthesized from transgenes in developing endosperm tissue. To examine the synthesis and accumulation of zein proteins in developing plant seeds, intact or modified zein genes have been transferred into dicot plant species such as petunia and tobacco by *Agrobacterium* -mediated transformation (84,114,135–137). Unfortunately, the results obtained are not very informative because of the inefficient regulation of zein promoters, the inefficient translation or the low stability of zein proteins, and the difference in the secretory pathway of storage proteins, namely, the secretion *via* the golgi into vacuole-derived protein bodies present in cotyledon in these dicot plants. For example, when intact α-class 19-kD and 22-kD zein genes were introduced into petunia or tobacco plants, their expressions were very low (and tissue-unspecific for the 19-kD zein gene). Consequently, very little zein protein was detected (84,114). To circumvent the inefficient regulation of zein promoters in these dicot plants, the promoters from a gene encoding a dicot seed storage protein, β-phaseolin (135–137) and the constitutive cauliflower mosaic virus 35S (CaMV 35S) (114) have been used to express intact or modified α-class zein coding sequences. In transgenic petunia and tobacco seeds, the maximal level of 19-kD zein mRNAs obtained from the β-phaseolin gene promoter were between 1% and 2.5% of the total seed polyA^+ RNA (136,137). Despite the detection of zein mRNAs, only a small amount (ranging up to 0.003 to 0.005% of the total seed protein) of zein proteins was detected (57,136,137). A large fraction of zein proteins synthesized in tobacco seeds was degraded quickly with a half-life of less than one hour (136). Only a small fraction of zein proteins was correctly processed (136,137). They accumulated in the reticulate network or the spherical inclusions at the periphery of cells as well as in small aggregates near or within the cell wall in the embryo of mature seeds (57). When the 19-kD and 22-kD zein genes were expressed under the control of CaMV 35S promoter in transgenic tobacco plants, a small fraction of zein proteins synthesized was properly processed. They stably accumulated in the leaf, root and endosperm tissue, but not in the embryo (114). Similarly, a β-class 15-kD zein gene was also expressed under the control of the β-phaseolin promoter in transgenic tobacco plants (135). In this case, a higher level (up to 1.6% of the total seed protein) of zein protein accumulation was observed in the transgenic seed. Although the 15-kD zein proteins synthesized in developing seeds were properly processed, they accumulated in both embryo and endosperm tissue in the crystalloid component of vacuolar protein bodies, rather than in the ER-derived protein bodies as in the maize endosperm tissue.

In addition to the modifications of zein protein structures, elevated syntheses of particular zein classes may also influence the overall amino acid composition of the total zein protein fraction. Particularly, the overexpression of the methionine-rich δ-class 10-kD and β-class 15-kD zeins would significantly increase the final methionine content of the maize seed proteins and consequently, improve the nutritive quality of maize seed protein as demonstrated in the naturally occurring variant, BSSS-53 (128). It would be easier to manipulate the expression of the genes encoding these two methionine-rich zeins since only one or two copies of genes are present in the genome. To achieve higher expression levels of these genes in developing endosperm, their promoters may require some modifications. In the absence of a clear understanding on the molecular mechanisms involved in the endosperm- and developmental stage-specific expression of zein genes, it is difficult to speculate on precise strategies for the manipulation of zein promoter sequences. However, it seems feasible to alter the expression mode of these methionine-rich genes by replacing their promoters with those from the α-class zein genes. Previous immuno-localization studies show that elevated syntheses of α-class zeins take place in the mature cells located in the central region of the endosperm whereas the β- and γ-class zein syntheses begin in the immature cells located at the subaleurone layer (52). If the spatially regulated α-class zein synthesis is controlled at the transcriptional level, the utilization of α-class zein promoters for the expression of the methionine-rich zein genes may lead to a higher accumulation of their proteins in the mature endosperm cells. Moreover, a high level of gene expression may also be attained through an extensive genome amplification process taking place in these centrally located mature endosperm cells. Another way to alter the expression mode of the β- and δ-class zein genes is to introduce a novel *cis*-acting mechanism into their promoters. As demonstrated recently, a specific interaction between O2 protein and a 10-bp *cis*-acting sequence element present in the 22-kD zein promoter mediates the 22-kD zein gene-specific transactivation mechanism of O2 protein (67). Introduction of this *cis*-acting element into the 27-kD zein promoter, which is not normally under the regulation of O2 protein, renders this modified promoter responsive to transactivation by O2 protein (68). Introduction of the O2 responsive *cis*-acting element into the methionine-rich zein promoters may allow these genes to be regulated by the O2 protein. Since *O2* gene is present in normal maize endosperm tissue, introduction of the O2 responsive *cis*-element into heterologous zein promoters may increase their transcriptional activity as well as their spatial expressed mode.

These are a few examples of potential approaches to modify zein gene structures. The maize genotype-specific effects of transcriptional and post-transcriptional mechanisms on the zein synthesis must be also taken into consideration when these modified genes are introduced into maize plants. Effectiveness of these approaches should be re-evaluated as we gain more knowledge on molecular regulations of zein gene expression and protein assembly. The qualitative effects of the genetically engineered zein proteins on the final amino acid balance of the maize seed protein remain to be examined in transgenic maize plants in the near future.

FUTURE PROSPECTS

Seed storage protein research is a remarkable example of how intricately basic science and genetic breeding are tied together. Coordinately regulated developmental and cellular processes attain accumulation of storage proteins in plant seeds. Thus, seed storage proteins provide a unique and valuable model system to study a wide range of biological processes, including gene regulation, protein assembly and protein secretion. In maize, the availability of numerous genetic regulatory mutations and variants affecting seed development further compliments such basic studies. In agriculture, these regulatory mutants have been successfully utilized in conventional breeding to generate various maize varieties with improved seed protein quality. However, the conventional breeding method cannot totally overcome the nutritive limitation of the maize seed protein. Identification and characterization of various *cis*- and *trans*-regulations involved in zein synthesis have been providing an enormous amount of information on how to maximize the amino acid balance of the maize seed protein. Genetic engineering of seed storage protein genes through the manipulation of these regulatory mechanisms is now feasible. Generation of genetically engineered storage proteins with desirable amino acid balance will increase the nutritive quality of maize seeds and open up a new dimension to animal nutrition which no conventional breeding method can achieve.

Acknowledgments. We are grateful to Kathy Ward and Lin Wu for their assistance in the preparation of this review article. This work was supported by U.S. Department of Energy Grant No. 84ER13367 to J.M.

REFERENCES

1 Gianazza, E., Viglienghi, V., Righetti, P.G., Salamini, F. and Soave, C. (1977) Phytochem. 16, 315–317.

2 Wilson, C.M. (1983) in Seed Proteins: Biochemistry, Genetics, Nutritive Value (Gottschalk, W. and Muller, H.P., eds.), pp. 271–307, Martinus Nijhoff/Junk, The Hague, Netherlands.

3 Shewry, P.R. and Kreis, M. (1991) in Biochemical Aspects of Crop Improvement (Khanna, K.R., ed.), pp. 225–253, CRC Press, Boca Raton, FL.

4 Mertz, E.T., Bates, L.S. and Nelson, O.E. (1964) Science 145, 279–280.

5 Nelson, O.E., Mertz, E.T. and Bates, L.S. (1965) Science 150, 1469–1470.

6 Messing, J. (1983) Trends Biotechnol. 1, 1–6.

7 Osborne, T.B. (1924) The Vegetable Proteins. 2nd ed., Longmans, Green and Co., London.

8 Lee, K.H., Jones, R.A., Dalby, A. and Tsai, C.Y. (1976) Biochem. Genet. 14, 641–650.

9 Tsai, C.Y. (1979) Biochem. Genet. 17, 1109–1119.

10 Esen, A. (1986) Plant Physiol. 80, 623–627.

11 Kirihara, J.A., Petri, J.B. and Messing, J. (1988) Gene 71, 359–370.

12 Heidecker, G. and Messing, J. (1986) Annu. Rev. Plant Physiol. 37, 439–466.

13 Esen, A.A. (1987) J. Cereal Sci. 5, 117–128.

14 Hagen, G. and Rubenstein, I. (1980) Plant Sci. Lett. 19, 217–223.

15 Righetti, P.G., Gianazza, E., Viotti, A. and Soave, C. (1977) Planta 136, 115–123.

16 Vitale, A., Soave, C. and Galante, E. (1980) Plant Sci. Lett. 18, 57–64.

17 Soave, C., Reggiani, R., DiFonzo, N. and Salamini, F. (1981) Genetics 97, 363–377.
18 Messing, J. (1987) in Genetic Engineering, vol. 6. (Rigby, P., ed.), pp. 1–45, Academic Press, London.
19 Burr, F.A. and Burr, B. (1981) J. Cell Biol. 90, 427–434.
20 Geraghty, D., Peifer, M.A., Rubenstein, I. and Messing, J. (1981) Nucl. Acids Res. 9, 5163–5174.
21 Marks, M.D. and Larkins, B.A. (1982) J. Biol. Chem. 257, 9976–9983.
22 Pedersen, K., Devereux, J., Wilson, D.R., Sheldon, E. and Larkins, B.A. (1982) Cell 29, 1015–1026.
23 Spena, A., Viotti, A. and Pirrotta, V. (1982) EMBO J. 1, 1589–1594.
24 Argos, P., Pedersen, K., Marks, M.D. and Larkins, B.A. (1982) J. Biol. Chem. 257, 9984–9990.
25 Pedersen, K., Argos, P., Naravana, S.V.L. and Larkins, B.A. (1986) J. Biol. Chem. 261, 6279–6284.
26 Prat, S., Pérez-Grau, L. and Puigdomènech, P. (1987) Gene 52, 41–49.
27 Prat, S., Cortadas, J., Puigdomènech, P. and Palau, J. (1985) Nucl. Acids Res. 13, 1493–1504.
28 Wang, S.-Z. and Esen, A. (1986) Plant Physiol 81, 70–74.
29 Altenbach, S.B., Pearson, K.W., Leung, F.W. and Sun, S.S.M. (1987) Plant Mol. Biol. 8, 239–250.
30 Paulis, J.W., Wall, J.S. and Sanderson, J. (1978) Cereal Chem. 55, 705–712.
31 Doggett, H. (1975) in High Quality Protein Maize (Bauman, L.F., Carballo, A., Mertz, E.T. and Sprague, E.W., eds.), pp. 371–373, Dowden, Hutchinson and Ross, Stroudsburg.
32 Kirihara, J.A., Hunsperger, J.P., Mahoney, W.C. and Messing, J.W. (1988) Mol. Gen. Genet. 211, 477–484.
33 Burr, B., Burr, F.A., St. John, T.P., Thomas, M. and Davis, R.W. (1982) J. Mol. Biol. 154, 33–49.
34 Gianazza, E., Righetti, P.G., Pioli, F., Galante, E. and Soave, C. (1976) Maydica 21, 1–17.
35 Hagen, G. and Rubenstein, I. (1981) Gene 13, 239–249.
36 Viotti, A., Sala, E., Marotta, R., Alberi, P., Balducci, C. and Soave, C. (1979) Eur. J. Biochem. 102, 211–222.
37 Viotti, A., Abildsten, D., Pogna, N., Sala, E. and Pirrotta, V. (1982) EMBO J. 1, 53–58.
38 Heidecker, G. Chaudhuri, S. and Messing, J. (1991) Genomics 10, 719–732.
39 Boronat, A., Martínez, M.C., Reina, M. Puigdomènech, P. and Palau, J. (1986) Plant Sci. 47, 95–102.
40 Das, O.P. and Messing, J.W. (1987) Mol. Cell. Biol. 7, 4490–4497.
41 Das, O.P., Levi-Minzi, S., Koury, M., Benner, M. and Messing, J. (1990) Proc. Nat. Acad. Sci. U.S.A. 87, 7809–7813.
42 Wilson, D.R. and Larkins, B.A. (1984) J. Mol. Evol. 20, 330–340.
43 Kridl, J.C., Vieira, J., Rubenstein, I. and Messing, J. (1984) Gene 28, 113–118.
44 Spena, A., Viotti, A. and Pirrotta, V. (1983) J. Mol. Biol. 169, 799–811.

45 Wandelt, C. and Feix, G. (1989) Nucl. Acids Res., 17, 2354.
46 Hu, N.-T., Peifer, M.A., Heidecker, G., Messing, J. and Rubenstein, I. (1982) EMBO J. 1, 1337–1342.
47 Burr, B. and Burr, F.A. (1976) Proc. Nat. Acad. Sci. U.S.A. 73, 515–519
48 Larkins, B.A., Bracker, C.E. and Tsai, C.Y. (1976) Plant Physiol. 57, 740–745.
49 Burr, B., Burr, F.A. Rubenstein, I. and Simon, M.N. (1978) Proc. Nat. Acad. Sci. U.S.A. 75, 696–700.
50 Larkins, B.A., Pedersen, K., Handa, A.K., Hurkman, W.J. and Smith, L.D. (1979) Proc. Nat. Acad. Sci. U.S.A. 76, 6448–6452.
51 Larkins, B.A. and Hurkman, W.J. (1978) Plant Physiol. 62, 256–263.
52 Lending, C.R. and Larkins, B.A. (1989) Plant Cell 1, 1011–1023.
53 Lending, C.R., Kriz, A.L., Larkins, B.A. and Bracker C.E. (1988) Protoplasma 143, 51–62.
54 Ludevid, M.D., Torrent, M., Martinez-Izquierdo, J.A., Puigdomènech, P. and Palau, J. (1984) Plant Mol. Biol. 3, 227–234.
54a Esen, A. and Stetler, D.A. (1992) Amer. J. Bot. 79, 243–248.
55 Hurkman, W.J., Smith, L.D., Richter, J. and Larkins, B.A. (1981) J. Cell Biol. 89, 292–299.
56 Wallace, J.C., Galili, G., Kawata, E.E., Cuellar, R.E., Shotwell, M.A. and Larkins, B.A. (1988) Science 240, 662–664.
57 Wallace, J.C., Ohtani, T., Lending, C.R., Lopes, M., Williamson, J.D., Shaw, K.L., Gelvin, S.B. and Larkins, B.A. (1990) in Plant Gene Transfer, UCLA Symposium on Molecular and Cellular Biology New Series vol. 129 (Lamb, C. and Beachy, R.N., eds.), pp. 205–216, Alan R. Liss, New York, NY.
58 Fontes, E.B.P., Shank, B.B., Wrobel, R.L., Moose, S.P., OBrian, G.R., Wurtzel, E.T. and Boston, R.S. (1991) Plant Cell 3, 483–496.
59 Galante, E., Vitale, A., Manzocchi, L., Soave, C. and Salamini, F. (1983) Mol. Gen. Genet. 192, 316–321.
60 Salamini, F., Bremenkamp, M., Di Fonzo, N., Manzocchi, L., Marotta, R., Motto, M. and Soave, C. (1985) in Molecular Form and Function of the Plant Genome (Vloten-Doting, L., Groot, G.S.P. and Hall, T.C., eds.), pp. 543–553, Plenum Press, New York, NY.
61 Boston, R.S., Fontes, E.B.P., Shank, B.B. and Wrobel, R.L. (1991) Plant Cell 3, 497–505.
62 Haas, I.G. and Wabl, M. (1983) Nature 306, 387–389.
63 Marocco, A., Santucci, A., Cerioli, S., Motto, M., Di Fonzo, N., Thompson, R. and Salamini, F. (1991) Plant Cell 3, 507–515.
64 Boston, R.S., Kodrzycki, R. and Larkins, B.A. (1986) in Molecular Biology of Seed Storage Proteins and Lectins (Shannon, L.M. and Chrispeels, M.J., eds.), pp. 117–126, University of California, Riverside, CA.
65 Kodrzycki, R., Boston, R.S. and Larkins, B.A. (1989) Plant Cell 1, 105–114.
66 Marks, M.D., Lindell, J.S. and Larkins, B.A. (1985) J. Biol. Chem. 260, 16445–16450.

67 Schmidt, R.J., Ketudat, M., Aukerman, M.J. and Hoschek, G. (1992) Plant Cell 4, 689–700.
68 Ueda, T., Waverczak, W., Ward, K., Sher, N., Ketudat, M., Schmidt, R.J. and Messing, J. (1992) Plant Cell 4, 701–709.
69 Zenke, M., Grundström, T., Matthes, H., Wintzerith, M., Schatz, C., Wildeman, A. and Chambon, P. (1986) EMBO J. 5, 387–397.
70 Brown, J.W.S., Wandelt, C., Feix, G., Neuhaus, G. and Schweiger, H.G. (1986) Eur. J. Cell Biol. 42, 161–170.
71 Brandt, A., Montembault, A., Camero-Mills, V. and Rasmussen, S.K. (1985) Carlsberg Res. Commun. 50, 333–345.
72 Colot, V., Rober, L.S., Kavanaph, T.A., Bevan, M.W. and Thompson, R.D. (1987) EMBO J. 6, 3559–3564.
73 DeRose, R.T., Ma, D.-P., Kwon, I.-S., Hasnain, S.E., Klassy, R.C. and Hall, T.C. (1989) Plant Mol. Biol. 12, 245–256.
74 Shotwell, M.A., Boyer, S.K., Chesnut, R.S. and Larkins, B.A. (1990) J. Biol. Chem. 265, 9652–9658.
75 Sugiyama, T., Rafalski, A., Peterson, D. and Soll, D. (1985) Nucl. Acids Res. 13, 8729–8737.
76 Thompson, G.A. and Larkins, B.A. (1989) BioEssays 10, 108–112.
77 Maier, U.-G., Brown, J.W.S., Toloczyki, C. and Feix, G. (1987) EMBO J. 6, 17–22.
78 Quayle, T. and Feix, G. (1992) Mol. Gen. Genet. 231, 369–374.
79 Grasser, K.D., Maier, U.G., Haass, M.M. and Feix, G. (1990) J. Biol. Chem. 265, 4185–4188.
80 Maier, U.G., Brown, J.W.S., Schmitz, L.M., Schwall, M., Dietrich, G. and Feix, G. (1988) Mol. Gen. Genet. 212, 241–245.
81 Maier, U.G., Grasser, K.D., Haass, M.M. and Feix, G. (1990) Mol. Gen. Genet. 221, 164–170.
82 So, J.S. and Larkins, B.A. (1991) Plant Mol. Biol. 17, 309–319.
83 Matzke, A.J.M., Stöger, E.M., Schernthaner, J.P. and Matzke, M.A. (1990) Plant Mol. Biol. 14, 323–332.
84 Ueng, P., Galili, G., Sapanara, V., Goldsbrough, P.B., Dube, P., Beachy, R.N. and Larkins, B.A. (1988) Plant Physiol. 86, 1281–1285.
85 Thompson, G.A., Boston, R.S., Lyznik, L.A., Hodges, T.K. and Larkins, B.A. (1990) Plant Mol. Biol. 15, 755–764.
86 Langridge, P. and Feix, G. (1983) Cell 34, 1015–1022.
87 Langridge, P., Eibel, H., Brown, J.W.S. and Feix, G. (1984) EMBO J. 3, 2467–2471.
88 Brown, J.W.S., Maier, U.G., Schwall, M., Schmitz, L., Wandelt, C. and Feix, G. (1988) Biochem. Physiol. Pflanzen 183, 99–106.
89 Langridge, P., Pintor-Toro, J.A. and Feix, G. (1982) Mol. Gen. Genet. 187, 432–438.
90 Langridge, P., Pintor-Toro, J.A. and Feix, G. (1982) Planta 156, 166–170.
91 Langridge, P., Brown, J.W.S., Pintor-Toro, J.A., Feix, G., Neuhaus, G., Neuhaus-Url, G. and Schweiger, H.-G. (1985) Eur. J. Cell Biol. 39, 257–264.
92 Quattrocchio, F., Tolk, M.A., Coraggio, I., Mol, J.N.M., Viotti, A. and Koes, R.E. (1990) Plant Mol. Biol. 15, 81–93.

93 Brown, J.W.S. and Feix, G. (1990) Nucl. Acids Res. 18, 111–117.
94 Giovinazzo, G., Manzocchi, L.A., Bianchi, M.W., Coraggio, I. and Viotti, A. (1992) Plant Mol. Biol. 19, 257–263.
95 Quayle, T.J.A., Brown, J.W.S. and Feix, G. (1989) Gene 80, 249–257.
96 Tsai, C.Y., Larkins, B.A. and Glover, D.V. (1978) Biochem. Genet. 16, 883–896.
97 Burr, F.A. and Burr, B. (1982) J. Cell Biol. 94, 201–206.
98 Jones, R.A., Larkins, B.A. and Tsai, C.Y. (1977) Plant Physiol. 59, 525–529.
99 Motto, M., Maddaloni, M., Ponziani, G., Brembilla, M., Marotta, R., Di Fonzo, N., Soave, C., Thompson, R. and Salamini, F. (1988) Mol. Gen. Genet. 212, 488–494.
100 Schmidt, R.J., Burr, F.A. and Burr, B. (1987) Science 238, 960–963.
101 Hartings, H., Maddaloni, M., Lazzaroni, N., Di Fonzo, N., Motto, M., Salamini, F. and Thompson, R. (1989) EMBO J. 8, 2795–2801.
102 Schmidt, R.J., Burr, F.A. Aukerman, M.J. and Burr, B. (1990) Proc. Nat. Acad. Sci. U.S.A. 87, 46–50.
103 Varagona, M.J., Schmidt, R.J. and Raikhel, N.V. (1991) Plant Cell 3, 105–113.
104 Lohmer, S., Maddaloni, M., Motto, M., Di Fonzo, N., Hartings, H., Salamini, F. and Thompson, R.D. (1991) EMBO J 10, 617–624.
105 Spena, A., Krause, E. and Dobberstein, B. (1985) EMBO J. 4, 2153–2158.
106 Phillips, R.L. and McClure, B.A. (1985) Cereal Chem. 62, 213–218.
107 Cruz-Alvarez, M., Kirihara, J.A. and Messing, J. (1991) Mol. Gen. Genet. 225, 331–339.
108 Kowles, R.V. and Phillips, R.L. (1988) Intern. Rev. Cytol. 112, 97–136.
109 Phillips, R.L., Kowles, R.V., McMullen, M.D., Enomoto, S. and Rubenstein, I. (1985) in Plant Genetics (Freeling, M., ed.), pp. 739–754, Alan R. Liss, New York, NY.
110 Fisk, E.L. (1927) Amer. J. Bot. 14, 53–75.
111 Randolph, L.F. (1936) J. Agric. Res. 53, 881–916.
112 Kowles, R.V. and Phillips, R.L. (1985) Proc. Nat. Acad. Sci. U.S.A. 82, 7010–6014.
113 Kowles, R.V., Srienc, F. and Phillips, R.L. (1990) Develop. Genet. 11, 125–132.
114 Schernthaner, J.P., Matzke, M.A. and Matzke, A.J.M. (1988) EMBO J. 7, 1249–1255.
115 Ueda, T. and Messing, J. (1991) Theor. Appl. Genet. 82, 93–100.
116 Motto, M., Di Fonzo, N., Hartings, H., Maddaloni, M., Salamini, F., Soave, C. and Thompson, R.D. (1989) Oxf. Surv. Plant Mol. Cell Biol. 6, 87–114.
117 Ortega, E.I. and Bates, L.S. (1983) Cereal Chem. 60, 107–111.
118 Paez, A.V., Helm, J.L. and Zuber, M.S. (1969) Crop Sci. 9, 251–252.
119 Bressani, R., Benavides, V., Acevedo, E. and Ortiz, M.A. (1990) Cereal Chem. 67, 515–518.
120 Magnavaca, R. (1991) in Proceedings of the International Conference on Sorghum Nutritional Quality (Ejeta, G., Kirleis, A. and Mertz, E., eds.), pp. 75–82, Purdue University Press, West Lafayette, IN.

121 Vasal, S.K., Villegas, E., Bjarnason, M., Gelaw, B. and Goertz, P. (1980) in Improvement of Quality Traits of Maize for Grain and Silage Use (Pollmer, W.G. and Phillips, R.H., eds.), pp. 37–73, Martinus Nijhoff, London, UK.
122 Geetha, K.B., Lending, C.R., Lopes, M.A., Wallace, J.C. and Larkins, B.A. (1991) Plant Cell 3, 1207–1219.
123 Gentinetta, E., Maggiore, F. and Salamini, F. (1975) Maydica 20, 145–164.
124 Paiva, E., Kriz, A.L., Peixoto, M.J.V.V.D., Wallace, J.C. and Larkins, B.A. (1991) Cereal Chem. 68, 276–279.
125 Wallace, J.C., Lopes, M.A., Paiva, E. and Larkins, B.A. (1990) Plant Physiol. 92, 191–196.
126 Lopes, M.A. and Larkins, B.A. (1991) Crop Sci. 31, 1655–1662.
127 Phillips, R.L., Morris, P.R., Wold, F. and Gengenbach, B.G. (1981) Crop Sci. 21, 601–607.
128 Messing, J. and Fisher, H. (1991) J. Biotechnol. 21, 229–238.
129 Schickler, H., Benner, M.S. and Messing, J. (1993) Plant J. (in press).
130 Benner, M.S., Phillips, R.L., Kirihara, J.A. and Messing, J.W. (1989) Theor. Appl. Genet. 78, 761–767.
131 Thompson, J.F., Madison, J.T., Waterman, M.A. and Muenster, A.-M.E. (1981) Phytochem. 20, 941–945.
132 Kermicle, J.L. (1970) Genetics 66, 69–85.
133 Fromm, M.E., Morrish, F., Armstrong, C., Williams, R., Thomas, J. and Klein, T.M. (1990) Bio/Tech. 8, 833–844.
134 Gordon-Kamm, W.J., Spencer, T.M., Mangano, M.L., Adams, T.R., Daines, R.J., Start, W.G., O'Brien, J.V., Chambers, S.A., Adams, W.R. Jr., Willetts, N.G., Rice, T.B., Mackey, C.J., Krueger, R.W., Kausch, A.P. and Lemaux, P.G. (1990) Plant Cell 2, 603–618.
135 Hoffman, L.M., Donaldson, D.D., Bookland, R., Rashka, K. and Herman, E.M. (1987) EMBO J. 6, 3213–3221.
136 Ohtani, T., Galili, G., Wallace, J.C., Thompson, G.A. and Larkins, B.A. (1991) Plant Mol. Biol. 16, 117–128.
137 Williamson, J.D., Galili, G., Larkins, B.A. and Gelvin, S.B. (1988) Plant Physiol. 88, 1002–1007.

INVESTIGATIONAL APPROACHES FOR STUDYING THE STRUCTURES AND BIOLOGICAL FUNCTIONS OF MYELOID ANTIMICROBIAL PEPTIDES

Michael E. Selsted

Departments of Pathology and
Microbiology and Molecular Genetics
College of Medicine
University of California
Irvine, CA 92717

INTRODUCTION

Macrophages and neutrophils are phagocytic leukocytes of myeloid lineage, arising from hematopoietic precursors in the bone marrow. Neutrophils are short-lived cells which are terminally differentiated as they enter the circulation, whereas macrophages derive from circulating monocytes which mature into specialized cells in various tissues including lung, liver and skin. The phagocytic and microbicidal functions of neutrophils and macrophages provide an essential component of the host's defense against colonization by a diverse spectrum of potential pathogens.

The antimicrobial functions of phagocytes rely in large part on their capacity to ingest and kill invading microorganisms. Microbial killing occurs via oxidative processes which entail the production of oxygen-derived metabolites, as well as by oxygen-independent processes (1). Killing by the latter pathway is dependent on phagolysosomal fusion whereby antimicrobial substances in the phagocyte's cytoplasmic granules are delivered to the microbe-containing phagosome (2).

The identification and characterization of antimicrobial components contained in phagocyte granules has been the subject of numerous investigations over the last four decades. In a series of publications beginning in the early 1960s, Zeya and Spitznagel reported on the *in vitro* antibacterial activities of several low molecular weight cationic proteins from granules of rabbit and guinea pig neutrophils (3–5). Subsequent biochemical investigations revealed these molecules to be disulfide-rich peptides now known to be members of a family of structurally-related antimicrobial molecules called

Genetic Engineering, Vol. 15, Edited by J.K. Setlow
Plenum Press, New York, 1993

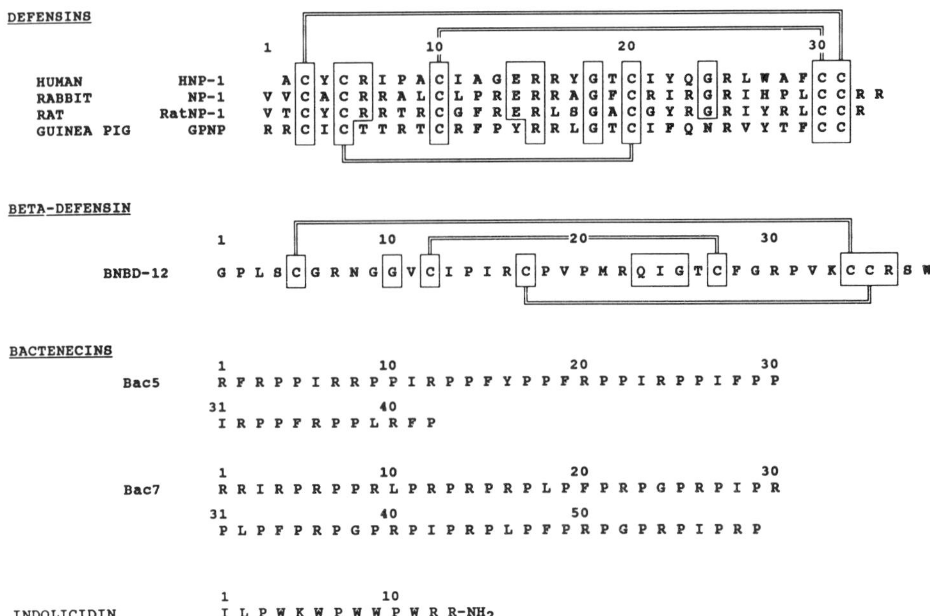

Figure 1. Covalent structures of representative myeloid antimicrobial peptides. The primary and disulfide structures of selected members of the defensin and beta-defensins families are shown with highly conserved (defensins) or invariant (beta-defensins) residues enclosed in boxes. The four defensins are representative of a myeloid peptide family which now has sixteen members; BNBD-12 is one of 13 bovine neutrophil beta-defensins. The cysteine connectivities, which differ between two families, are indicated with double lines. Bac5 and Bac7 are bectenecins from bovine neutrophils and are characterized by their abundance of arginine and proline, and the absence of disulfides. Indolicidin is among the shortest antimicrobial peptides, is highly cationic and uniquely rich in tryptophanyl residues.

defensins (6). It is now well established that the cytoplasmic granules of mammalian phagocytes contain a complex array of antibacterial proteins and peptides (2,7–10). Among the larger polypeptides of phagocyte granules are proteins such as bactericidal/permeability increasing factor (BPI), CAP 37/azurocidin, cathespsin G, lactoferrin and lysozyme (2,7–9). The antimicrobial peptides which have been isolated and characterized in recent years constitute a diverse and rapidly expanding list which includes the bactenecins (10), defensins (6), beta-defensins (11,12) and indolicidin (13). The covalent structures of representative members of these peptide families are shown in Figure 1.

The aim of this chapter is to introduce several methodological approaches which have been developed for investigating the biochemical and functional properties of antimicrobial peptides from neutrophils and macrophages. Emphasis is given to techniques for the purification, characterization, synthesis and functional analysis of the peptides; the reader is encouraged to consult several recent reports on the molecular genetics of these molecules (14–17), as these studies will not be discussed.

PURIFICATION OF PHAGOCYTE ANTIMICROBIAL PEPTIDES

Cellular Fractionation

Since all of the known peptides in neutrophils and macrophages are packaged in lysosome-like cytoplasmic granules, an initial enrichment of these molecules can be achieved by preparation of a granule-enriched subcellular fraction. Freshly-obtained macrophages or neutrophils may be homogenized in a Dounce or Potter-Elvejhem homogenizer, or by nitrogen cavitation in a Parr bomb (18). The latter method is preferred in this laboratory because it is rapid and cell breakage appears to be nearly quantitative (13). The granule-rich fraction is harvested by differential centrifugation to yield a pellet which is stored at -70° C (13).

Polypeptide Extraction

Efficient extraction of peptide from granules or intact cells can be performed by suspending tissue pellets directly into ice cold 30% acetic or formic acid. Sonication or homogenization is sometimes required to generate a homogeneous suspension, and extraction is typically allowed to proceed by stirring in a melting ice bath for 8 to 18 hours. Repeated extraction of the centrifuged pellet will occasionally increase the yield by 5 to 10%. The acid extract, containing granule proteins and trace amounts of cellular ions, is then taken to a dry powder by lyophilization.

Peptide Fractionation

Because the sizes of the known antimicrobial peptides are in the range of 1.5 to 7 kD (10), gel filtration is an efficient step for generating relatively large scale (1 to 100 milligram) quantities of low molecular weight material. For example, in the purification of antimicrobial peptides from 1.3×10^{10} bovine neutrophils, an acid extract of the granules was fractionated on a 4.8 × 110 cm column of BioGel P-60 equilibrated in 5% acetic acid (13; Figure 2). Eluent fractions were analyzed on 10 to 20% gradient SDS polyacrylamide gels, revealing that peaks 5 and 6 contained peptides of 4 to 5 kD (beta-defensins) and approximately 2 kD (indolicidin) respectively (Figure 2). Both peaks eluted well after the solvent-included volume of the column, indicating that the peptides are retarded by interaction with the gel matrix, a phenomenon also observed for defensins. Recognition of this anomalous chromatographic behavior is important for two reasons: 1) it may be exploited to provide substantially purified material in a single step (e.g., see Figure 2); and 2) if

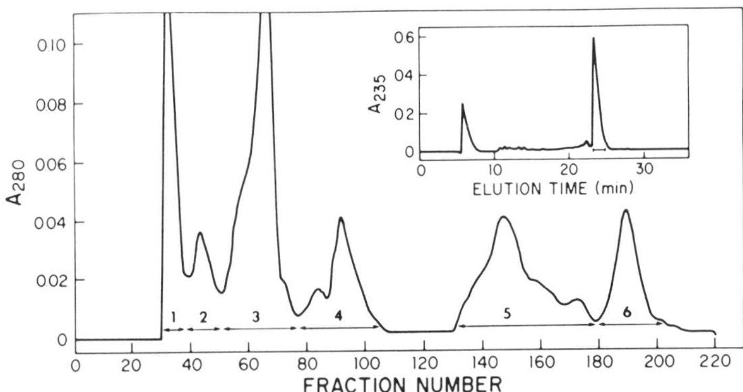

Figure 2. Gel filtration purification of antimicrobial peptides from acid extract of bovine neutrophil granules. Beta-defensins (~ 12 mg; peak 5) and indolicidin (~ 1.5 mg; peak 6) were purified from larger granule constituents on a column of Bio-Gel P-60. Subsequent reversed phase HPLC purification of indolicidin (inset) revealed that the peptide in peak 6 was nearly pure (from reference 13, with permission).

collection of fractions is terminated at the included volume of the column, biologically active peptides may be inadvertently lost.

Screening of Fractions for Antimicrobial Activity

To screen for antimicrobial activity in whole cell or granule extracts, or from chromatographically resolved fractions, microbicidal or microbial inhibition assays are employed. The former, described in detail later in this chapter, allows for the detection of compounds that kill microbes whereas the latter method does not distinguish between microbial killing and inhibition. Since the microbicidal assay is relatively arduous, the number of samples which can be evaluated in a single determination is limited. Alternatively, large numbers of samples can be screened with a microbial inhibition method such as the plate diffusion assays described by Hultmark et al. (19) and modified by Lehrer et al. (20). In these assays, nutrient-containing agar (or agarose) plates are seeded with a selected target microorganism, and peptide samples (5 to 10 μl) are then placed into small wells formed in the solid medium or onto discs placed on the surface. Following an appropriate incubation interval, microbial growth inhibition may be visualized and quantitated by measuring the clear zones around each well or disc. If the seeded medium contains limited nutrient (20), it may be necessary to overlay the agar(ose) with a layer of enriched solid medium in order to support microbial growth outside the perimeter of inhibition. Hultmark (19) and Lehrer (20) and their coworkers have also described overlay methods for detecting antimicrobial activities in polyacrylamide gels.

As noted previously, the granules of phagocytes contain a diverse collection of microbicidal polypeptides, and it is not surprising, therefore, that the bioassay of gel filtration chromatography fractions discloses many antimicrobial fractions. For example, each of the peaks in Figure 2 contained proteins which were antibacterial against *E. coli* and *S. aureus* in both inhibition and microbicidal assays (11,13).

Purification of Homogeneous Peptide Preparations

A number of methods have been used to obtain highly purified antimicrobial peptides from material which has been partially purified by gel filtration chromatography. In some of our earliest defensin purifications, six rabbit neutrophil peptides were separated from each other by preparative acid-urea PAGE (21). Individual bands were localized by brief immersion of the gel in an alkaline solution of eosin Y (22), and following excision of the gel segments, the peptides were electroeluted. Subsequently, an ion-exchange HPLC method was developed as an alternative, high capacity procedure for separating each rabbit defensin (23). Invariably the final step in peptide purification is reversed-phase HPLC with volatile solvents. In many cases, trifluoroacetic acid (0.1% vol/vol) has been used as the ion pairing reagent in water/acetonitrile gradients. In the separation of beta-defensins, 0.13% (vol/vol) heptafluorobutyric acid was required for one dimension of RP-HPLC which provided an additional measure of selectivity for purifying this large (n = 13) family of highly homologous peptides (11).

For assessing purity, we routinely use four criteria: 1) homogeneous behavior on RP-HPLC, 2) single bands on SDS-PAGE and acid-urea PAGE, 3) a unique aminoterminal residue upon sequence analysis and 4) an amino acid composition consistent with sequence. The utility of acid-urea PAGE deserves special mention. Peptides as small as 10 amino acids can be separated as a combined function of charge and size in this electrophoretic system, and detection of 1 μg is easily accomplished with formalin-containing solutions of Coomassie Brilliant Blue (13). The ability of acid-urea gels to resolve closely related peptides can be demonstrated by the resolution of 15 myeloid defensins (24–27; Figure 3) and 13 recently purified beta-defensins from bovine neutrophils (11; Figure 4).

STRUCTURAL CHARACTERIZATION

Covalent Structural Analysis

The early work of Zeya and Spitznagel revealed that guinea pig and rabbit neutrophil "cationic proteins" were rich in arginine, and the abundance of this amino acid is a feature of virtually every myeloid antimicrobial peptide isolated to date (see Figure 1). As sequence analysis became routine in the 1980s, it was possible to determine the primary structures of many of these molecules. The first sequences of myeloid antimicrobial peptides were of two defensins from rabbit lung macrophages (28), followed by determinations of the primary

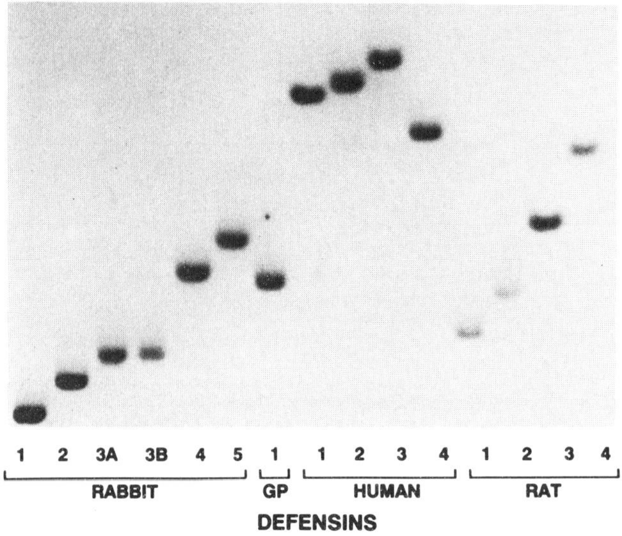

Figure 3. Acid-urea PAGE of defensins. 1 to 3 μg of purified defensins from rabbit, guinea pig (GP), human and rat neutrophils were resolved on this 12.5% acid-urea gel. The numbering of individual peptides corresponds to the nomenclature used in the original reports on these defensins.

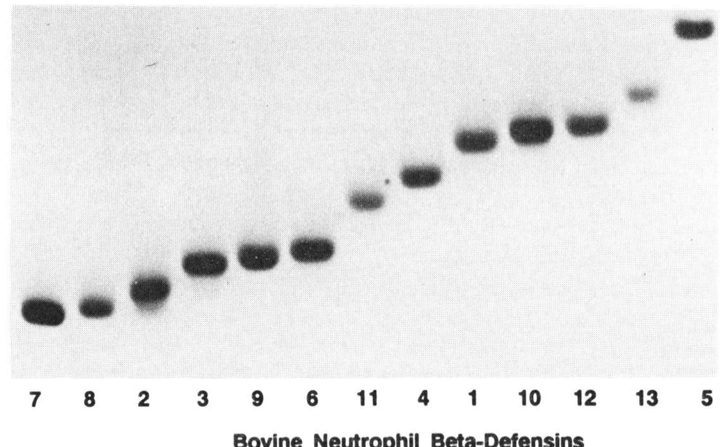

Figure 4. Acid-urea PAGE of bovine neutrophil beta-defensins. Thirteen highly homologous peptides were purified by gel filtration and RP-HPLC methods, and assessed for purity on a 12.5% acid-urea gel. Two μg of each beta-defensin (BNBD 1–13), loaded in order of decreasing cationicity, were electrophoresed for 4 hr and stained with formalin-containing Coomassie blue. The numbering of the peptides is based on their order of elution on RP-HPLC.

structures of six rabbit neutrophil defensins (24). These determinations were performed with the use of the relatively inefficient spinning cup sequencers which often required milligram quantities of material for sequencing. With the advent of gas-phase automated sequencing, it has become possible to sequence at the tens-of-picomoles level in a fully automated mode. Instruments employing these technologies have been used for sequencing the majority of the myeloid antimicrobial peptides characterized to date.

Though the great majority of residues identified in the antimicrobial peptide sequences (e.g., Figure 1) were established with modern automated Edman degradation, several special problems have been encountered which merit mention. The carboxytermini of five rabbit defensins end with the dipeptide Arg-Arg (24). To ascertain whether one or two arginine residues occurred at this position required amino acid analysis of carboxyterminal fragments released following chymotryptic or staphylococcal protease digestion (24,28). Among the 13 beta-defensins are 7 peptides in which the aminoterminus is blocked with a pyroglutamyl residue (11). Sequencing of these peptides required enzymatic removal of the aminoterminal residue with pyroglutamyl aminopeptidase prior to automated Edman degradation. Further, the carboxyterminal tryptophans of several beta-defensins were not detected by automated sequence determinations, and required analysis by amino acid analysis of carboxypeptidase A-treated samples (11). The presence of the carboxamidated C-terminus of indolicidin was established by fast atom bombardment mass spectrometry (13).

The defensin and beta-defensin families are characterized by conserved cystine arrays which are unique to each family (Figure 1). To determine the cysteine connectivities, representative members of each family (defensin: HNP-2; beta-defensin: BNBD-12) were digested with selected proteases, and the disulfide-containing fragments were purified by RP-HPLC and characterized by amino acid analysis (13,29). In both HNP-2 and BNBD-12, a Cys-Cys dipeptide occurs near the carboxyterminus. In order to assign the disulfide bond involving each of the cysteines in the respective dipeptides, a partially digested oligopeptide was subjected to a single step of Edman degradation, cleaving the Cys-Cys peptide bond and liberating two disulfide-containing fragments (Figure 5). Amino acid analysis of the fragments allowed assignment of the tri-disulfide motif in the defensin and beta-defensin families as shown in Figurc 1.

Secondary and Tertiary Structures

Several defensins have been characterized by circular dichroism spectrometry. Though quite small, these peptides are rather uniformly rich in beta-stranded secondary structure and lack measurable helical content. This feature was confirmed by two-dimensional NMR experiments on three defensins: NP-5 (30), NP-2 and HNP-1 (31,32) which show that all three peptides contain similar highly-folded conformations composed predominantly of beta-sheet. The three-dimensional solution structures determined for these defensins were the first to demonstrate the amphiphilicity of the folded peptide chain. In

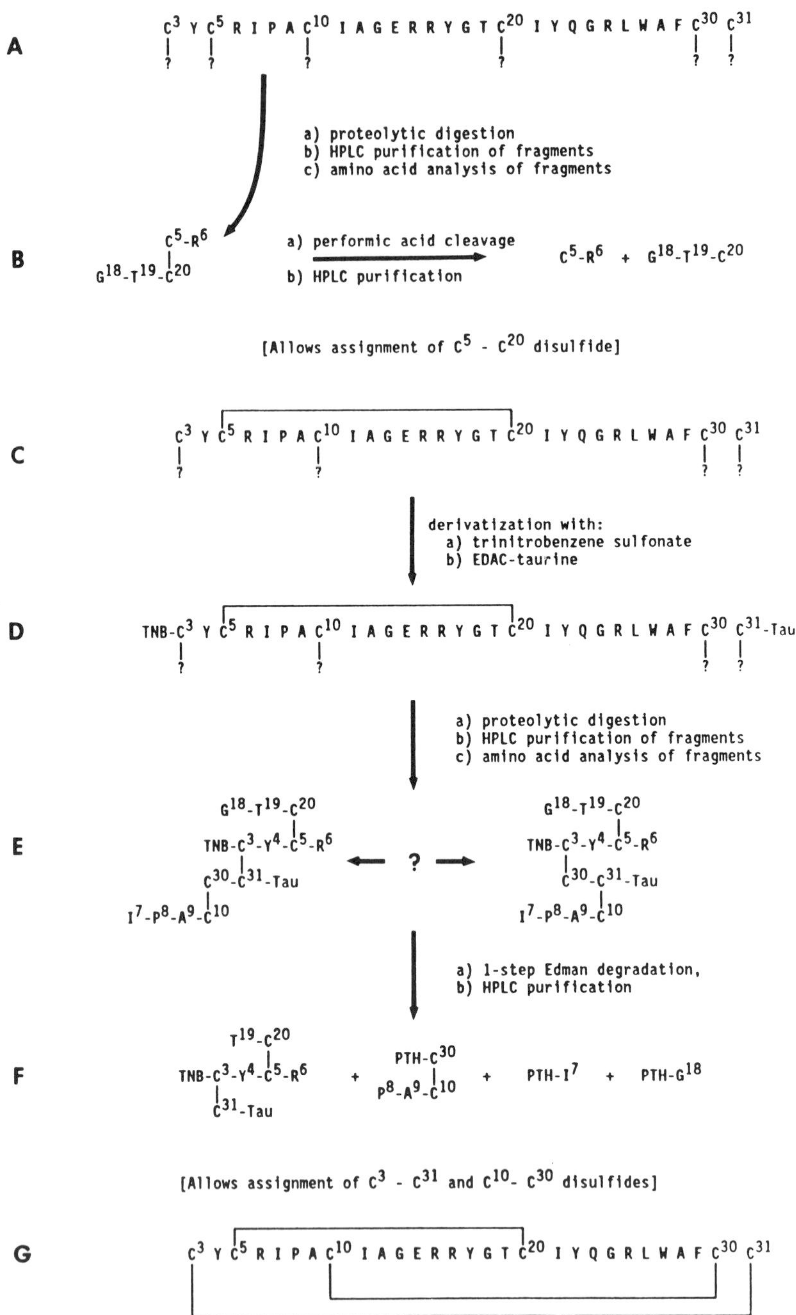

Figure 5. The strategy for determination of the disulfide bonds in human defensin HNP-2 (from reference 29, with permission).

addition, they provided confirmatory data on the disulfide assignments (Figures 1 and 5).

Defensins are rather easily crystallized at low pH and, to date, crystals of NP-1, NP-2, NP-4, NP-5, HNP-1 and HNP-3 have been grown by vapor diffusion in hanging drops. Interestingly, HNP-3 is the only defensin which has been readily derivatized with heavy atoms (33). This enabled collection of 2.4 to 2.5 Å diffraction data on platinum salt derivatives of HNP-3 crystals. The crystal structure of HNP-3 was solved by the multiple isomorphous replacement method (33), revealing that the structure of the HNP-3 monomer is very similar to that of HNP-1 determined by 2D-NMR in solution (31). The monomeric structures of defensins, determined in crystals or in solution, reveal that they are dominated by a triple-stranded beta-sheet which is highly constrained by three disulfides. As will be discussed below, defensins are membrane active, and as such are structurally unique compared to other known lytic peptides, most of which are amphiphilic helices.

Quaternary Relationships

The crystal structure of HNP-3 showed that the peptide is dimerized through several hydrogen bonds (33), a feature which is consistent with the presence of several slowly-exchanging protons measured in NMR experiments on HNP-1 (31). The dimer is extremely stable, and confers to the structure a high degree of amphiphilicity (33). NMR experiments demonstrate that higher-order aggregates of HNP-1 exist in solution, a finding which was also confirmed by equilibrium sedimentation studies of HNP-3 (33). As discussed below, additional assembly of defensin monomers and dimers appears to take place as the peptide interacts with lipid bilayers. Several models have been proposed to explain the cytocidal interaction of the peptide with its cellular target, each requiring the disruption of the cytoplasmic membrane. This may involve a nonspecific deformation ("wedge hypothesis"; Figure 6) of the membrane or may result from the formation of membrane pores (33).

Solid Phase Peptide Synthesis (SPPS)

A number of the investigational approaches described here have benefited from the development of protocols for chemical synthesis of myeloid antimicrobial peptides. In particular, the synthetic production of tens to hundreds of milligrams of several peptides has provided material for crystallographic and NMR investigations, and as antigens for production of anti-peptide antibodies. Anti-defensin polyclonal IgG has proven to be a valuable reagent for immunohistochemical localization of defensins in neutrophils (25), macrophages (34) and intestinal Paneth cells (35). It should be noted that production of antibodies which recognize native defensins has been achieved only with folded peptide preparations in which the cysteines are correctly paired, an indication that the epitope pattern is dependent upon peptide tertiary structure. SPPS additionally allows for the introduction of isotopically-labelled amino acids into the peptide chain, thereby circumventing structural alterations which occur with most chemical methods of radiolabelling. The resulting isotope-

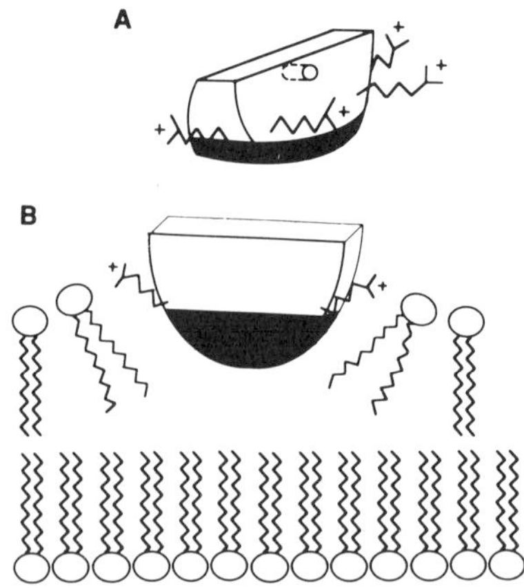

Figure 6. Possible interaction of human defensins with lipid bilayers. A. Cartoon representation of the amphiphilic dimer of human defensin HNP-3 based on its atomic crystal structure. The base of the dimer (shaded) is hydrophobic, six arginine side chains extend outward, and at the top of the dimer are located the amino- and carboxytermini of both monomers. B. The "wedge hypothesis" of destabilization of the bilayer involves the electrostatic interaction of the arginine side chains with negatively-charged phospholipid head groups, resulting in the insertion of the hydrophobic base into the hydrocarbon core (from reference 33, with permission).

containing peptide is a valuable reagent for numerous applications including radioimmunoassay, binding studies, and for tracing the metabolic fate of the labelled molecule.

Initial attempts to utilize SPPS for peptide production were directed toward the synthesis of rabbit defensin NP-2 with the use of tertiary butoxycarbonyl (tBoc) protection schemes. Though the conditions for folding and disulfide formation were established with natural peptide, preparation of an equivalent synthetic peptide with tBoc protocols met with limited success. The major pitfall was the inability to generate quantitatively intact sulfhydryl groups in the six cysteine residues in NP-2. Our SPPS protocols were subsequently adapted to fluoromethoxycarbonyl (Fmoc) protection schemes which employ base-labile N^{α}-protection, and relatively mild conditions for final cleavage of the peptide-resin and side-chain removal steps. Equally important, the cysteine side chains can be efficiently restored to the SH form, allowing for the subsequent formation of disulfides.

A representative example of this approach is the Fmoc solid phase synthesis of RatNP-1 (Figure 1), one of four neutrophil defensins isolated from Sprague-Dawley rats. The RatNP-1 peptide chain was synthesized on an Eppendorf Synostat instrument with N^{α}-Fmoc protection based on methods described by Kent (36) and Barany et al. (37). Cleavage and deprotection utilized Reagent R (38), a TFA-based cocktail, followed by extraction of

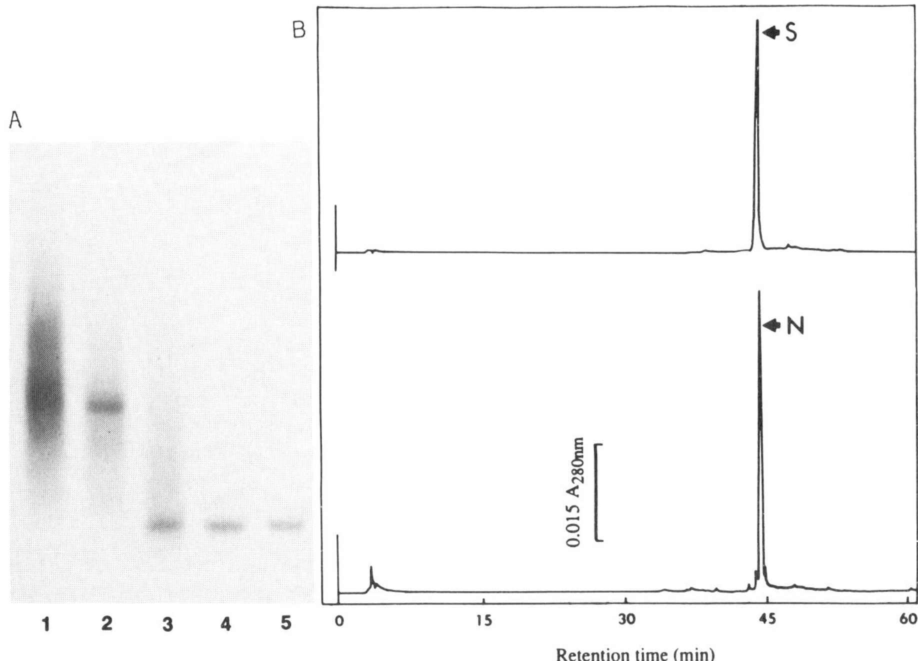

Figure 7. Solid phase peptide synthesis, oxidation/folding and purification of a rat neutrophil defensin. A. RatNP-1, synthesized with the use of an Fmoc-based protocol, was analyzed at each stage of synthesis on a 12.5% acid-urea gel. Lane 1, 6 μg of crude peptide (reduced form); lane 2, 2 μg of reduced peptide after RP-HPLC of the crude peptide; lane 3, 6 μg of crude peptide following 24 hr of air oxidation/folding at pH 7.0; lane 4, 2 μg of oxidized/folded peptide following RP-HPLC purification; lane 5, 2 μg of natural RatNP-1 purified from neutrophils. B. RP-HPLC elution of natural (N) and synthetic (S) RatNP-1.

scavengers from the crude peptide with methylene chloride. Conditions for folding and disulfide formation were air oxidation in ammonium acetate buffer, pH 7.0. The folded synthetic product eluted during preparative RP-HPLC with a retention time identical to natural RatNP-1 (Figure 7), and the natural and synthetic peptides behaved identically on acid-urea PAGE (Figure 7). Similar approaches have been used for the synthesis of other defensins including guinea pig (GPNP) and rabbit (NP-2) defensins.

A somewhat different application of SPPS was utilized to identify a defensin peptide expressed in the epithelium of the mouse small intestine. Ouellette and colleagues previously characterized a cDNA clone from intestinal Paneth cells which encoded a defensin-like precursor (39). In an attempt to produce a "standard" for use in purifying the predicted peptide from intestinal extracts, the putative mature peptide was synthesized, folded and oxidized as described above. Termed cryptdin-C, the synthetic peptide was used both as a biochemical marker for identifying the authentic intestinal defensin (cryptdin) in acid-urea gels, and as an immunogen for generating antibody. The strategy was successful in allowing for the identification of five intestinal defensins (cryptdins; 35) in crude extracts of intestinal tissue, and for production of anti-cryptdin antibody. The antibody recognized the natural peptide both in an enzyme-linked immunoassay and in intestinal tissue sections (35).

The utility of SPPS in studies of antimicrobial peptides was further demonstrated by the synthesis of indolicidin, a unique tryptophan-rich peptide from bovine neutrophils (13; Figure 1). Since the peptide contains five tryptophanyl residues, great care was taken in developing conditions for cleavage, deprotection, and purification. More than two grams of indolicidin were synthesized with Fmoc protocols, purified, and by several criteria were determined to be equivalent natural peptide (40). Attempts to crystallize synthetic indolicidin are under way.

BIOLOGICAL ACTIVITIES

Most experimental studies have suggested that the predominant biological roles of phagocyte-derived antimicrobial peptides involve host defense functions carried out by inflammatory cells. *In vivo* , the activities of the peptides appear to function as antimicrobial and cytotoxic agents, and they may also act as chemoattractants for other inflammatory cells. Briefly described below are experimental methods of use in analyzing the role of myeloid antimicrobial peptides as mediators of these biological activities.

Antimicrobial Activities

As mentioned above, the *in vitro* antimicrobial properties of the myeloid peptides have been determined with microbicidal or microbial inhibition assays. The most widely used method, employed for the majority of such studies on purified defensins, entails the incubation of a standard suspension of bacterial or fungal test organism (Table 1) with peptide in buffers of varied composition (41,42). Cultures are grown to mid log phase in an appropriate medium, harvested, washed and resuspended to 1 to 2 $\times 10^7$ colony forming units (CFU) per ml. To conserve peptide, the incubation volume is usually 0.05 ml, with the final cell concentration being 1 to 2 $\times$ 10^6 CFU/ml. Peptide stock solutions, usually made up in 0.01% acetic acid, are diluted in the incubation buffer to a final concentration of 1 to 100 μg/ml, and the incubation is initiated by addition of an appropriate volume of the bacterial or fungal stock suspension to the prewarmed (37° C) peptide-buffer mixture. At timed intervals, 50 or 100 μl samples are removed and diluted serially prior to plating on nutrient-containing agar plates. Killing activity is quantitated by determining the reduction in CFU relative to appropriate control incubations. The alternative plate diffusion method, described above, has the advantage of greater sample throughput, but is limited by its inability to distinguish killing from inhibitory activities, and does not provide information on the time course of antimicrobial action.

Table 1
Antimicrobial Spectrum of Defensins

Gram Positive Bacteria	Obligate Anaerobes
Staphylococcus aureus	*Bacteroides fragilis*
Group B streptococci	*Clostridium difficile*
Listeria monocytogenes	
	Fungi
	Candida albicans
Gram Negative Bacteria	*Aspergillus*
Pseudomonas aeruginosa	*Candida parapsilosis*
Acinetobacter spp.	*Cryptococcus neoformans*
Escherichia coli	*Coccidioides immitis*
Proteus mirabilis	
Haemophilus influenzae	**Viruses**
Salmonella typhimurium	Herpes simplex 1 and 2
	Vesicular stomatitis
	Influenza A

Antiviral Activity

Early experiments revealed that certain defensins were potently antiviral against herpes simplex viruses 1 and 2 in addition to other enveloped viruses (43,44). Two methods for evaluating antiviral activity have been employed. The first measures inhibition of cytopathic effect of peptide-treated virus suspensions on cultured monolayers, e.g., Vero monkey kidney cells, and provides a qualitative measure of inhibition of viral replication. To measure direct virucidal activity, the reduction in viral titer after incubation with peptide is determined by quantitative plaque assay on cell monolayers. Such studies with human (25), rabbit (44) and guinea pig defensins (26) demonstrated a potent, direct antiviral effect against HSV-1 and -2, vesicular stomatitis virus, and to a lesser degree against strains of influenza virus and cytomegalovirus. Of the strains tested, all viruses shown to be susceptible to defensins are enveloped; naked viruses such as echovirus and reovirus were unaffected, suggesting that the membranous viral envelope is the site of defensin virucidal activity.

Cytotoxicity of Myeloid Antimicrobial Peptides

Most studies suggest that the contents of peptide-containing (i.e., azurophil) granules are normally destined for delivery to intracellular phagosomes, as opposed to the contents of "specific" granules which are secreted extracellularly. However, in acute or chronic inflammatory processes, cell turnover and other stimulatory processes may cause antimicrobial peptides to accumulate in extracellular compartments where they may contribute to injury and even death of host cells. The cytotoxic effects of human and rabbit defensins have been evaluated in studies by Lichtenstein and co-workers, in which tumor cells lines were incubated with defensins under a variety of conditions (23,45). Lysis was evaluated by measuring chromium-51 release and by loss of trypan blue exclusion. As compared to antimicrobial effects, cytolytic activity against mammalian targets is much slower, requiring 4 to 8 hours for maximal activity. In the presence of serum, cytotoxicity was dramatically inhibited, suggesting that there may be protective components in serum which function to reduce cellular damage resulting from defensin released extracellularly. More recent studies demonstrate that defensin-mediated cytotoxic activity requires target cell metabolism and possibly endocytosis (45). An early event associated with cytotoxicity is an increase in membrane permeability of small molecules (ions, trypan blue), which by itself is inadequate to cause lysis as measured by release of chromium-labelled macromolecules (46). Subsequent to their cellular uptake, defensins may induce single-strand breaks in DNA, possibly explaining the relatively long incubation periods required for cytotoxicity *in vitro* (47).

Interactions of Myeloid Antimicrobial Peptides with Lipid Bilayers

Numerous lines of evidence implicate the cell membrane as an important site of activity of defensins (6). To facilitate the study of these interactions, Kagan et al. evaluated the effect of defensins on planar lipid bilayers (48). One side of a voltage-clamped membrane bilayer was exposed to increasing concentrations of rabbit or human defensin while the current was continuously measured. Conductance measurements revealed that both peptides form voltage-dependent currents across the bilayer, and that reversal of the system's polarity causes the current to cease. Additional experiments suggest that membrane pores may form as 2 to 4 peptide molecules assemble to form multimers. The observation that several bacteria and fungi are most sensitive to defensins if the microorganism is growing is consistent with the voltage dependence of channel formation in phospholipid bilayers. Additional experiments will be required to elucidate the orientation and position of the peptide assembly in the target membrane.

Other Activities

A possible role for defensins as chemoattractants for mononuclear phagocytes was demonstrated by Territo and co-workers (49). At 5×10^{-9} M, the human defensin HNP-1 elicited selective recruitment of monocytes in multiwell chambers, suggesting that the late influx of mononuclear cells which occurs *in vivo* may be triggered by extracellular release of these peptides from neutrophils.

Solomon and coworkers have demonstrated that certain defensins act as inhibitors of the adrenal cortical response to adrenocorticotropin (ACTH) by binding to the ACTH receptor; based on this inhibitory activity, the peptides were proposed to be corticostatins (50). Though the binding of certain defensins to the ACTH receptor is binding is clearly demonstrable, the physiological significance of this interaction *in vivo* is not known.

The peptide effector molecules of phagocytic leukocytes constitute a structurally diverse system of endogenous antibiotics which have evolved to protect against invasive microorganisms. Until very recently the expression of these molecules appeared limited to cells of myeloid lineage. With the recent discovery that defensins (35) and beta-defensins (51) are expressed in non-myeloid specialized epithelial cells, it would not be surprising to find a much broader distribution of these molecules in tissues of higher vertebrates.

Acknowledgments: I am grateful to Drs. Tomas Ganz and Robert Lehrer for their invaluable collaboration, and to the many students and research associates who have contributed to this work, especially Sylvia Harwig and Wendy Morris. This work was supported by NIH grants AI 22931 and AI 31696, the Office of Naval Research NOOO14-90-1310, and a University of California Biotechnology Training Grant. This is publication #4 of the NCDDG-OI program, cooperative agreement UO1-AI31696.

REFERENCES

1 Thomas, E.L., Lehrer, R.I. and Rest, R.F. (1988) Rev. Infect. Dis. 10 (suppl. 2), S450–S456.
2 Lehrer, R.I. and Ganz, T. (1990) Blood 76, 2169–2181.
3 Zeya, H.I. and Spitznagel, J.K. (1963) Science 142, 1085–1087.
4 Zeya, H.I. and Spitznagel, J.K. (1966) J. Bacteriol. 91, 750–754.
5 Zeya, H.I. and Spitznagel, J.K. (1968) J. Exp. Med. 127, 927–939.
6 Ganz, T., Selsted, M.E. and Lehrer, R.I. (1990) Eur. J. Haematol. 44, 1–8.
7 Gabay, J.E., Scott, R.W., Campanelli, D., Griffith, J., Wilde, C., Marra, M.N., Seeger, M. and Nathan, C.F. (1989) Proc. Nat. Acad. Sci. U.S.A. 86, 5610–5614.
8 Spitznagel, J.K. (1990) J. Clin. Invest. 86, 1381–1386.
9 Elsbach, P. (1990) TibTech 8, 26–30.
10 Gennaro, R., Romeo, D., Skerlavaj, B. and Zanetti, M. (1989) in Blood Cell Biochemistry (Harris, J.R., ed.), pp. 335–368, Plenum Press, New York, NY.
11 Selsted, M.E., Tang, Y.-Q., Morris, W.L., McGuire, P.A., Novotny, M.J., Smith, W., Henschen, A.H. and Cullor, J.S. (1992) (in press).
12 Tang, Y.-Q. and Selsted, M.E. (1992). Characterization of the disulfide motif in BNBD-12, an antimicrobial beta-defensin peptide from bovine neutrophils (in press).
13 Selsted, M.E., Novotny, M.J., Morris, W.L., Smith, W. and Cullor, J.S. (1992) J. Biol. Chem. 267, 4292–4295.
14 Ganz, T., Rayner, J.R., Valore, E.V., Tumolo, A., Talmadge, K. and Fuller, F. (1989) J. Immunol. 143, 1358–1365.
15 Daher, K.A., Lehrer, R.I., Ganz, T. and Kronenberg, M. (1988) Proc. Nat. Acad. Sci. U.S.A. 85, 7327–7331.
16 Sparkes, R.S., Kronenberg, M., Heinzmann, C., Daher, K.A., Klisak, I., Ganz, T. and Mohandas, T. (1989) Genomics 5, 240–244.
17 Michaelson, D., Rayner, J., Couto, M. and Ganz, T. (1992) J. Leuk. Biol. 51, 634–639.

18 Borregard, N., Heiple, J.M., Simons, E.R. and Clark, R.A. (1983) J. Cell. Biol. 97, 52–61.
19 Hultmark, D., Engstrom, A., Andersson, K., Steiner, H., Bennich, H. and Boman, H.G. (1983) EMBO J. 2, 571–576.
20 Lehrer, R.I., Rosenman, M., Harwig, S.S.L., Jackson, R. and Eisenhauer, P. (1991) J. Immunol. Meth. 137, 167–173.
21 Selsted, M.E., Szklarek, D. and Lehrer, R.I. (1984) Infect. Immun. 45, 150–154.
22 Selsted, M.E. and Becker, H.W., III (1986) Anal. Biochem. 155, 270–274.
23 Lichtenstein, A., Ganz, T., Selsted, M.E. and Lehrer, R.I. (1986) Blood 68, 1407–1410.
24 Selsted, M.E., Brown, D.M., DeLange, R.J. and Lehrer, R.I. (1985) J. Biol. Chem. 260, 4579–4584.
25 Ganz, T., Selsted, M.E., Szklarek, D., Harwig, S.S.L., Daher, K. and Lehrer, R.I. (1985) J. Clin. Invest. 76, 1427–1435.
26 Selsted, M.E. and Harwig, S.S.L. (1987) Infect. Immun. 55, 2281–2286.
27 Eisenhauer, P.B., Harwig, S.S.L., Szklarek, D., Ganz, T., Selsted, M.E. and Lehrer, R.I. (1989) Infect. Immun. 57, 2021–2027.
28 Selsted, M.E., Brown, D.M., DeLange, R.J. and Lehrer, R.I. (1983) J. Biol. Chem. 258, 14485–14489.
29 Selsted, M.E. and Harwig, S.S.L. (1989) J. Biol. Chem. 264, 4003–4007.
30 Pardi, A., Hare, D.R., Selsted, M.E., Morrison, R.D., Bassolino, D.A. and Bach, A.C., (1988) J. Mol. Biol. 201, 625–636.
31 Pardi, A., Selsted, M.E., Yip, P.F., Skalicky, J.J. and Zhang, X.L. (1992) Biochemistry (in press).
32 Zhang, X.L., Selsted, M.E. and Pardi, A. (1992) Biochemistry (in press).
33 Hill, C.P., Yee, J., Selsted, M.E. and Eisenberg, D. (1991) Science 251, 1481–1485.
34 Ganz, T., Sherman, M.P., Selsted, M.E. and Lehrer, R.I. (1985) Amer. Rev. Respir. Dis. 132, 901–904.
35 Selsted, M.E., Miller, S.I., Henschen, A.H. and Ouellette, A.J. (1992) J. Cell. Biol. 118, 929–936.
36 Kent, S.B.H. (1988) Annu. Rev. Biochem. 57, 957–989.
37 Barany, G., Sole, N.A., Van Abel, R.J., Albericio, F. and Selsted, M.E. (1992) in Innovation and Perspectives in Solid Phase Synthesis and Related Technologies: Peptides, Polypeptides and Oligonucleotides (Epton, R., ed.), pp. 29–38, Intercept, Andover, England.
38 Albericio, F., Kneib-Cordonier, N., Biancalana, S., Gera, L., Masad, R.I., Hudson, D. and Barany, G. (1990) J. Org. Chem. 55, 3730–3743.
39 Ouellette, A.J., Greco, R.M., James, M., Frederick, D., Naftilan, J. and Fallon, J.T. (1989) J. Cell. Biol. 108, 1687–1695.
40 Selsted, M.E., Levy, J.N., Van Abel, R.J., Cullor, J.S., Bontems, R.J. and Barany, G. (1992) in Peptides: Chemistry and Biology (Smith, J.A. and Rivier, J.E., eds.), pp. 905–907, ESCOM, Leiden.
41 Selsted, M.E., Szklarek, D., Ganz, T. and Lehrer, R.I. (1985) Infect. Immun. 49, 202–206.
42 Lehrer, R.I., Barton, A., Daher, K.A., Harwig, S.S.L., Ganz, T. and Selsted, M.E. (1989) J. Clin. Invest. 84, 553–561.

43 Daher, K., Selsted, M.E. and Lehrer, R.I. (1986) J. Virol. 60, 1068–1074.
44 Lehrer, R.I., Daher, K., Ganz, T. and Selsted, M.E. (1985) J. Virol. 54, 467–472.
45 Lichtenstein, A.K., Ganz, T., Nguyen, T., Selsted, M.E. and Lehrer, R.I. (1988) J. Immunol. 140, 2686–2694.
46 Lichtenstein, A. (1991) J. Clin. Invest. 88, 93–100.
47 Gera, J.F. and Lichtenstein, A. (1991) Cell. Immunol. 138, 108–120.
48 Kagan, B.L., Selsted, M.E., Ganz, T. and Lehrer, R.I. (1990) Proc. Nat. Acad. Sci. U.S.A. 87, 210–214.
49 Territo, M.C., Ganz, T., Selsted, M.E. and Lehrer, R.I. (1989) J. Clin. Invest. 84, 2017–2020.
50 Solomon, S., Hu, J., Zhu, Q., Belcourt, D., Bennett, H.P.J., Bateman, A. and Antakly, T. (1991) J. Steroid Biochem. Molec. Biol. 40, 391–398.
51 Diamond, G., Zasloff, M., Eck, H., Brasseur, M., Maloy, W.L. and Bevins, C.L. (1991) Proc. Nat. Acad. Sci. U.S.A. 88, 3952–3956.

PROGRESS IN THE CLONING OF GENES FOR PLANT STORAGE LIPID BIOSYNTHESIS

Vic C. Knauf

Calgene, Inc.
1920 Fifth Street
Davis, CA 95619

INTRODUCTION

Functional membranes are essential for plant life, indeed for all life. Plant cells rely on glycerolipids to constitute a large proportion of these membranes. Moreover, the observation of almost exclusively 16 and 18 carbon fatty acids esterified to the glycerol backbone of glycerolipids in membranes indicates that strong functional requirements define the composition of these polar structural lipids. Many plants, however, also use neutral glycerolipid (primarily triacylglycerols, TAG) as an energy-rich storage compound. This chapter will focus on the synthesis of storage lipids in the seeds of plants although the reader will be reminded that the biosynthetic pathways for storage and structural glycerolipids largely overlap. Several recent reviews (1–3) cover plant lipid biosynthesis; however, there have been rapid advances in cloning genes encoding lipid biosynthesis enzymes since their publication.

TAGs accumulate within intracellular oil bodies. Some plants, such as soybean, sunflower and peanut, store TAG in the cotyledons of seed embryos themselves. Others, such as castor bean and tobacco, store TAG within endosperm cells. Finally, mesocarp tissues of fruit from oil palm and avocado are very rich in TAG. Storage lipids typically accumulate in very discrete phases of development (4). When seeds germinate, oxidation of TAG provides energy and carbon for the young plantlet.

Discussion here of cloned genes encoding factors in the biosynthesis of seed storage lipids is organized in terms of the biochemical pathways: supply of energy and substrates; synthesis of basic fatty acid components in the proplastid (or chloroplast) organelle; transport to the endoplasmic reticulum (ER); and synthesis of the glycerolipids themselves.

Genetic Engineering, Vol. 15, Edited by J.K. Setlow
Plenum Press, New York, 1993

SUBSTRATE SUPPLY

A triacylglycerol consists of a glycerol molecule with a fatty acid esterified to each of the three glycerol carbons. The fatty acids are synthesized from acetate in the proplastid organelle and are linked to glycerol in the endoplasmic reticulum.

Glycerol is presumably provided by the glycolytic pathway in the cytoplasm. These reactions have been studied in oilseeds but not extensively. Supply of glycerol for TAG is considered not to be rate-limiting in view of the general opinion that glycolysis of imported sucrose is the primary fuel for seed development and metabolism. As a one-step derivative of dihydroxyacetone phosphate in glycolysis, the simplest model is the glycerol-3-phosphate is readily available as needed when acyl-CoA is available for the glycerol-3-phosphate acyltransferase enzyme. However, the presence of glycerol phosphate dehydrogenase has not been detected in seeds from several species of oilseeds (5,6). Glycerol fed to tissue slices is readily incorporated into TAG, suggesting a possible *in vivo* path of dephosphorylation of glyceraldehyde-3-phosphate, followed by reduction to glycerol, and then phosphorylation by glycerol kinase (5,7). Glycerol kinase has been suggested as a rate-limiting step. However, if the supply of glycerol-3-phosphate is rate-limiting in TAG synthesis, either fatty acid biosynthesis is uncoordinated and acyl-CoA pools can accumulate, or synthesis of fatty acids within the proplastid somehow recognizes the degree of glycerol-3-phosphate availability in the cytoplasm. The simplest model is that phosphoglycerate phosphatase is negatively regulated by free glycerol pools, allowing available acyl-CoA substrates for TAG synthesis to pull glycerol-3-phosphate as needed from glycolysis.

Fatty acid biosynthesis in the proplastid occurs by serial elongations of two-carbon increments. Each elongation step requires a malonyl-ACP (ACP = acyl carrier protein) molecule and two molecules of NADH or NADPH. In "white" seed tissues such as castor bean endosperm or sunflower embryo, the reducing power presumably comes from the catabolism of sucrose imported into the cell. Notably, lipid-storing cells in developing seeds of plants such as *Brassica* spp. are green and likely active photosynthetically (8). If so, undoubtedly some of the reductant necessary for fatty acid biosynthesis is supplied by the chloroplast itself. The organelle which synthesizes fatty acids in the leaf tissue is the chloroplast. The genetically identical organelle in the seed is variously called a leucoplast, a proplastid, or a chloroplast. Even amyloplasts, in pea seeds as an example, are the likely site of fatty acid biosynthesis. For simplicity's sake, I will use "plastidial compartment" to describe the double-membrane-bounded organelle which may or may not contain thylakoids in developing seed.

Malonyl-ACP derives directly from malonyl-CoA which derives from acetyl-CoA. Acetyl-CoA, of course, is central to many catabolic and anabolic pathways (9). Multiple sources of acetyl-CoA have been proposed for fatty acid biosynthesis during storage lipid deposition in seeds. These include: a) acetate from mitochondrial respiration (10) which passively diffuses to the cytoplasm and then to the plastidial compartment where acetyl-CoA synthetase (11) converts acetate to acetyl-CoA; an acetylcarnitine intermediate which may have a role in transport (12,13); b) acetyl-CoA from plastidial pyruvate dehydrogenase activity (14,15) following plastidial glycolysis (16); c) malic enzyme activity on imported malate (17).

Perhaps some combination of these are active and collectively contribute to fatty acid biosynthesis. Perhaps different plant species recruit acetyl-CoA from these or other sources differentially.

SYNTHESIS OF FATTY ACIDS IN PLASTIDIAL ORGANELLES

Fatty acids are synthesized very similarly in plant cells as they are in animal, fungal and bacterial cells. That is, from a two or three carbon "initiator," fatty acids are elongated in two-carbon increments by a repeating series of four enzymatic reactions (Figure 1): a) a condensing enzyme, or ketoacyl synthase (KAS), that condenses an activated acetyl-ACP derived from a malonyl-ACP onto the thioester carbon of the elongating fatty acid covalently linked to the synthase enzyme to create a ketoacyl-thioester of ACP; b) a ketoacyl-ACP reductase which uses NADH or NADPH (depending on isozyme) to reduce the acyl chain to a hydroxyacyl-ACP; c) a β-hydroxyacyl-ACP dehydrase which removes water to create a *trans* double bond enoyl-ACP group; followed by d) an enoyl-ACP reductase activity which uses NADH or NADPH (depending on isozyme) to hydrogenate carbons at the double bond to regenerate a saturated acyl-ACP two carbons longer than the acyl-ACP starting the four-enzyme cycle.

In animal cells, these four enzyme activities (plus three others and ACP itself) correspond to specific domains within a single large multifunctional protein (18) which is active in a dimer formation. In yeast, these four enzyme activities (plus two others and ACP) are split up as domains in two different polypeptide subunits (19,20) which assemble into an α6β6 complex to form active enzyme. Although other bacteria may have different architectures, the best characterized system, in *E. coli,* resembles that found in plant plastidial extracts in that the fatty acid synthesis (FAS) enzymes are separate and can be purified away from each other and retain activity (21–23). Also similar to *E. coli,* the FAS cofactor ACP is a relatively small protein post-translationally modified with a pantetheine prosthetic group.

Kuo and Ohlrogge (24) purified spinach leaf ACP and determined its primary structure. Using the amino acid sequence and synthetic oligonucleotides, Scherer and Knauf (25) cloned and sequenced a cDNA for spinach leaf ACP-I. Reports of cDNA and genomic clones for other ACPs soon followed (26–28). Ohlrogge and colleagues (29,30) have characterized different ACP isoforms including ACP-I, ACP-II and ACP-III isoforms from spinach containing highly conserved domains and other peptide regions less conserved at the amino acid sequence level. These are apparently expressed differentially in different parts of the spinach plant. Some *in vitro* experiments suggested that some FAS enzymes may discriminate between acyl-thioesters of different ACP isoforms, but this has not yet been shown to have significance *in vivo.* In deed, there may be as many as 35 ACP genes in *Brassica napus* (28). Different ACP gene family members which express in different plant tissues may have

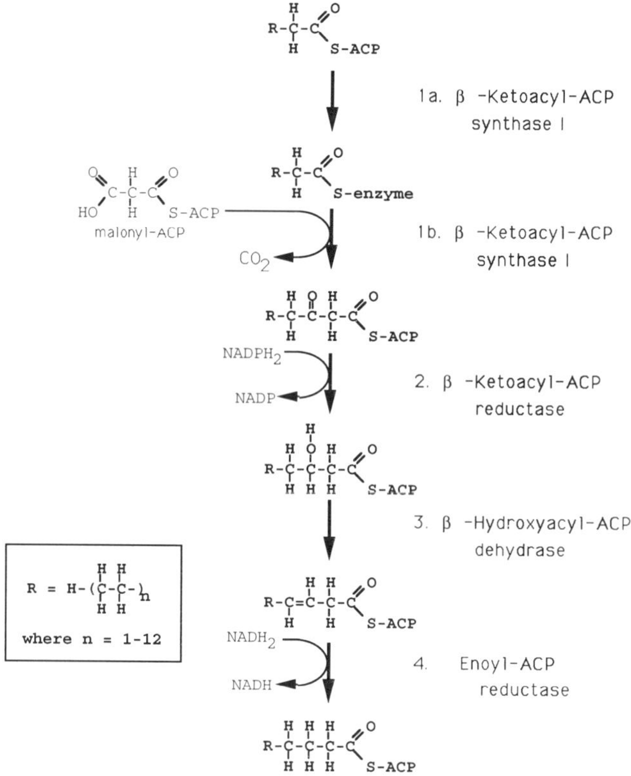

Figure 1. Elongation of acyl-ACP by FAS enzymes. When palmitoyl-ACP is elongated (n = 14), reactions 1a and 1b are catalyzed by β-ketoacyl-ACP synthase II.

diverged without resulting in functional differences. Recently, Shintani and Ohlrogge (31) have identified an *Arabidopsis thaliana* ACP analogue apparently targeted to the mitochondrion. Brody and colleagues (32) have studied mitochondrial ACPs in *Neurospora,* but the function of such proteins remains unknown. Clearly, however, ACP has at least two radically different functions in *E. coli* (33).

Shimakata and Stumpf characterized two separate β-ketoacyl-ACP synthase activities in spinach leaf (23). KAS-I prefers substrates from acetyl-ACP through myristoyl-ACP with significantly lower activity on palmitoyl-ACP and longer substrates (34). KAS-II exhibits significant activity on palmitoyl-ACP (with less but significant activity on myristoyl-ACP, and considerably less affinity for shorter substrates). Thus, the work of Shimakata and Stumpf explained the almost complete inhibition of cerulenin on elongation of short chains while some long chain elongating ability remained. Moreover, from a physiological perspective, the results suggested that a separate activity for the elongation of C16:0-ACP to C18:0-ACP would be a mechanism to control the relative ratio of 16 and 18 carbon fatty acids used for structural lipid biosynthesis. C x:y is a commonly used abbreviation for fatty acids wherein x = the number of carbons and y = the number of double bonds.

Siggard-Anderson et al. (35) reported the cloning of a β-ketoacyl-ACP synthase cDNA from barley using polymerase chain reactions (PCR) based on amino acid sequence of a protein tagged with radioactive cerulenin. Recently, Thompson and colleagues at Calgene (36) have cloned cDNAs from *Brassica campestris* rapeseed embryos and from *Ricinus communis* (castor bean) endosperm. Highly purified KAS-I preparations from castor bean endosperm primarily contained a single protein band corresponding to 50 kD as estimated by SDS PAGE. Analogous preparations for KAS-II contained polypeptides of apparent molecular size 46 kD and 50 kD. Monoclonal antibodies raised against the KAS-II 50 kD protein band cross-reacted with the 50 kD protein from the KAS-I preparation but not with the 46 kD polypeptide in the KAS-II preparation. The assembled data including near identical primary structure for 50 kD bands in KAS-I and KAS-II preparations suggest they are highly similar if not functionally interchangeable proteins. DNA sequence analyses of the castor bean cDNAs corresponding to the 50 kD and 46 kD polypeptides reveal clear DNA and amino acid homology shared with each other and with corresponding rapeseed cDNAs. Rapeseed embryos also appear to have "50 kD" and "46 kD" class mRNAs. Expression analyses in *E. coli* by Genez et al. (37) suggest that both proteins may be required for KAS-II activity.

β-ketoacyl-ACP reductase has been purified from *Brassica napus* embryos and partially sequenced (38). Klein et al. (39) subsequently used the amino acid sequence to design PCR reactions leading to the cloning of a cDNA for this FAS enzyme from *Cuphea lanceolata* .

β-hydroxyacyl-ACP dehydrase remains somewhat elusive at this time. Amy et al. (18) identified most of the FAS enzyme domains in the multifunctional rat protein, and assigned that for the β-hydroxyacyl-ACP dehydrase to a large peptide region otherwise unaccounted for. The *E. coli* gene for an analogous enzyme, β-hydroxydecanoyl thioester dehydrase, has been cloned by complementation of a mutant but this is not the central enoyl-ACP dehydratase involved in the synthesis of saturated fatty acids (40).

Enoyl-ACP reductase has been purified from *Brassica napus* rapeseed embryos and a partial amino acid sequence published (41). A cDNA characterized by Nijkamp and colleagues (42) as encoding a protein that binds ACP showed good homology with the reported amino acid sequence and thus appears to encode enoyl-ACP reductase. Notably, however, the enzymology suggests there may be isozymes of enoyl-ACP reductase (22,41). As in the case of ACP, it is not clear yet if isozymes with functionally significant differences (such as substrate chain length specificity) exist or if the enzymes are encoded by a multigene family with slight evolutionary divergence among the various genes.

The four enzyme serial repeat initially requires an acyl group acceptor onto which the additional two-carbon increment is bound by the β-ketoacyl-ACP synthase. In both yeast and animal FAS, the initial cycle in the synthesis of a fatty acid begins with acetyl-ACP. This initiating nucleus is provided by the activity of an enzyme domain which effectively switches the thioester bound CoA in acetyl-CoA for a similarly pantetheine-thioester-bound ACP.

Until recently, acetyl-ACP was also thought to be the initiating nucleus in *E. coli* and in plants (43). There are measurable pools of acetyl-ACP and measurable acetyl-CoA ACP transacylase activity in plant and in *E. coli* extracts. However, a third type of β-ketoacyl synthase activity (KAS-III) has been identified as cerulenin-resistant, with the use of acetyl-CoA as an initiating nucleus, and with malonyl-ACP will form a β-ketobutyryl-ACP, which is then further elongated by the four enzyme series reactions described above (44–46). Aside from the use of acetyl-CoA as the condensing acceptor, the KAS-III activity apparently also accepts short chain acyl-ACPs as well. The *E. coli* gene sequence was identified while another gene was being sequenced that happened to be adjacent in the *E. coli* genome; no cloned plant gene has yet been reported.

The role(s) of acetyl-ACP pools in plants and *E. coli* has become speculative. Data from thiolactomycin-resistant mutants (47) suggest that the observed acetyl-CoA ACP transacylase activity of *E. coli* may be the same enzyme as KAS-III. Indeed, in the absence of malonyl-ACP, KAS enzymes may act as transacylases between ACP and CoA.

Malonyl-ACP is derived from malonyl-CoA which in turn is the product of carboxylation of acetyl-CoA *via* acetyl-CoA carboxylase. Malonyl-CoA ACP transacylase is easily detected in the plant extracts and generally not considered as a regulating step for fatty acid biosynthesis. It is instructive to recall that malonyl-CoA is utilized by several biosynthetic pathways in plants whereas malonyl-ACP represents the first true commitment of an acetate group for fatty acid elongation. Although the enzyme has been reported as purified to homogeneity (48), cloned genes for plant malonyl-CoA ACP transacylase have not yet been reported.

The structure of acetyl-CoA carboxylase in *E. coli* was described in the 1970s as four separate proteins: a biotin carboxylase, a biotin carboxyl carrier protein (BCCP), and two dissimilar subunits comprising the acetyl-CoA transcarboxylase. The last two genes to be cloned, those encoding the transcarboxylase, have recently been reported (49). Animal (50,51) and yeast (52) acetyl-CoA carboxylase cDNAs have been described encoding a single large multifunctional protein including recognizable domains for BCCP, the biotin carboxylase, and the terminal transcarboxylase.

The data for plants are somewhat less clear. Older reports describe biotin-containing components of acetyl-CoA carboxylase as peptides of 60 kD or less. Other workers have identified large MW enzymes from plants with acetyl-CoA carboxylase activity (53). It now seems clear that other biotin-requiring carboxylases such as 3-methylcrotonyl-CoA carboxylase (54) may be present in plants. In addition, the requirement for malonyl-CoA in several pathways (including endoplasmic reticulum elongation of fatty acids discussed below) suggests that isozymes of acetyl-CoA carboxylase are likely. Identification of herbicides that specifically inhibit monocotyledonous plant acetyl-CoA carboxylase (55,56) and the availability of streptavidin-bound detection systems that recognize the biotin prosthetic group of acetyl-CoA carboxylase suggest that genes for higher plant acetyl-CoA carboxylase will likely be reported soon.

Once a growing acyl-ACP chain reaches 16 carbons in length (C16:0-ACP), several possibilities exist: a) KAS-II begins an elongation cycle on C16:0-ACP to generate C18:0-ACP; b) an acyltransferase may transfer the acyl group from ACP directly to a glycerol-3-phosphate or to lysophosphatidic acid as part of plastid structural lipid biosynthesis including those lipids found in chloroplast thylakoid membranes; or c) an acyl-ACP thioesterase cleaves the ACP from the C16:0 fatty acid. Free fatty acid then diffuses into the plastid cell wall where an acyl-CoA synthetase in the outer membrane may effectively direct it into the cytoplasm by forming soluble C16:0-CoA in the cytosol. The glycerol-3-phosphate acyltransferase in plastids is soluble and has been cloned as a cDNA from cotyledons from squash (57). An homologous clone from cucumber was subsequently cloned (58).

Acyl-ACP thioesterase activity partially purified from plants (59,60) shows an apparent preference for C18:1-ACP as a substrate over C18:0-ACP which in turn appears to be preferred over C16:0-ACP (activity on shorter acyl-ACP substrates is very low *in vitro*). Whether such plant extract activity really represents more than one enzyme with perhaps differing substrate specificities remains to be determined. Notably, most plant tissues and most seed oils contain less C18:0 fatty acids than either C16:0 or C18:1. Exactly how plants use the KAS-II elongation cycle in combination/competition with acyl-ACP thioesterases to control the apparently important ratios of C16 and C18 fatty acids is an intriguing area for experimentation with cloned genes.

Oleoyl-ACP thioesterase of safflower embryos has been partially purified and two distinct clones were obtained from an embryo cDNA bank (61). Expression of each cDNA clone in *E. coli* suggested slightly different specificity patterns on C16:0-ACP, C18:0-ACP and C18:1-ACP but the physiological significance of this is not known.

Finally, for the basic FAS enzymes found in the plastid which provide the bulk of fatty acids stored in TAG, there is the stearoyl-ACP Δ9 desaturase which, in combination with ferredoxin as an electron transfer cofactor, converts 18:0-ACP to 18:1-ACP. cDNA clones were first reported for safflower (62) and castor bean (63,64), and subsequently for other sources (65,66) including rapeseed (67).

An interesting seed-specific desaturation is found in parsley, carrot and a few other plants which store petroselenic acid which is an 18:1 fatty acid but with the double bond at the #6 carbon. Cahoon et al. have cloned a cDNA which when expressed in tobacco callus generates small amounts of petroselenic acid presumably *via* a stearoyl-ACP Δ6 desaturase (68). This cloned cDNA shares DNA homology with the stearoyl-ACP Δ9 desaturase.

Other seed-specific plastid FAS variants of some note include those which generate shorter chain fatty acids by intervention of enzymes such as the lauroyl-ACP thioesterase found in *Umbellaria californica* (69). *U. californica* (California bay) is a tree producing a seed oil with 60% C12:0 fatty acids. Expression of the cloned cDNA for a seed-expressed lauroyl-ACP thioesterase behind a seed-specific promoter in both *Arabidopsis thaliana* and *Brassica napus* generated transgenic seed oils dramatically enriched in TAG containing C12:0 fatty acids (70). Thioesterase enzyme activities from other plants with different compositions of shorter chain fatty acids have also been identified, including plant seed sources of *Cuphea hookeriana, Ulmus americana* (elm), coconut and Chinese camphor (71). Interestingly, the *Umbellaria* lauroyl-ACP thioesterase does show DNA and amino acid sequence homology with oleoyl-ACP thioesterases.

In all the cases of cloned FAS genes described above for plants, the genes are encoded in the nuclear genome. Excepting the mitochondrial form of ACP, then, all of the cDNAs correspond to preproteins containing a transit peptide which effects uptake into the plastidial organelle. Transit peptides among the FAS genes encoding different enzymes show little if any homology. This is observed even when the two homologous KAS-I and KAS-II cDNAs from castor bean are compared. Considerable divergence can be seen when comparing the transit regions of Δ9 stearoyl-ACP desaturase from safflower, castor bean and jojoba.

All of the FAS enzymes described above recognize an acyl-ACP substrate. However, no kind of consensus ACP binding site has suggested itself. The availability of a series of acyl-ACP thioesterases with differing chain length specificities may provide a basis to recognize features of enzyme structure that recognize the acyl group of the relevant substrate.

EXPORT OF FATTY ACIDS FROM THE PLASTID AND TRANSPORT TO ENDOPLASMIC RETICULUM

As mentioned above, plastids contain acyl-ACP thioesterases which presumably provide pools of free fatty acids (FFA). Certainly, C16:0 and longer FFA will be very hydrophobic and likely partition to the plastid membranes. An acyl-CoA synthetase has been identified in the outer membrane of the chloroplast (72); presumably a similar activity is present in seed plastids producing FFA for TAG synthesis.

An operating hypothesis would be that this enzyme synthesizes a soluble acyl-CoA "into" the cytoplasm, effectively pumping FFA out of the plastidial compartment. It is interesting to note that the transgenic plants expressing the lauroyl-ACP thioesterase appear to have no problem transporting C12:0 out of the plastidial organelle even though cruciferous plant seeds and their endogenous acyl-CoA synthetase do not naturally see C12:0 as a FFA.

So-called lipid transfer protein genes expressed in leaf tissue have been cloned but what, if any, role such proteins have in storage lipid biosynthesis is not clear. Another protein of poorly understood significance is an acyl-CoA thioesterase (73) found in the same outer-membrane-enriched fractions in which Joyard and colleagues identified the acyl-CoA synthetase. There are data from *Arabidopsis* mutants suggesting that fatty acids from ER lipids can somehow make their way back into plastidial lipids of entirely different structures. Perhaps the acyl-CoA thioesterase represents a mechanism to pump FFAs from certain cytoplasmic acyl-CoA pools back into the chloroplast.

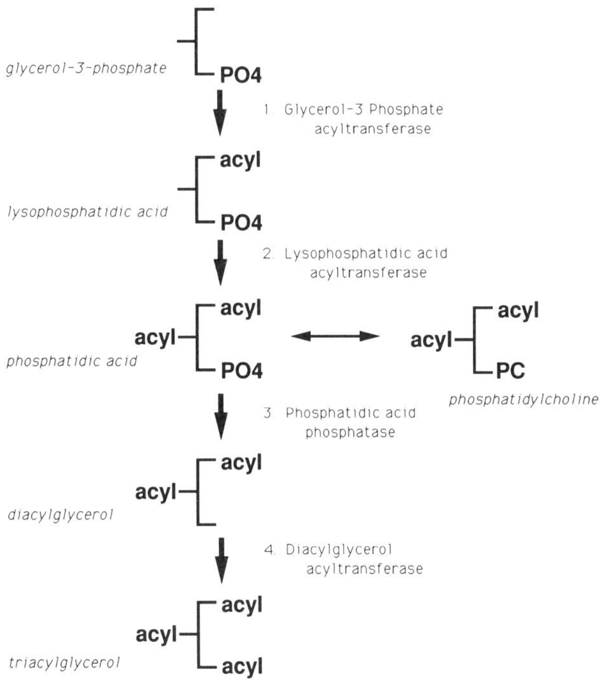

Figure 2. Synthesis of triacylglycerol. Desaturation of 18:1 and 18:2 acyl groups likely occurs when acylated at the #2 position of phosphatidyclcholine; some acyl group exchange presumably occurs to explain 18:2 and 18:3 fatty acids found at the #1 and #3 positions of triacylglyerols in seed oils.

STORAGE LIPID BIOSYNTHESIS IN THE ER AND OIL BODY FORMATION

Triacylglyerol biosynthesis in plants is generally believed to resemble the Kennedy pathway established for animal TAG synthesis. The first step in the sequence (Figure 2) is the acylation of glycerol-3-phosphate by 3-phosphoglycerol acyltransferase. The acyl-group is drawn from the cytoplasmic pool of acyl-CoAs. Isolated enzyme activity specificities for different fatty acids appear to be rather non-discriminatory, reflecting acyl-CoA pool composition (74).

A cDNA for the squash cotyledon plastidial glycerol-3-phosphate acyltransferase has been cloned (57). This, however, is a soluble enzyme unlike the ER enzyme which is believed to be microsomal and membrane-bound. There have been reports of a soluble glycerol-3-phosphate acyltransferase from cacao involved in TAG synthesis; further work is necessary to validate this conclusion (75).

The second acylation by another microsomal membrane-bound enzyme, lysophosphatidyl acyltransferase (LPAT, but also known as LPAAT), generally appears more discriminating in substrate specificity (76). In most temperate zone plants, the acyl groups in the second position of TAGs are almost exclusively unsaturated; indeed, the same rule holds for the acyl group in the second position of structural lipids, No cloned genes have yet been described despite considerable attention in recent years (77–80).

According to the Kennedy pathway, the enzymatic steps leading up to diacylglycerol (DAG) therefore are also found in the synthesis of structural lipids like phosphatidylcholine (PC) and phosphatidylethanolamine (PE). Moreover, *in vitro* enzymatic data from isolated microsomes suggest that the synthesis of linoleic (18:2) and linolenic (18:3) fatty acids occurs as a desaturation of oleoyl- and linoleoyl-acyl groups in the second position of PC (81). This observation is consistent with the LPAT specificity which preferentially places 18:1 in the 2nd position as substrate to synthesize the proportionally large amounts of 18:2 and 18:3 fatty acids found in structural lipids. *Arabidopsis* mutant lines have been described that appear to have lesions in the terminal 18:1 and 18:2 desaturases (82,83). A modified chromosomal walking protocol was used to identify a rapeseed seed cDNA encoding the 18:2 desaturase (84); in addition, a T-DNA insertion mutant has led to the direct cloning of the genomic locus encoding the 18:2 desaturase (85). This desaturase is also known as the Δ15 desaturase since the third double bond is placed at the #15 carbon of linoleic acid (and similarly, Δ12 for the 18:1 desaturase). Since 18:2 and 18:3 fatty acids can be found in all three positions of stored TAG, there is presumably some transfer mechanism which remains poorly characterized at present. These enzyme reactions require an electron carrier, cytochrome b5, which has now been cloned from cauliflower (86,87).

Phosphatidic acid phosphatase converts phosphatidic acid to diacylglycerol. Although there is a suggestion in animal cells that this enzyme is rate-limiting in the formation of TAG, this remains speculative in plants. Again, diacylgycerol is an intermediate in structural lipid synthetic pathways as well.

Diacylglycerol acyltransferase (DAGAT) which places the third fatty acid on the glycerol backbone is believed to be the last and perhaps the unique enzyme in TAG synthesis (88–91). This enzyme has a very hydrophobic substrate and is itself membrane-bound. No clones for genes encoding this enzyme from any living organism have been reported so far.

Notably, plants that store unusual fatty acids suggest that DAG synthesis may be catalyzed by a similar but separate set of enzymes for TAG compared to structural lipids. For example, TAG in castor bean endosperm may approach 95% ricinoleic fatty acids (92). Since ricinoleic fatty acids are not found in structural lipids, it seems that mono- or diacylglycerols with ricinoleate at the number 1 and 2 positions are fated to become TAG. Slabas and colleagues observed a similar phenomenon with *Cuphea procumbens* which contains 80% short-chain fatty acids (93).

The high levels of ricinoleic fatty acid in castor bean TAG, or short-chain fatty acids in *Cuphea* TAG, also indicate that either LPAT or some so far uncharacterized transfer or switching mechanism exists to place the seed-specific fatty acids into the number two position of triacylglycerides. That the enzymatic base for this appear unique is evident in the observations that many common temperate zone oilseeds do maintain the discriminatory base for which fatty acids are found in the number two position of TAG. For example, erucic

(22:1) fatty acids naturally found in rapeseed TAG (up to 55% by weight) are nonetheless not found in the number two position; however, nasturtium seed containing TAG at 80% 22:1 suggests that some plant species can synthesize TAG with 22:1 at all three positions. *In vitro* experiments indicate that rapeseed TAG-synthesizing extracts discriminate against short-chain fatty acids in the TAG. *In vivo* transgenic rapeseed oils with 12:0 may yet tell a different story.

In these regards, perhaps of some note is the report of a diacylglycerol synthesizing mechanism which can be disassociated from microsomal membranes (94). If the Kennedy pathway enzymes do account for structural lipid biosynthesis in seeds as other tissues, it may be possible to identify the enzyme activities in plant extracts whether or not they represent the major synthetic pathway for TAG. Two separate systems would be one rational explanation for the ability of castor bean and other crops with uncommon fatty acids in TAG to keep those fatty acids out of structural lipids. However, to date, no genes encoding TAG synthesis enzymes from plants have been isolated.

As TAG concentrates in developing seeds, oil bodies with a boundary monolayer membrane can be found within the cell. Typically, these proliferate as a population of lipid droplets. The unproven but most convincing model for oil body formation is that the TAG products of DAGAT, being very hydrophobic, accumulate within the bilayer membranes of the ER. This accumulation leads to a swelling, physically separating each half of the bilayer membrane, until a small oil body "buds" off surrounded by a monolayer of polar lipids and structural membrane proteins—and perhaps with some lipid synthesis enzymes in the monolayer in an active conformation. The non-enzymatic structural protein present in large amounts in that "half-unit" membrane is termed oleosin (95). First cloned by Huang and colleagues using protein sequence to identify cDNA clones (96), numerous cDNA and genomic clones have now been identified from carrot, rapeseed, soybean and other plants (97–102). Structure-function analyses have been attempted with an attractive hypothesis for the abundance and conservation of oleosins in oilseeds, suggesting that oleosins help keep the numerous oil bodies from coalescing into huge oil bodies. During seed germination when stored lipids must be mobilized for oxidation, smaller oil bodies would provide a larger surface area, which in turn allows more efficient access by cytosolic enzymes.

Mention of erucic and ricinoleic fatty acids above indicates another aspect of seed TAG synthesis. The plastidial compartment typically exports 16:0, 18:0 and 18:1 fatty acids. Some of the monounsaturated fatty acids are further modified at some stage of TAG synthesis. For instance, ricinoleic acid appears to be a hydroxylation of 18:2 fatty acids, presumably while esterified at the number two position of PC by an enzyme either evolved from an 18:2 desaturase or perhaps a P450-type activity (92). Other analogous enzyme activities seem likely in plant seeds that store TAG containing epoxy fatty acids and cyclopropane fatty acids.

A seed-specific elongation pathway from 18:1 to 20:1 to 22:1 acyl-CoAs is found in disparate plants such as rapeseed, jojoba, nasturtium and meadowfoam (103). The elongation is believed to go through the same cycle of reactions requiring acetate and malonyl-CoA as for FAS in the plastidial compartment and other organisms. Paul Stumpf has suggested that since the plant plastidial FAS is like the prokaryotic type II system of *E. coli,* then the ER erucic acid FAS may be "eukaryotic" in having the activities represented by one or two multifunctional proteins. This proposal is consistent with the general reliance of FAS systems on ACP-bound substrates; the absence of any small cytoplasmic ACPs found to date implies that the pantetheinylated ACP function in erucic acid biosynthesis may be part of a larger protein. *Arabidopsis* mutants with apparent blocks in the elongation pathway have been isolated but no cloned genes have yet been reported (82,83).

And finally, *Simmondsia chinensis* (jojoba) appears unique among higher plants in that its seeds store liquid wax esters rather than TAG. The wax molecules consist of a fatty alcohol esterified to a fatty acid. Fatty acids synthesized in jojoba seeds for wax esters are almost entirely 20:1, 22:1 and 24:1 synthesized from 18:1 produced in the plastid compartment and elongated in the ER. A membrane-bound acyl-CoA reductase converts monounsaturated acyl-CoAs to the corresponding monounsaturated fatty alcohol. A second membrane-bound enzyme ligates the alcohol to the acyl group of an acyl-CoA. Putative cDNAs for jojoba acyl-CoA reductase and for the wax synthase enzyme have been identified (104).

SUMMARY

Most of the plant lipid storage synthesis genes reported to date have been obtained with the use of protein sequence data from purified enzyme or antibodies raised against purified enzyme extracts. DNA and peptide homology appear to be the general rule for genes from different plants encoding the same enzyme. Thus once a cloned gene sequence is available from one species, finding analogous genes from other species is straightforward. Indeed, peptide homology with a Δ9 stearoyl-ACP desaturase helped to identify a Δ6 stearoyl-ACP desaturase. Similarly, peptide sequence from a cyanobacterial "prokaryotic pathway" Δ12 desaturase helped identify the rapeseed ER Δ15 desaturase.

Notably, the rapeseed Δ15 desaturase was independently cloned by two groups using genetics rather than purifying the enzyme activity sufficiently for direct sequencing or raising antibodies. Predictably, the availability of Δ12 desaturase mutants in *Arabidopsis* suggests that this gene will soon be reported as will certain other genes involved in structural lipid biosynthesis.

It should not be long before all of the basic FAS and TAG genes (and many unique lipid genes like those for jojoba wax synthesis enzymes) are cloned and described in the literature. This is quite remarkable since the first, spinach leaf ACP-I, was reported only five years ago (25). Two reports in 1992 described successful engineering of two dramatically different and unique storage lipids in canola with antisense of an endogenous gene in one case (67) and expression of transgene from a wild species in the other (70). The applied uses of having these cloned genes should help spur further gene isolation and characterization for the enzymes of a complex and fascinating pathway specific to seed tissues of plants.

REFERENCES

1 Harwood, J.L. (1988) Annu. Rev. Plant Physiol. Plant Mol. Biol. 39, 101–138.
2 Slabas, A. and Fawcett, T. (1992) Plant Mol. Biol. 19, 169–191.
3 Ohlrogge, J.B., Browse, J. and Somerville, C.R. (1991) Biochim. Biophys. Acta 1082, 1–26.
4 Ohlrogge, J.B. and Kuo, T.-M. (1984) Plant Physiol., 74, 622–625.
5 Ghosh, S. and Sastry, P.S. (1986) Indian J. Biochem. Biophys. 23, 100–104.
6 Gurr, M.L., Blades, J., Appleby, R.S., Smith, C.G., Robinson, M.P. and Nichols, B.W. (1974) Eur. J. Biochem. 43, 281–290.
7 Sangwan, R., Gauthier, D., Turpin, D., Pomeroy, M. and Plaxton, W. (1992) Planta 187, 198–202.
8 Murphy, D.J. and Cummings, I. (1989) J. Plant Physiol. 135, 63–69.
9 Liedvogel, B. (1986) J. Plant Physiol. 124, 211–222.
10 Murphy, D. and Stumpf, P. (1981) Arch. Biochem. Biophys. 212, 730–739.
11 Zeiher, C. and Randall, D. (1991) Plant Physiol. 96, 382–389.
12 Masterson, C., Wood, C. and Thomas, D. (1990) Plant Cell and Environ. 13, 755–765.
13 Roughan, P., Post-Beittemiller, J., Ohlrogge, J. and Browse, J. (1992) in Metabolism, Structure and Utilization of Plant Lipids (Cherif, A., ed.) , pp. 177–180, Centre National Pédagogique, Tunis, Tunisia.
14 Camp, P.J. and Randall, D.D. (1985) Plant Physiol. 77, 571–577.
15 Springer, J. and Heise, K. (1989) Planta 177, 417–421.
16 Simcox, P., Reid, E., Canvin, D. and Dennis, D. (1977) Plant Physiol. 59, 1228–1232.
17 Smith, R., Gauthier, D., Dennis, D. and Turpin, D. (1992) Plant Physiol. 98, 1233–1238.
18 Amy, C.M., Witkowski, A., Naggert, J., Williams, B., Randhawa, Z. and Smith, S. (1989) Proc. Nat. Acad. Sci. U.S.A. 86, 3114–3118.
19 Mohamed, A.H., Chirala, S.S., Mody, N.H., Huang, W.-Y. and Wakil, S.J. (1988) J. Biol. Chem. 263, 12315–12325.
20 Chirala, S.S., Kuziora, M.A., Spector, D.M. and Wakil, S.J. (1987) J. Biol. Chem. 262, 4231–4240.
21 Shimakata, T. and Stumpf, P. (1982) Arch. Biochem. Biophys. 217, 144–154.
22 Shimakata, T. and Stumpf, P. (1982) Arch. Biochem. Biophys. 218, 77–91.
23 Shimakata, T. and Stumpf, P.K. (1982) Proc. Nat. Acad. Sci. U.S.A. 79, 5808–5812.
24 Kuo, T.M. and Ohlrogge, J.B. (1984) Arch. Biochem. Biophys. 230, 110–116.
25 Scherer, D.E. and Knauf, V.C. (1987) Plant Mol. Biol. 9, 127–134.
26 Rose, R.E., DeJesus, C.E., Moylan, S.L., Ridge, N.P., Scherer, D.E. and Knauf, V.C. (1987) Nucl. Acids Res. 15, 7197.
27 Hansen, L. (1987) Carlsberg Res. Commun. 52, 381–392.
28 de Silva, J., Loader, N.M., Jarman, C., Windust, J.H.C., Hughes, S.G. and Safford, R. (1990) Plant Mol. Biol. 14, 537–548.

29 Ohlrogge, J.B. and Kuo, T.M. (1985) J. Biol. Chem. 260, 8032–8037.
30 Battey, J.F. and Ohlrogge, J.B. (1990) Planta 180, 352–360.
31 Shintani, D.K. and Ohlrogge, J.B. (1992) Plant Physiol. 99, 88.
32 Brody, S. and Mikolajczyk, S. (1988) Eur. J. Biochem. 173, 353–359.
33 Therisod, H., Weissborn, A.C. and Kennedy, E.P. (1986) Proc. Nat. Acad. Sci. U.S.A. 83, 7236–7240.
34 Mackintosh, R., Hardie, D. and Slabas, A. (1989) Biochim. Biophys. Acta 1002, 114–124.
35 Siggard-Andersen, M., Kauppinen, S. and von Wettstein-Knowles, P. (1991) Proc. Nat. Acad. Sci. U.S.A. 88, 4114–4118.
36 Nelsen, J., Genez, A., Stalker, D.M., McCarter, D., Bleibaum, J. and Thompson, G.A. (1992) (unpublished data).
37 Genez, A., Nelsen, J., Thompson, G.A. and Stalker, D.M. (1992) (unpublished data).
38 Sheldon, P.., Kekwick, R., Sidebottom, C., Smith, C. and Slabas, A. (1990) Biochem. J. 271, 713–720.
39 Klein, B., Pawlowski, K., Horicke-Grandpierre, C., Schell, J. and Topfer, R. (1992) Mol. Gen. Genet. 223, 122–128.
40 Cronan, J.E., Li, W.-B., Coleman, R., Narasimhan, M., de Mendoza, D. and Schwab, J.M. (1988) J. Biol. Chem. 263, 4641–4646.
41 Slabas, A.R., Cottingham, I., Austin, A., Fawcett, T. and Sidebottom, C.M. (1991) Plant Mol. Biol. 17, 911–914.
42 Kater, M., Koningstein, G., Nijkamp, J. and Stuitje, A. (1991) Plant Mol. Biol. 17, 895–909.
43 Jaworski, J., Clough, R. and Barnum, S. (1989) Plant Physiol. 90, 41–44.
44 Jaworski, J., Post-Beittenmiller, D. and Ohlrogge, J. (1992) in Metabolism, Structure and Utilization of Plant Lipids (Cherif, A., ed.) pp. 113–120, Centre National Pédagogique, Tunis, Tunisia.
45 Walsh, M., Kloppenstein, W. and Harwood, J. (1990) Phytochemistry 29, 3797–3799.
46 Jackowski, S. and Rock, C.O. (1987) J. Biol. Chem. 262, 7927–7931.
47 Jackowski, S., Murphy, C.M., Cronan, J.E. and Rock, C.O. (1989) J. Biol. Chem. 264, 7624–7629.
48 Caughey, I. and Kekwick, R.G.O. (1982) Eur. J. Biochem. 123, 553–561.
49 Li, S.J. and Cronan, J.E. (1992) J. Biol. Chem. 267, 16841–16847.
50 Takai, T., Yokoyama, C., Wada, K. and Tanabe, T. (1988) J. Biol. Chem. 263, 2651–2657.
51 Lopez-Casillas, F., Bai, D., Luo, X., Kong, I., Hermondson, M. and Kim, K. (1988) Proc. Nat. Acad. Sci. U.S.A. 85, 5784–5788.
52 Al-Feel, W., Chirala, S.S. and Wakil, S.J. (1992) Proc. Nat. Acad. Sci. U.S.A. 89, 4534–4538.
53 Slabas, A. and Hellyer, A. (1985) Plant Sci. 39, 177–182.
54 Baldet, P., Alban, C., Axiotis, S. and Douce, R. (1992) Plant Physiol. 99, 450–455.
55 Secor, J. and Cseke, C. (1988) Plant Physiol. 86, 10–12.
56 Burton, J.D., Gronwald, J.W., Somers, D.A., Connelly, J.A., Gengenbach, B.G. and Wyse, D.L. (1987) Biochem. Biophys. Res. Commun. 148, 1039–1044.

57 Ishizaki, O., Nishida, I., Agata, K., Eguchi, G. and Murata, N. (1988) FEBS Lett. 238, 424–430.
58 Johnson, T.C., Schneider, J.C. and Somerville, C. (1992) Plant Physiol. 99, 771–772.
59 Hellyer, A. and Slabas, A. (1990) in Plant Lipid Biochemistry Structure and Utilization (Quinn, P.J. and Harwood, J.C., eds.), pp. 157–159, Portland Press, London.
60 Imai, H., Nishida, I. and Murata, N. (1992) Plant Mol. Biol. (in press).
61 Knutzon, D.S., Bleibaum, J.L., Nelsen, J., Kridl, J.C. and Thompson, G.A. (1992) Plant Physiol. 100 (in press).
62 Thompson, G.T., Scherer, D.E., Foxall-Van Aken, S., Kenney, J.W., Young, H.L., Shintani, D.K., Kridl, J.C. and Knauf, V.C. (1991) Proc. Nat. Acad. Sci. U.S.A. 88, 2578–2582.
63 Shanklin, J. and Somerville, C. (1991) Proc. Nat. Acad. Sci. U.S.A. 88, 2510–2514.
64 Knutzon, D.K., Scherer, D.E. and Schreckengost, W.E. (1991) Plant Physiol. 96, 344–345.
65 Sato, A., Becker, C. and Knauf, V. (1992) Plant Physiol. 99, 362–363.
66 Shanklin, J., Mullins, C. and Somerville, C. (1991) Plant Physiol. 97, 467–468.
67 Knutzon, D.S., Thompson, G.A., Radke, S.E., Johnson, W.B., Knauf, V.C. and Kridl, J.C. (1992) Proc. Nat. Acad. Sci. U.S.A. 89, 2624–2628.
68 Cahoon, E.B., Shanklin, J. and Ohlrogge, J.B. (1992) Proc. Nat. Acad. Sci. U.S.A. 89, 11184–11188.
69 Pollard, M., Anderson, L., Fan, C., Hawkins, D. and Davies, H. (1991) Arch. Biochem. Biophys. 284, 306–312.
70 Voelker, T., Worrell, A., Anderson, L., Bleibaum, J., Fan, C., Hawkins, D., Radke, S. and Davies, H. (1992) Science 257, 72–73.
71 Davies, M. (1992) (personal communication).
72 Joyard, J. and Stumpf, P.K. (1981) Plant Physiol. 67, 250–256.
73 Block, M.A., Dorne, A.J., Joyard, J. and Douce, R. (1983) FEBS Lett. 153, 377–381.
74 Sun, C., Cao, Y. and Huang, A. (1988) Plant Physiol. 88, 56–60.
75 Frtiz, P.J., Kauffman, J.M., Robertson, C.A. and Wilson, M.R. (1986) J. Biol. Chem. 261, 194–199.
76 Bafor, M., Jonsson, L., Stobart, A.K. and Stymne, S. (1990) Biochem. J. 272, 31–38.
77 Oo, K. and Huang, A., (1989) Plant Physiol. 91, 1288–1295.
78 Peterek, G., Schmidt, V., Wolter, F. and Frentzen, M. (1992) in Metabolism, Structure and Utilization of Plant Lipids (Cherif, A., ed.) , pp. 401–408, Centre National Pédagogique, Tunis, Tunisia.
79 Berneth, R. and Frentzen, M. (1990) Plant Sci. 67, 21–28.
80 Cao, Y.Z., Oo, K.C. and Huang, A.H.C. (1990) Plant Physiol. 94, 1199–1206.
81 Stymne, S., Tonnet, M. and Green, A. (1992) Arch. Biochem. Biophys. 294, 557–563.
82 James, D. and Dooner, H. (1990) Theor. Appl. Genet. 80, 241–245.

83 Lemieux, B., Miquel, M., Somerville, C. and Browse, J. (1990) Theor. Appl. Genet. 80, 234–240.
84 Arondel, V., Lemieux, B., Hwang, I., Gibson, S., Goodman, H.M. and Somerville, C.R. (1992) Science 258, 1353–1355.
85 Pierce, J. (1992) (personal communication).
86 Kearns, E., Hughly, S. and Sommerville, C. (1991) Arch. Biochem. Biophys. 284, 431–436.
87 Kearns, E., Keck, P. and Somerville, C. (1992) Plant Physiol. 99, 1254–1257.
88 Ichihara, K., Takahashi, T. and Fujii, S. (1988) Biochim. Biophys. Acta 958, 125–129.
89 Oo, K.C. and Chew, Y.H. (1992) Plant and Cell Physiol. 33, 189–195.
90 Cao, Y.Z. and Huang, A.H.C. (1986) Plant Physiol. 84, 762–765.
91 Weselake, R., Taylor, D., Pomeroy, M., Lawson, S. and Underhill, E. (1991) Phytochemistry 30, 3533–3538.
92 Moreau, R. and Stumpf, P. (1981) Plant Physiol. 67, 672–676.
93 Slabas, A.R., Roberts, P.A., Ormesher, J. and Hammond, E.W. (1982) Biochim. Biophys. Acta 711, 411–420.
94 Murphy, D.J. (1988) Lipids 23, 157–163.
95 Huang, A.H.C. (1992) Annu. Rev. Plant Physiol. Plant Mol. Biol. 43, 177–200.
96 Qu, R. and Huang, A.H.C. (1990) J. Biol. Chem. 265, 2238–2243.
97 Qu, R., Wang, S.M., Lin, Y.H., Vance, V.B. and Huang, A.H.C. (1986) Biochem. J. 234, 57–65.
98 Hatzopoulos, P., Franz, G., Choy, L. and Sung, R. (1990) Plant Cell 2, 457–467.
99 Cummins, I. and Murphy, D.J. (1992) Plant Mol. Biol. 19, 873–876.
100 Kalinski, A., Leor, D., Weisenann, J., Matthews, B. and Herman, E. (1991) Plant Mol. Biol. 17, 1095–1098.
101 Rooijen, G., Terning, L. and Moloney, M. (1992) Plant Mol. Biol. 18, 1177–1179.
102 Keddie, J., Hubner, G., Slocombe, S., Jarvis, R., Cummins, I., Edwards, E., Shaw, C. and Murphy, D. (1992) Plant Mol. Biol. 19, 1079–1083.
103 Fehling, E., Lessire, R., Cassagne, C. and Mukherjee, K.D. (1992) Biochim. Biophys. Acta 1126, 88–94.
104 Metz, J. and Lassner, M. (1992) (personal communication).

GENES FOR CROP IMPROVEMENT

John Bennett

Division of Plant Breeding, Genetics and Biochemistry
International Rice Research Institute
Los Banos, Philippines

INTRODUCTION

Between 1987 and 1991, the United States Department of Agriculture received well over 100 proposals for field trials of transgenic plants. The majority of trials were designed to test for enhanced resistance to biotic and abiotic stresses (Table 1). Other trials were concerned with alterations in product quality: delayed fruit ripening, modified seed proteins, increased carbohydrate content of tubers, increased sterol content, modification of oil content to reduce rancidity, and alteration of flower pigmentation. The common theme was crop improvement and the desired phenotype was conferred by transformation, usually with a single foreign gene.

Plant breeders depend on variation as the raw material for crop improvement. The three main sources of variation for conventional plant breeding are: (a) segregation and recombination following hybridization, (b) chemical or physical mutagenesis, and (c) germplasm collections of related wild varieties and species. The advent of transformation gives plant breeders access to a new and broader gene pool. Plants have been transformed with foreign genes from bacteria, viruses, animals, and of course other plants (1), and chemically synthesized genes have also been transferred to plants (2,3). This illustrates the power of genetic engineering to expand the range of gene transfer beyond that obtainable by conventional breeding.

This review of genes for crop improvement begins with a summary of the current status of plant transformation, especially the recent progress in transformation of cereals. There follows a survey of the potentially useful foreign genes that have been introduced into plants by transformation to enhance stress resistance or alter product quality. Finally, I consider the structural modifications that may be needed in foreign genes to ensure their efficient expression in plants. I shall not discuss in any depth the important and

Genetic Engineering, Vol. 15, Edited by J.K. Setlow
Plenum Press, New York, 1993

rapidly expanding applications of transformation as a tool to understand the regulation of plant growth and metabolism (4–7).

RECENT DEVELOPMENTS IN PLANT TRANSFORMATION

The transfer of foreign DNA to plants was first reported in 1983, when three groups described the recovery of kanamycin-resistant calli from tobacco and petunia plants (8–10). The transferred gene was a chimera composed of the coding sequence of the neomycin phosphotransferase (*neo*) gene from the bacterial transposon Tn*5* and the promoter and transcriptional terminator of the nopaline synthase (*nos*) gene from the tumor-inducing (Ti) plasmid of *Agrobacterium tumefaciens*. The *neo* gene provided for inactivation of the antibiotic kanamycin, while the *nos* regulatory sequences permitted expression of the gene in plant cells. During induction of the crown gall disease, *A. tumefaciens* naturally transforms its host, and the *nos* gene is expressed as part of the redirection of host metabolism toward the supply of C and N to the bacterium in the form of nopaline. Thus, the first successes in plant transformation were achieved by commandeering some of the machinery involved in the only known example of routine gene transfer between kingdoms of cellular organisms.

Agrobacterium-mediated transformation is now used to introduce foreign genes into the nucleus of many dicots (11,12). For some species such as tobacco, tomato and potato, specific varieties have been found to be extremely readily transformed by *Agrobacterium,* that is, transgenic calli are produced efficiently and are regenerated with high frequency into viable, fertile plants. However, for other dicotyledonous plants *Agrobacterium*-mediated transformation may proceed but regeneration may be infrequent, or transformation itself may prove a stumbling block. For this reason, tobacco, tomato and potato feature prominently in the list of plants engineered for crop improvement.

Many attempts have been made to determine whether *Agrobacterium*-mediated transformation of cereals is feasible. Two groups have reported *Agrobacterium*-mediated transformation of rice. Raineri et al. (13) presented Southern and western blots as evidence for stable transformation with the β-glucuronidase (*gus*) reporter gene but did not recover fertile plants. The frequency of transformation was highest with 2- to 4-day-old embryos. Chan et al. (14) reported expression of the *gus* and *neo* genes. Southern blots and enzyme assays were included in their evidence for success. Transformation was best with 3- to 4-day-old seedlings but was dependent on the prior exposure of rice tissue to medium conditioned by a potato suspension culture. The key substance secreted by potato is most likely to be a flavonoid capable of triggering the *Agrobacterium*-plant interaction (11,12).

Over the last 6 to 7 years, the most successful methods for transformation of cereals involved the use of protoplasts (15). DNA was introduced into protoplasts via electroporation (16,17) or PEG-mediated uptake (18). After transformation, the protoplasts were allowed to form callus and then to regenerate into plantlets (19–21).

Table 1
Applications to USDA to Field Test Genetically Engineered Plants
1987–1991

Crop	Type of Study						
	Virus Resistance	Bacterial Resistance	Fungal Resistance	Insect Resistance	Herbicide Resistance	Freezing Resistance	Product Quality
Potato	+	+		+	+		+
Tomato	+		+	+		+	+
Corn				+	+		+
Tobacco	+			+			+
Cotton				+	+		
Rapeseed				+			+
Rice				+			+
Alfalfa	+						
Cantaloupe	+						
Chrysanthemum							+
Cucumber	+						
Soybean					+		
Sunflower							+

Table 2
History of Rice Transformation

Year	Advance	Ref.
1986	Transient expression of foreign gene in japonica protoplasts:	
	- after electroporation	17
	- after PEG-mediated uptake	18
1988	Regeneration of transgenic japonica plants after transformation of protoplasts	19,20,21
	Transient expression after biolistic transformation	25
1989	Recovery of fertile transgenic plants from japonica protoplasts	22
1990	Transformation of indica protoplasts	23
	Recovery of fertile transgenic plants from indica protoplasts	24
1991	Recovery of fertile transgenic plants from biolistic transformation of indica and japonica callus and embryos	27

Table 2 summarizes the transformation history of rice, the first cereal to be regenerated from protoplasts to give fertile, transgenic progeny (22). The main problem with protoplast-based methods is the low frequency with which most varieties regenerate viable, fertile plants. While certain japonica varieties show a high capacity for regeneration, indica varieties are more difficult. Even so, several elite indica cultivars have now been transformed as protoplasts and regenerated to give fertile plants (23,24).

In the last five years, direct delivery of DNA into plant cells by the biolistic method has come to prominence (25). Gold or tungsten particles coated with DNA are fired into intact plants, tissue explants, callus or cell suspensions. Acceleration of the particles has been achieved by such methods as gunpowder (25), helium gas (26) and electric discharge (27,28). The biolistic approach is successful for transient expression studies (25,29) and for nuclear transformation of rice (27,28) and maize (29–31), together with sugarcane, wheat and sorghum (28). The biolistic method appears to be independent of variety and has been successful with both japonica and indica varieties of rice (27,28).

Plants contain three distinct genomes (Table 3). In addition to the nuclear genome, genetically functional DNA exists in the chloroplast and the mitochondrion. Most transformation studies attempt to integrate foreign genes into the nuclear genome, but there is increasing interest in organellar transformation. Chloroplasts have been transformed in the green alga *Chlamydomonas* (32,33) and in tobacco (34). So far, no report of successful transformation of the mitochondrial genome has been reported for plants, although it has been successful for yeast (35,36). The biolistic method is

Table 3
Transformation of Nuclear, Chloroplast and Mitochondrial Genomes of Plants (mid-1992)

Genome	*Agrobacterium*	Protoplasts		Biolistic	
	(stable)	(stable)	(transient)	(stable)	(transient)
Nucleus	+	+	+	+	+
Chloroplast	-	-	-	+	+
Mitochondrion	-	-	-	-	-

currently the only successful method for chloroplast transformation. It is also the method by which the mitochondria of yeast cells were transformed.

In principle, foreign genes could exist in host cells in three states: transiently as the introduced plasmid, on a longer term basis as an autonomously replicating plasmid, or permanently as a segment of DNA integrated into one of the three genomes by way of homologous or nonhomologous recombination. The third state is clearly preferable but depends on a recombination event. Most nuclear transformation studies to date have relied on nonhomologous recombination to integrate the foreign gene into nuclear DNA. Since the site of integration of a foreign gene could affect its own expression and also inactivate a host gene at or near that site, there is growing interest in homologous recombination or targeted integration (37). Should this procedure become routine, gene replacement would become possible and would be of considerable academic and applied interest.

MARKER GENES AND REPORTER GENES FOR TRANSFORMATION

Foreign genes are widely exploited in the plant transformation process itself (Table 4). Since very few cells in a target population become transformed, selection of transformants demands the use of selectable markers. Selectable marker genes usually encode enzymes which inactivate either an antibiotic or a herbicide. Only transformed cells survive and grow on media containing an appropriate concentration of the antibiotic or herbicide. The *neo* gene is used during transformation (10) to detoxify antibiotics such as kanamycin and G418, whereas the bacterial hygromycin phosphotransferase gene (38) inactivates hygromycin. The use of herbicides as selectable markers is discussed below.

Colorimetric, fluorometric, luminometric, or radiometric monitoring of transformation also exploits certain foreign genes known collectively as reporter genes. The most commonly used reporter genes are those encoding β-glucuronidase (39), luciferase (40) and chloramphenicol acetyl transferase (CAT) (10). Their principal uses are in facilitating the development of transformation protocols and in the characterization of plant promoters.

Table 4
Genes of Use for Crop Improvement

Transformation protocols	Abiotic stress
Antibiotic resistance:	Heavy metal:
neomycin phosphotransferase (B)	Metallothionen II (A)
hygromycin phosphotransferase (B)	Salt:
	Mannitol-1-phosphate dehydrogenase (B)
Herbicide resistance:	Freezing:
phosphinothricin acetyl transferase (B)	Fish antifreeze protein (A)
EPSP synthase (B,P).	Oxidation:
	Mn-Superoxide dismutase (P)
Reporter genes:	
chloramphenicol acetyl transferase (B)	Modified Product Quality
β-glucuronidase (B)	Fatty acid composition:
luciferase (A)	ACP thioesterase (P)
	Amino acid composition:
Pest & Disease Resistance	Methionine-rich protein (P)
Virus:	Dihydropicolinate synthase (B)
coat protein (V)	Delayed fruit ripening:
RNA-binding protein (V, antisense)	Polygalacturonase (P, antisense)
Satellite RNA (V)	ACC oxidase (P, antisense)
AL1 virus replication gene (V, antisense)	ACC synthase (P, antisense)
	ACC deaminase (B)
	Flower pigments
Fungus:	Chalcone synthase (P, antisense)
Chitinase (P)	
Ribosome inactivating protein (P)	Other Properties
	Polyhydroxybutyrate formation:
Insect:	Acetoacetyl CoA reductase (B)
Bt Cry 1A(b) (B,S)	PHB synthase (B)
Bt Cry 1A(c) (B,S)	Male sterility:
Cowpea trypsin inhibitor (P)	Barnase (B) and Barstar (B)
Potato inhibitor II (P)	Antibodies:
	Heavy chain (A)
	Light chain (A)

(Source of genes: A = animal, B = bacterium, P = plant, S = synthetic, V = virus)

Herbicide Tolerance

Transgenic mechanisms of herbicide resistance have obvious commercial significance in addition to their use as selectable markers for transformation. Considerable progress has been made in devising strategies for increasing herbicide tolerance in plants (41). One of these strategies envisions the transformation of plants with foreign genes which detoxify herbicides. An example is provided by the herbicide Basta/phosphinothricin, which inhibits glutamine synthetase, leading to accumulation of ammonia. DeGreef et al. (42) obtained field resistance to Basta through use of the *bar* gene from *Streptomyces hygroscopicus* (38). The *bar* gene encodes an acetyltransferase which acetylates the active component of Basta, and renders it inactive as a herbicide. This strategy is reminiscent of the mechanism to which certain corn lines owe their natural tolerance of atrazine: detoxification of the herbicide through conjugations with glutathione-S-transferase (43). In the context of selectable marker genes, the *bar* gene is probably the most popular of the herbicide resistance genes.

The second strategy for obtaining transgenic resistance to herbicides is to overexpress the target protein of the herbicide, while the third strategy is to express a desensitized target protein. Both of these strategies may be illustrated by reference to the herbicide glyphosate [N-(phosphonomethyl)glycine]. This herbicide inhibits aromatic amino acid biosynthesis in plants and bacteria by binding to the enzyme 5-enolpyruvylshikimate-3-phosphate synthase (EPSPS). Resistance to glyphosate has been achieved in plants both by overexpression of EPSPS (so that more herbicide is required to achieve a given level of killing) (44) and by expression of a modified EPSPS that is relatively insensitive to the herbicide (45,46).

Basta and glyphosate belong to the new generation of relatively benign herbicides with very low mammalian toxicity. However, their post-emergence use has been limited because of their nonselective (broad spectrum) mode of action. The development of Basta- and glyphosate-resistant crop plants will permit the use of these herbicides in new situations and encourage the phasing-out of the highly toxic herbicides currently in use. Of course, this strategy would eventually be rendered ineffective if there were gene transfer between the transgenic crop and wild relatives. To reduce the chance that weeds would acquire herbicide tolerance through cross-pollination, it would be desirable to integrate genes for herbicide resistance into the chloroplast DNA of the transgenic plant. In most crop plants, chloroplast DNA is inherited maternally, i.e., not transmitted through pollen (47). Since some progress toward chloroplast transformation has already been made (29,32–34), it is likely that the transfer of herbicide resistance genes to the chloroplast genome will be a focus of future work.

GENES FOR RESISTANCE TO PESTS AND DISEASES

Virus Resistance

It has been known for many years that plants can be protected against a virulent strain of certain viruses by prior infection with an attenuated strain of the same virus or a related virus. This phenomenon is known as cross-protection (48). In the case of positive-strand RNA viruses, a similar protective effect is observed in transgenic plants expressing the coat protein gene of the virus (49–52). The genes used in this type of transformation are double-stranded cDNAs derived from the viral RNA. Although the mechanism of cross-protection is not fully understood, it does not always appear to require high levels of expression of the coat protein itself (53,54).

Genetically-engineered resistance has also been obtained for a negative-strand RNA virus (55). Tomato spotted wilt virus is an enveloped virus containing an internal RNA-binding protein known as the nucleocapsid protein. Double-stranded cDNA derived from the nucleocapsid gene was introduced into tobacco plants by the *Agrobacterium* route and was found to be expressed and to confer considerable protection from a later challenge by the virus itself. This approach may be useful for producing plants resistant to infection by other negative-strand viruses. This mechanism of cross-protection is also not understood but could involve anti-sense inhibition of viral transcription or replication.

Virus resistance has been enhanced through transformation of plants with double-stranded DNA derived from satellite RNAs (56,57). Again, the mechanism of cross-protection is unclear but may involve competition between infective and satellite RNA sequences for cellular components.

A transgenic cross-protection method has been developed for a DNA-containing virus, tomato golden mosaic virus, one of the gemini viruses (58). Expression of an antisense construct derived from the viral AL1 gene conferred resistance in transgenic tobacco. The AL1 gene is involved in DNA replication. It is to be expected that many other viral genes can be used to give cross-protection.

Fungal Resistance

Many plants, animals and microorganisms produce proteins which are toxic to fungi. The genes for some of these proteins have been cloned and will presumably be employed in transgenic approaches to enhance fungal resistance in plants. Two examples relate to plants showing enhanced resistance to the important soil-borne fungal pathogen *Rhizoctonia solani.* In one case, transgenic plants expressed bean chitinase under the control of the CaMV 35S promoter (59), while in the other case plants expressed a ribosome inactivating protein (RIP) from barley under the control of the *wun* 2 promoter from potato (60).

R. solani causes disease symptoms on roots, stems and leaf sheaths of a wide range of plant species. For example, it causes leaf scurf in potato and sheath blight in rice. Fungal cell walls depend on chitin (a β-1,4-polymer of N-acetylglucosamine) for mechanical strength; treatment of fungal mycelia with chitinase leads to the bursting of the cells at the growing hyphal tip (61). Broglie et al. (59) found that, following transformation with the bean chitinase gene, tobacco seedlings expressed elevated levels of chitinase and showed better survival rates than control plants in soil heavily infested with *R. solani.* In contrast, transgenic plants showed no protection against *Pythium aphanidermatum,* a pathogen lacking a chitin-containing cell wall.

Canola plants (*Brassica napus*) transformed with the chitinase gene also showed enhanced survival in the presence of *R. solani* (59). The extent of disease resistance observed in the transgenic tobacco or canola plants varied with the amount of fungal inoculum, a property characteristic of quantitative resistance. However, the delay in the appearance of symptoms as well as the lower severity of disease may enable young seedlings to survive the critical period during stand establishment in the field when they are most susceptible to attack by soil-borne pathogens.

Logemann et al. (60) transformed tobacco with the barley RIP protein, which inactivates some eukaryotic ribosomes by hydrolyzing a N-glycosidic bond in 28S rRNA. There is considerable specificity in the action of RIP because its expression in transgenic tobacco does not impair tobacco growth but does retard growth of *R. solani.* Since *in vitro* studies (62) with *Trichoderma reesei* and *Fusarium sporotrichioides* demonstrated that combinations of barley RIP and barley chitinase inhibit growth more efficiently that either enzyme does alone, it is possible that the access of RIP to fungal cells is impeded by chitin. It would be interesting to know whether the simultaneous expression of both chitinase and RIP in transgenic plants leads to a synergistic enhancement of resistance to *R. solani.*

Insect Resistance

Insects are a major source of yield loss, especially in the countries of the humid and sub-humid tropics. The potential attractiveness of the transgenic approach to insect control arises from five considerations: (a) the economic and human health costs of insecticide use, (b) the development of insecticide resistance in pests, (c) the counter-productive effects of insecticides on many of the natural enemies of crop pests, (d) the absence of effective host plant resistance to many insect pests, and (e) the tendency of effective host plant resistance, when it does exist, to break down in the face of adaptive changes in the pest population.

Resistance to several insects has been enhanced through expression in plants of *Bacillus thuringiensis* (BT) toxin genes (2,3,63,64) and also genes encoding proteinase inhibitors (65,66). To be effective, these two types of inhibitor must be expressed in tissue consumed by the insect. The inhibitors interfere with aspects of insect digestion. BT toxins bind to epithelial glycoproteins of the intestine, especially the midgut, and cause fatal leakage of fluids between the intestine and the hemocoel (67). BT toxin genes of the *Cry*IA class have been effective against certain lepidopteran insects in transgenic tobacco, tomato and cotton but ineffective against other lepidopterans. However, their effectiveness has been enhanced 10- to 100-fold by chemical synthesis of the gene sequence to eliminate many of the adenine-thymine (AT)-rich sequence motifs which cause instability in the mRNA of transgenic plants (2,3) (see below for a fuller discussion of this problem).

In transgenic tobacco, the cowpea trypsin inhibitor enhances resistance to *Heliothis virescens* (65), and potato inhibitor II enhances resistance to *Manduca sexta* (66). It seems likely that proteinase inhibitors act by sequestering digestive proteinase but this remains to be established; they might act in a more subtle manner.

GENES FOR RESISTANCE TO ABIOTIC STRESS

Heavy Metal Tolerance

The first example of stable transgenic resistance to an abiotic stress was provided by the *Agrobacterium*-mediated introduction of the human metallothionen-II gene into *Brassica napus* and tobacco (68). The CaMV 35S promoter was used in conjunction with the *nos* terminator. The growth of root and shoot of transformed seedlings was unaffected by up to 0.1 mM $CdCl_2$, whereas control seedlings showed severe inhibition of root and shoot growth and chlorosis of leaves. The tolerance phenotype segregated as a dominant, single-locus Mendelian character.

Salt Tolerance

Many low molecular weight substances have been found to accumulate in living cells to provide protection from salt stress. In the case of plants, various species accumulate glycine betaine, proline and sugar alcohols such as mannitol and sorbitol. Tarczynski et al. (69) introduced into tobacco the *E. coli* gene encoding mannitol-1-phosphate dehydrogenase under the control of the CaMV 35S promoter and the *nos* terminator. In *E. coli* this reversible enzyme acts primarily to oxidize mannitol-1-phosphate to fructose-6-phosphate as part of growth on mannitol. In transgenic tobacco, however, this sequence of reactions is driven in reverse by excess fructose-6-phosphate, the mannitol-1-phosphate is hydrolyzed by a nonspecific phosphatase, and mannitol accumulates to more than 6 µmol/g fresh weight. When transgenic and control tobacco plants were compared for tolerance to 25 mM NaCl, mannitol accumulation provided significant protection to mature transgenic plants, enabling them to flower and set seed, whereas control plants died before flowering (H. Bohnert, personal communication). Protection was not observed in younger seedlings.

Freezing Tolerance

Many fruits and vegetables suffer considerable damage when exposed even for one night to freezing conditions. One transgenic approach to achieving freezing tolerance in plants was initiated by Hightower et al. (70), who introduced into tobacco the gene encoding an alanine-rich "anti-freeze" protein from fish. Although this class of protein has not yet been detected in plants, its action in fish is well established. Such proteins depress the freezing-point of water and thereby reduce the probability of cellular damage. The protein is expressed in tobacco but whether it depresses the freezing point of water in plant tissues and prevents ice formation has not been reported.

Oxidative Stress

Atmospheric oxygen and oxygen produced by photosynthesis can be reduced in plant cells to highly toxic chemicals known collectively as active oxygen species. Hydrogen peroxide, the hydroxyl radical and superoxide are examples of this group. Plants have evolved enzymatic mechanisms (catalases, peroxidases and superoxide dismutases) to deal with these molecules, which can be produced under a variety of stress situations, particularly those that block photosynthetic electron transport under moderate to high light intensities (e.g., cold, herbicides, presence of gaseous pollutants such as SO_2 and O_3, and certain fungal toxins). Bowler et al. (71) found that overexpression of manganese superoxide dismutase and targeting of the enzyme to the chloroplast led to significant protection from paraquat-induced damage in the light. Targeting of the same enzyme to the mitochondrion did not lead to protection, a result consistent with the fact that in the light paraquat-induced superoxides are formed in the chloroplast as a result of electron transfer from photosystem I.

GENES FOR MODIFIED PRODUCT QUALITY

Alteration of Fatty Acid Composition

Voelker et al. (72) have redirected the synthesis of fatty acid chains in *Arabidopsis* away from C_{16} or C_{18} molecules to the C_{12} laurate, which is of industrial importance. They transferred to *Arabidopsis* the gene for 12:0-acyl-carrier protein thioesterase from the oilseed plant, California bay (*Umbellularia californica*). The thioesterase, by prematurely hydrolyzing the growing acylthioesters, is thought to play a crucial role in the production of medium-chain fatty acids. To ensure that the thioesterase was active in seeds of *Arabidopsis* at the time of triacylglycerol production, it was fused with the promoter from the napin gene. Napin is a seed storage protein of *Brassica napus*. The transgenic plants were found to produce seeds which contained C_{12} laurate as their major fatty acid in triacylglycerols.

Alteration of Amino Acid Quantity

Humans and other animals utilize a full complement of 20 amino acids for protein synthesis. Although animals can synthesize some of these compounds from the citric acid cycle, there are 10 so-called essential amino acids (arginine, histidine, isoleucine, leucine, lysine, methionine, phenylalanine, threonine, tryptophan and valine) which must be obtained in the diet, that is, directly or indirectly from plant sources. Since methionine and lysine are often deficient in grains or seeds, it is interesting to note two recent studies on these amino acids: a chimeric gene encoding a methionine-rich seed protein from Brazil nut has been used to enhance by 30% the methionine content of tobacco seed proteins (73), and the lysine content of tobacco plants has been increased by expressing bacterial dihydropicolinate synthase in their chloroplasts (74). Dihydropicolinate synthase is the first enzyme of the lysine biosynthetic pathway; since the bacterial enzyme is not subject to the same regulatory mechanisms as the endogenous plant enzyme, its presence leads to a deregulation of lysine biosynthesis.

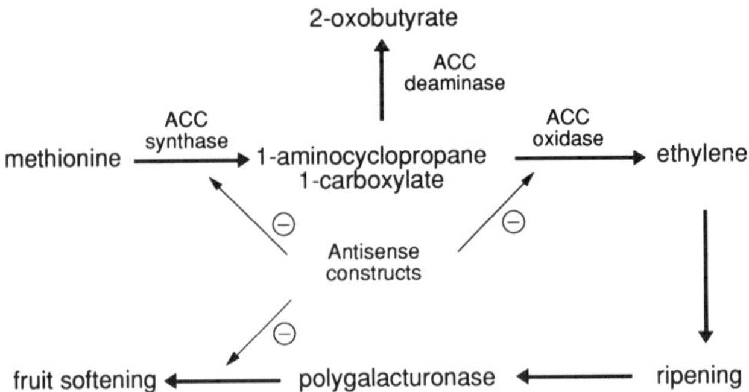

Figure 1. Transgenic modifications of the ethylene biosynthetic pathway to prolong the shelf life of tomato fruit.

Delayed Fruit Ripening in Tomato

Polygalacturonase is one of the key enzymes in the softening of fruit during ripening (Figure 1). Sheehy et al. (75) and Smith et al. (76,77) demonstrated that the antisense gene approach could lower the polygalacturonase levels in transgenic tomato fruit and prolong shelf life.

Hamilton et al. (78) also used an antisense gene approach to delay tomato fruit softening. However, in their case, the antisense construct produced RNA that was complementary to mRNA encoding ACC oxidase, the enzyme responsible for oxidizing 1-aminocyclopropane-1-carboxylate (ACC) to ethylene (Figure 1). It has been known for decades that ethylene is the plant hormone which regulates ripening. Oeller et al. (79), in a third antisense approach, succeeded in lowering the ethylene level in ripening tomato fruit by 99.5% by using the antisense construct from the gene encoding the first enzyme in the ethylene biosynthetic pathway, ACC synthase (Figure 1).

In a further attack on the ethylene biosynthetic pathway in tomato fruit, Klee et al. (80) employed a bacterial gene encoding ACC deaminase. Expression of this gene in transgenic tomato reduced the ethylene content of fruit by as much as 90%. The fruit showed delays in softening of up to 6 weeks.

Flower Pigmentation

The antisense approach was used by van der Krol et al. (81) to alter flower pigmentation patterns in petunia. The construct contained in the reverse orientation the gene for chalcone synthase, one of the early enzymes in the production of flavonoids from phenylalanine.

OTHER GENES FOR ADDED VALUE

Polyhydroxybutyrate Production

Polyhydroxybutyrate (PHB), a high molecular weight polyester of industrial importance, is accumulated as a storage form of carbon in many species of bacteria and is a biodegradable thermoplastic. Poirier et al. (82) have used genes from the bacterium *Alcaligenes eutrophus* that encode two enzymes required to convert acetoacetyl-coenzyme A to PHB (acetoacetyl-CoA reductase and PHB synthase). These genes have been placed under the control of the CaMV 35S promoter and introduced into *Arabidopsis.* Transgenic plant lines that contained both genes accumulated PHB as electron-translucent granules. This illustrates the potential of using plants for the production of bioplastics and other biopolymers.

Male Sterility

Hybrid vigor has proved of immense value in increasing the yield of several outbreeding crops. The commercial importance of maize hybrid seed production is based entirely on this phenomenon. Its exploitation in inbreeding crops such as rice is relatively recent. Cytoplasmic male sterility (cms) is a key element in the intensive commercial production of hybrids. A maize line showing cms is employed as the female parent. If the cms is stable, no selfing of the female parent will occur and all seed produced by it after cross-pollination by a male-fertile line will be hybrid seed. Since hybrid seed can provide large increases in yield, farmers are often prepared to buy such seed each year in preference to multiplying lower-yielding, male-fertile inbred lines.

Commercial producers of hybrid maize propagate cms lines by crossing them with maintainer lines. They can also restore male fertility by crossing the cms line with a restorer line. Much of the effort being expended in establishing hybrid seed production in crops other than maize is devoted to the search for stable cms lines and for maintainer and restorer lines.

An alternative approach is the development of artificial systems of reversible nuclear male sterility through genetic engineering. In one such approach, a bacterial ribonuclease (Barnase) is expressed under the control of a promoter that is specific for the tapetal cells that feed the developing pollen sacs. The resultant death of the tapetal cells induces male sterility (83). To complete this system, it would be necessary to find a mechanism of reversing sterility. Reversal has been achieved through expression of Barstar, an inhibitor of Barnase (84). Expression of an anti-sense Barnase construct might also prevent accumulation of the mRNA for Barnase. An intriguing alternative way of artificially down-regulating expression of the Barnase gene could be to use genes encoding ribozymes. The latter degrade specific mRNA molecules by RNA-catalyzed splicing reactions (85). It is not yet clear whether the antisense approach or the ribozyme approach would give sufficiently marked and stable down-regulation to act as restorers of male fertility.

The advantage of artificial nuclear male sterility is that it gives the breeder much more flexibility in his breeding program. He is not restricted to a few lines showing stable cms and responsiveness to restorers and maintainers.

Antibodies

One of the most intriguing possibilities in plant genetic engineering is to exploit the selectivity of antibodies to inhibit or interrupt specific processes. Assembly of functional antibodies within transgenic plants has already been demonstrated (86,87). Since antibodies are composed of two types of subunit, light chains and heavy chains, it is necessary to express both proteins in plants to obtain a functional antibody. Hiatt et al. (86), prepared two transgenic lines of tobacco, one transformed with a light chain gene and the other transformed with a heavy chain gene. Both lines produced their respective immunoglobulin chain but only hybrid plants assembled functional antibody containing both chains. Düring et al. (87), transformed tobacco with a single construct containing both genes under the control of separate promoters; functional antibody accumulated within cells. By analogy with experiments conducted in other systems (88), it should be possible to fuse the heavy and light chain genes into a single open reading frame and still recover active antibody.

MODIFICATION OF GENES FOR ENHANCED EXPRESSION

Promoters and Terminators

Among the principal requirements for foreign gene expression in plant cells are a suitable promoter and terminator (Figure 2). The promoter most commonly used in plant transformation is the cauliflower mosaic virus 35S promoter. This viral promoter controls the synthesis of a 35S RNA during infection of plants by CaMV and is highly expressed in many tissues of monocotyledonous and dicotyledonous plants (39,89,90). Another generally expressed promoter is the actin promoter (91). Such promoters are useful for initial transgenic studies when the basic efficacy of the foreign gene is to be assessed. However, subsequent studies would in all likelihood require more specific expression of the foreign gene: expression in specific tissues, at specific times or in response to specific plant hormones (92–94) or specific environmental cues, such as heat shock (95), light (96), wounding (97) and/or fungal elicitors (98). Many such promoters are known and there are methods available for identifying additional promoters with interesting properties (99). Promoters frequently have complex structures; it may be possible to obtain various patterns of expression with the use of different permutations and combinations of regions of a single promoter. This is seen clearly with the CaMV 35S promoter (88). Promoter strength may also be enhanced or modulated by placing two or more copies of a promoter in tandem or by combining promoters (100,101).

Transcriptional terminators in plants contain at least one polyadenylation signal (eg., AATAAA, AATTAA or AACCAA) (102,103) but may also need

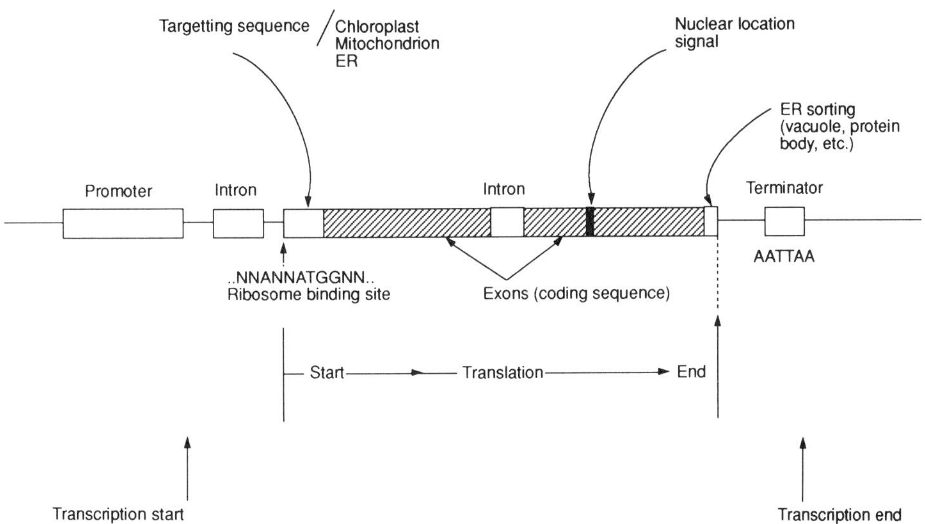

Figure 2. Modifications of a foreign gene for optimal expression in plants.

other 3′ sequences as well (104,105). The two most commonly-used terminators for plant transformation are the CaMV 35S terminator and the *nos* terminator (from the nopaline synthase gene of the T-DNA of *Agrobacterium*). The polyA tail may stabilize mRNAs against degradation (106).

Ribosome-binding Site

Genes may have to be modified to provide for appropriate ribosome binding sites in mRNA if high levels of expression are desired. The bacterial ribosome binding site (the Shine-Dalgarno sequence) is quite well defined but plant ribosome binding sites are rather different from their bacterial counterparts and less conserved. From a study of 75 published genomic DNA sequences from several higher plants, Joshi (107) concluded that the consensus context of the initiating ATG of plant mRNAs is TAAACA<u>ATG</u>GCT. It appears that the most important residue is the third residue before the initiating ATG. This residue should be a purine (usually A) for an abundantly expressed protein (107–109). Translational efficiency can be enhanced by inclusion of a viral untranslated leader sequence between the promoter and the coding region of the gene (110,111).

Introns

An intriguing phenomenon is the effect of introns on foreign gene expression in monocots. Callis et al. (112) were the first to report that introns increase gene expression in cultured maize cells. Tanaka et al. (113) reported as much as 90-fold higher levels of GUS expression in transgenic rice when an intron from the castor bean catalase gene is inserted into the GUS coding sequence. Enhanced expression was correlated with an increased level of mRNA and efficient splicing of the intron. This enhancement was not observed in transgenic tobacco. A stimulation of gene expression was observed in transgenic rice when the first intron of the *ActI* actin gene was inserted between the CaMV 35S promoter and the *gus* gene (114). Maas et al. (115), studied the effects of both exon 1 and intron 1 of the maize *Shrunken-1* gene on expression of CAT gene in transient expression experiments in rice and maize. The exon alone stimulated expression 10-fold, the intron stimulated 100-fold, and the two elements in combination stimulated 1000-fold when placed between the 35S promoter and the CAT gene. The exon is also stimulatory in tobacco protoplasts but the intron is inhibitory. It is not clear why monocots and dicots differ in their response to the presence of introns in transformation constructs.

In Figurc 2, two introns are included in the gene to indicate that the intron may be located in the coding sequence itself, which is then split into two exons, or in the upstream region between the transcriptional start site and the translational start site. The former location is the more usual in plant genes but the latter location appears to be equally effective and has the distinct advantage of providing for easier construction of transformation vectors. It can be quite tricky to place an intron within a coding sequence without altering the coding properties of the gene.

Potential Glycosylation Sites

It may also be important to make other specific modifications of genes to assist in their successful expression in plant cells. Many proteins that enter the endoplasmic reticulum undergo glycosylation at specific short amino acid sequences (Asn-X-Ser/Thr) which are recognized as glycosylation sites by glycosyl transferases. Similar sequences may also be present in proteins which never enter the endoplasmic reticulum. Should these proteins be forced to enter the endoplasmic reticulum (e.g., by transformation of plants with appropriate constructs), those previously unglycosylated sites might become glycoslyated and the protein might be inactivated. This problem was actually foreseen for the β-glucuronidase of *E. coli* (116). Based on the DNA sequence of the *gus* gene, two putative N-linked glycosylation sites were predicted: N^{358}LS and N^{423}IS. The *gus* gene is satisfactorily expressed in plants if the enzyme is directed to the cytosol (39), the chloroplast (117) or the mitochondrion (118), but if the protein is directed to the endoplasmic reticulum it becomes glycosylated, resulting in loss of activity and poor staining of tissue with X-gluc (119). The problem was solved through alteration of both sites, N^{358} being mutated to serine, threonine or proline and N^{423} being eliminated through a spontaneous alteration in sequence (12). The modified GUS protein can be directed through the endoplasmic reticulum to the vacuole and recovered with high activity.

AT/GC Bias

Modification on a larger scale has been necessary for BT toxin genes (2,3). *Cry* IA(b) and *Cry* IA(c) BT toxin genes were found to be expressed very poorly in transgenic tobacco and tomato. This was attributed to three main problems: codon bias, out-of-position polyadenylation/termination signal sequences and mRNA instability. In fact, all three problems were consequences of the high AT content of BT toxin genes. Some AT-rich codons are commonly used in the toxin genes but are rarely used in plant genes (e.g., TTA coding for leucine). Thus, in plants the translation of a foreign gene rich in TTA codons would lead the ribosome to pause, thereby increasing the probability of premature termination of the polypeptide. Premature transcriptional termination would be expected also in such a situation because of the high frequency within the BT toxin gene of sequences such as AATTAA resembling polyadenylation signal sequences (103). Finally, ATTTA sequences are suspected to destabilize mRNA. Direct evidence for the instability of *cry* IA(b) mRNAs in transgenic carrot plants and protoplasts has been obtained by Murray et al. (121). By changing 21% of the residues in the 1845 bp gene through chemical synthesis (without changing the predicted amino acid sequence), Perlak et al. (3) were able to increase expression of the BT toxin by 10-fold for *cry* IA(b) and 100-fold for *cry* IA(c), and thereby increase the toxicity of the transgenic plants to insects.

Subcellular Targeting

The site of function of every plant protein is quite specific. Most proteins are found in the cytosol. Such proteins are usually synthesized on free cytoplasmic ribosomes and need no special signals to remain in the cytosol. Many other proteins that are synthesized on free cytoplasmic ribosomes enter the chloroplast, the mitochondrion or the nucleus. Specific signals are required for this targeting (122–124). The signals of chloroplast and mitochondrial targeting are 35 to 60 amino acids in length, are located at the N-terminus of proteins and are usually removed after entry into the organelle. Suborganellar targeting such as between the different compartments of the chloroplast require additional targeting signals (123,125).

For nuclear proteins too large to diffuse through nuclear pores (> 60 kD), a short internal targeting sequence may be required. The sequence Pro-Lys-Lys-Lys-Arg-Lys-Val of the large T-antigen of Simian Virus 40 has been shown to be sufficient to target several proteins to the mammalian nucleus (126). The same sequence is able to transport the bacteriophage T7 RNA polymerase (100 kD) (124) and the GUS enzyme (homotetramer of 68 kD subunits) (127) to the nucleus of transgenic tobacco. The highly basic DNA-binding domains of three nuclear proteins of plants were found to facilitate the import of GUS into nuclei of transgenic tobacco (127), suggesting a close association or overlap of the DNA binding and nuclear targeting domains of B-ZIP proteins.

Many proteins are synthesized on cytoplasmic ribosomes that are initially free but rapidly become bound to the endoplasmic reticulum as the N-terminus of the growing polypeptide emerges from the ribosome. The N-terminus of these proteins forms the signal peptide for entry of the protein into the ER (128–131). The signal peptide (20 to 25 residues) consists of three regions designated n, h and c. The n-region of 3 to 5 residues contains a single positively charged amino acid residue, the h-region contains 7 to 15 hydrophobic residues, and the c-region of 3 to 5 small neutral residues is important for recognition by a processing protease which removes the signal peptide very soon after it enters the lumen of the ER. Proteins which contain no other targeting sequence follow a so-called default pathway through the ER system and are secreted from the cell (132). Proteins with additional specific targeting sequences are deposited in protein bodies, lysosomes, vacuoles, peroxisomes and other membrane-bound derivates of the ER (133–136). A minority of chloroplast and mitochondrial proteins is encoded and synthesized in the organelles themselves (137,138).

Scaffold-associated DNA Sequences

Plant nuclear genes operate within a complex and highly organized chromosomal structure (139). Specific DNA sequences contribute to this structure by providing binding sites for scaffolding proteins involved in the bending and packing of chromosomes (140). Studies on animal cells indicate that inclusion of DNA sequences (about 3 kb) of scaffold-associated regions (SARs) in transfection vectors increases the transcription of foreign genes and dampens the position effects that arise from the uncontrolled insertion of foreign genes into different sites within the genome (141,142). It is likely that future plant transformation vectors will include SARs to achieve these same benefits.

CONCLUDING COMMENTS

Genetic engineering will lead to the improvement of many key characteristics of crop plants. It is already evident that the resistance of plants to several biotic stresses such as weeds, viruses, insects and microorganisms can be significantly enhanced by this approach. Less progress has been made towards resistance to abiotic stress but this situation will improve as we learn more about the biochemistry of stress responses. Improvements are also being made in the quantity and quality of plant products, and it has become possible to introduce a form of nuclear male sterility and thereby assist the production of higher-yielding hybrid plants in species where this has not yet been achieved.

Much concern has been expressed about the potential dangers of plant genetic engineering. Relatively few countries have adopted regulatory policies to deal with this problem. It is important that both industrialized and developing countries have in place clear and specific guidelines that cover containment for laboratory work, confinement for field-testing and certification for release to national agricultural research systems or the private sector. The guidelines should allow for rapid approval in non-controversial cases and thorough analysis in more difficult cases. The definition of what is non-controversial will itself evolve over time (143).

Of special concern is the spread of genes from transgenic crops to weedy wild relatives following cross-pollination. As we have seen, in those crop species which show maternal inheritance of plastids, the problem might be solved through integration of foreign genes into chloroplast DNA. However, before such a possibility can be made a reality, it will be necessary to develop efficient mechanisms of chloroplast transformation. Furthermore, a chloroplast location for foreign genes will make biochemical sense in only a minority of cases, e.g., some herbicide resistance genes. The dumping of foreign genes in the chloroplast should not be seen as a substitute for a realistic assessment of likely dangers.

Another vexing question is the impact of large-scale cultivation of transgenic crops on the genetic constitution of populations of field pests and diseases (144). Intensive cultivation of insect-resistant plants can be expected to lead to the selection of resistant strains. This problem is of course not unique to transgenic plants. It is seen also for insect-resistant plants developed by classical breeding techniques and finds a parallel in the development of insecticide-resistance in many pest populations following the excessive use of pesticides. Considerable insights will be needed into the population dynamics of insects and their mechanisms of genetic change, if appropriate strategies for deployment of engineered plants are to be formulated. Are weak resistance mechanisms likely to provide less selection pressure than strong mechanisms? Does security and sustainability lie in the deployment of several diverse resistance mechanisms, and should these be deployed simultaneously or sequentially? It should be noted that the best answers to these questions can be provided by small-scale field experimentation with transgenic plants.

Will transgenic plants display yield depression or reduction in fertility as a result of passage through tissue culture? Careful agronomic studies will have to be undertaken to determine whether transgenic plants suffer genetic changes as a result of the transformation protocol and whether the losses resulting from these changes outweigh the benefits accruing from the presence of the foreign gene.

Acknowledgments. I thank Dr. Pia Bennett, Mrs. Shirley Raymundo, and Ms. Minnie Bernardino for assistance in the preparation of the article.

REFERENCES

1 Uchimiya, H., Handa, T. and Brar, D.S. (1989) J. Biotechnology 12, 1–20.
2 Perlak, F.J., Deaton, R.W., Armstrong, T.A., Fuchs, R.L., Sims, S.R., Greenplate, J.T. and Fischhoff, D.A. (1990) Bio/Technology 8, 939–943.
3 Perlak, F.J., Fuchs, R.L., Dean, D.A., McPherson, S.L. and Fischhoff, D.A. (1991) Proc. Nat. Acad. Sci. U.S.A. 88, 3324–3328.
4 Keller, J.M., Shanklin, J., Vierstra, R.D. and Hershey, H.P. (1989) EMBO J. 8, 1005–1012.

5 Nagatani, A., Kay, S.A., Deak, M., Chua, N.H. and Furuya, M. (1991) Proc. Nat. Acad. Sci. U.S.A. 88, 5207–5211.
6 Lagrimini, L.M., Bradford, S. and Rothstein, S. (1990) Plant Cell 2, 7–18.
7 Schmulling, T., Schell, J. and Spena, A. (1988) EMBO J. 7, 2621–2629.
8 Bevan, M.W., Flavell, R.B. and Chilton, M.D. (1983) Nature 304, 184–187.
9 Fraley, R.T., Rogers, S.G., Horsch, R.B., Sanders, P.R., Flick, J.S., Adams, S.P., Bittner, M.L., Brand, L.A., Fink, C.L., Fry, J.S., Galluppi, G.R., Goldberg, S.B., Hoffman, N.L. and Woo, S.C. (1983) Proc. Nat. Acad. Sci. U.S.A. 80, 4803–4807.
10 Herrera-Estrella, L., De Block, M., Messens, E., Hernalsteens, J.P., Van Montagu, M. and Schell, J. (1983) EMBO J. 2, 987–995.
11 Klee, H., Horsch, R. and Rogers, S. (1987) Annu. Rev. Plant Physiol. 38, 467–486.
12 Weising, K., Schell, J. and Kahl, G. (1988) Annu. Rev. Genet. 22, 421–477.
13 Raineri, D.M., Bottino, P., Gordon, M.P. and Nester, E.W. (1990) Bio/Technology 8, 33–38.
14 Chan, M.T., Lee, T.M. and Chang, H.H. (1992) Plant Cell Physiol. 33, 577–583.
15 Potrykus, I. (1991) Annu. Rev. Plant. Physiol. Plant Mol. Biol. 42, 205–225.
16 Fromm, M., Taylor, L.P. and Walbot, V. (1985) Proc. Nat. Acad. Sci. U.S.A. 82, 5824–5828.
17 Ou-Lee, T.M., Turgeon, R. and Wu, R. (1986) Proc. Nat. Acad. Sci. U.S.A. 83, 6815–6819.
18 Uchimiya, H., Fushimi, T., Hashimoto, H.M., Harada, H., Syon, K. and Sugawara, Y. (1986) Mol. Gen. Genet. 204, 204–207.
19 Toriyama, K., Arimoto, Y., Uchimiya, H. and Hinata, K. (1988) Bio/Technology 6, 1072–1074.
20 Zhang, W. and Wu, R. (1988) Theor. Appl. Genet. 76, 835–850.
21 Zhang, H.M., Yang, H., Rech, E.L., Golds, T.J., Davis, A.S., Mulligan, B.J., Cocking, E.C. and Davey, M.R. (1988) Plant Cell Rep. 7, 379–384.
22 Shimamoto, K., Terada, R., Izawa, T. and Fujimoto, H. (1989) Nature 338, 274–276.
23 Datta, S.K., Peterhans, A., Datta, K. and Potrykus, I. (1990) Bio/Technology 8, 736–740.
24 Peng, J., Lyznik, L.A. and Hodges, T.K. (1990) Plant Cell Rep. 9, 168–172.
25 Wang, Y.C., Klein, T.M. Fromm, M., Cao, J., Sanford, J.C. and Wu, R. (1988) Plant Mol. Biol. 11, 433–439.
26 Sanford, J.C., Devit, M.J., Russell, J.A., Smith, F.D., Harpending, P.R., Roy, M.K. and Johnston, S.A. (1991) Technique J. Meth. Cell Molec. Biol. 3, 3–16.
27 Christou, P., Ford, T.L. and Kofron, M. (1991) Bio/Technology 9, 957–962.
28 Christou, P. (1992) Plant J. 2, 275–281.

29 Daniell, H., Vivekananda, J., Nielsen, B.L., Ye, G.N., Tewari, K.K. and Sanford, J.C. (1990) Proc. Nat. Acad. Sci. U.S.A. 87, 88–92.
30 Fromm, M.E., Morrish, F.M., Armstrong, C., Williams, R., Thomas, J. and Klein, T.M. (1990) Bio/Technology 8, 833–839.
31 Spenser, T.M., Gordon-Kam, W.J., Daines, R.J., Start, W.G. and Lemaux, P.G. (1990) Theor. Appl. Genet. 79, 625–631.
32 Boynton, J.E., Gillham, N.W., Harris, E.H., Hosler, J.P., Johnson, A.M., Jones, A.R., Randolph-Anderson, B.L., Robertson, D., Klein, T.M., Shark, K.B. and Sanford, J.C. (1988) Science 240, 1534–1538.
33 Blowers, A.D., Bogorad, L., Ye, G.N., Shark, K.B. and Sanford, J. (1989) Plant Cell 1, 123–132.
34 Svab, Z., Hajdukiewicz, P. and Maliga, P. (1990) Proc. Nat. Acad. Sci. U.S.A. 87, 8526–8530.
35 Butow, R.A. and Fox, T.D. (1990) Trends Biochem. Sci. 15, 465–468.
36 Fox, T.D., Sanford, J.C. and McMullin, T.W. (1988) Proc. Nat. Acad. Sci. U.S.A. 85, 7288–7292.
37 Yoder, J.I. and Kmiec, E. (1991) in Genetic Engineering Vol. 13 (Setlow, J.K., ed.) pp. 265–278, Plenum Press, New York, NY.
38 Van den Elzen, P., Townsend, J., Lee, K.Y. and Bedbrook, J.R. (1985) Plant Mol. Biol. 5, 299–302.
39 Jefferson, R.A., Kavanagh, T.A. and Bevan, M.W. (1987) EMBO J. 6, 3901–3907.
40 Ow, D.W., Wood, K.V., Deluca, M., De Wet, J.R., Helinski, D.R. and Howell, S.H. (1986) Science 234, 856–869.
41 Mazur, B.J. and Falco, S.C. (1989) Annu. Rev. Plant Physiol. Plant Mol. Biol. 40, 441–470.
42 DeGreef, W., Delon, R., De Block, M., Leemans, J. and Botterman, J. (1989) Bio/Technology 7, 61–64.
43 Shimabukuro, R.H., Frear, D.S., Swanson, H.R. and Walsh, W.C. (1971) Plant Physiol. 47, 10–14.
44 Shah, D.M., Horsch, R.B., Klee, H.J., Kishore, G.M., Winter, J.A., Tumer, N.E., Hironaka, C.M., Sanders, P.R., Gasser, C.S., Aykent, S., Siegel, N.R., Rogers, S.G. and Fraley, R.T. (1986) Science 233, 478–481.
45 Comai, L., Facciotti, D., Hiatt, W.R., Thompson, G., Rose, R.E. and Stalker, D.M. (1985) Nature 317, 741–744.
46 Della-Cioppa, G., Bauer, S.C., Taylor, M.L., Rochester, D.E., Klein, B.K., Shah, D.M., Fraley, R.T. and Kishore, G.M. (1987) Bio/Technology 5, 579–584.
47 Tilney-Bassett, R.A.E. (1984) in Chloroplast Biogenesis (Ellis, R.J., ed.) pp. 13–50, Cambridge University Press, Cambridge.
48 Fulton, R.W. (1986) Annu. Rev. Phytopathol. 24, 67–81.
49 Powell-Abel, P., Nelson, R.S., De, B., Hoffmann, N., Rogers, S.G., Fraley, R.T. and Beachy, R.N. (1986) Science 232, 738–743.
50 Stark, D.M. and Beachy, R.N. (1989) Bio/Technology 7, 1257–1262.
51 Lawson, C., Kaniewski, W., Haley, L., Rozmann, R., Newell, C., Sanders, P. and Tumer, N.E. (1990) Bio/Technology 8, 127–137.
52 Van der Wilk, F., Willink, D.P.L., Huisman, M.J., Huttinga, H. and Goldbach, R. (1991) Plant Mol. Biol. 17, 431–439.

53 Hemenway, C., Fang, R.-X., Kaniewski, W.K., Chua, N.H. and Tumer, N.E. (1988) EMBO J. 7, 1273–1280.
54 Beachy, R.N., Loesch-Fries, L.S. and Tumer, N. (1990) Annu. Rev. Phytopath. 28, 451–474.
55 Gielen, J.J.L., de Haan, P., Kool, A.J., Peters, D., van Grinsven, M.Q.J.M. and Goldbach, R.W. (1991) Bio/Technology 9, 1363–1367.
56 Gerlach, W.L., Llewellyn, D. and Haselhoff, J. (1987) Nature 328, 802–805.
57 Harrison, B.D., Mayo, M.A. and Baulcombe, D.C. (1987) Nature 328, 799–802.
58 Day, A.G., Bejarano, E.R., Buck, K.W., Burrell, M. and Lichtenstein, C.P. (1991) Proc. Nat. Acad. Sci. U.S.A. 88, 6721–6725.
59 Broglie, K., Chet, I., Holliday, M., Cressman, R., Biddle, P., Knowlton, S., Mauvais, C.J. and Broglie, R. (1991) Science 254, 1194–1197.
60 Logemann, J., Jach, G., Tommerup, H., Mundy, J. and Schell, J. (1992) Bio/Technology 10, 305–308.
61 Schlumbaum, A., Mauch, F., Vogeli, U. and Boller, T. (1986) Nature 324, 365–367.
62 Leah, R., Tommerup, H., Svendsen, I. and Mundy, J. (1991) J. Biol. Chem. 266, 1564–1573.
63 Fischhoff, D.A., Bowdish, K.S., Perlak, F.J., Marrone, P.G., McCormick, S.M., Niedermeyer, J.G., Dean, D.A., Kusano-Kretzmer, K., Mayer, E.J., Rochester, E.E., Rogers, S.G. and Fraley, R.T. (1987) Bio/Technology 5, 807–813.
64 Vaeck, M., Reynaerts, A., Hofte, H., Jansens, S., DeBeuckeleer, M.D., Dean, C., Zabeau, M., Van Montagu, M.V. and Leemans, J. (1987) Nature 328, 33–37.
65 Hilder, V.A., Gatehouse, A.M.R., Sheerman, S.E., Baker, R.F. and Boulter, D. (1987) Nature 330, 160–163.
66 Johnson, R., Narvaez, J., An, G. and Ryan, C. (1989) Proc. Nat. Acad. Sci. U.S.A. 86, 9871–9875.
66 Höfte, H. and Whiteley, H.R. (1989) Microbiol. Rev. 53, 242–255.
68 Misra, S. and Gedamu, L. (1989) Theor. Appl. Genet. 78, 161–168.
69 Tarczynski, M.C., Jensen, R.C. and Bohnert, H.J. (1992) Proc. Nat. Acad. Sci. U.S.A. 89, 2600–2604.
70 Hightower, R., Baden, C., Penzes, E., Lund, P. and Dunsmuir, P. (1991) Plant. Mol. Biol. 17, 1013–1021.
71 Bowler, C., Slooten, L., Vandenbranden, S., De Rycke, R., Botterman, J., Sybesma, C., Van Montagu, M. and Inze, D. (1991) EMBO J. 10, 1723–1732.
72 Voelker, T.A., Worrell, A.C., Anderson, L., Bleibaum, J., Fan, C., Hawkins, D.J., Radke, S.E. and Davies, H.M. (1992) Science 257, 72–74.
73 Altenbach, S., Pearson, K. Meeker, G., Staraci, L. and Sun, S. (1990) Plant. Mol. Biol. 13, 513–522.
74 Shaul, O. and Galili, G. (1992) Plant J. 2, 203–209.
75 Sheehy, R.E., Kramer, M. and Hiatt, W.R. (1988) Proc. Nat. Acad. Sci. U.S.A. 85, 8805–8809.

76 Smith, C.J.S., Watson, C.F., Ray, J., Bird, C.R., Morris, P.C., Schuch, W. and Grierson, D. (1988) Nature 334, 724–726.
77 Smith, C.J.S., Watson, C.F., Morris, P.C., Bird, C.R., Seymour, G.B., Gray, J.E., Arnold. C., Tucker, G.A. and Schuch, W. (1990) Plant Mol. Biol. 14, 369–380.
78 Hamilton, A.J., Lycett, G.W. and Grierson, D. (1990) Nature 346, 284–287.
79 Oeller, P.W., Min-Wong, L., Taylor, L.P., Pike, D.A. and Theologis, A. (1991) Science 254, 437–439.
80 Klee, H.J., Hatford, M.B., Kretzmer, K.A., Barry, G.F. and Kishore, G.M. (1991) Plant Cell 3, 1187–1193.
81 Van der Krol, A.R., Lenting, P.E., Veenstra, J., van der Meer, I.M., Koes, R.E., Gerats, A.G.M., Mol, J.M.N. and Stuitje, A.R. (1988) Nature 333, 866–869.
82 Poirier, Y., Dennis, D.E., Klomparens, K. and Somerville, C. (1992) Science 256, 520–523.
83 Mariani, C., De Beuckeleer, M., Truettner, J., Leemans, J. and Goldberg, R.B. (1990) Nature 347, 737–741.
84 Mariani, C., Gosele, V. DeBeuckeleer, M., DeBlock, M., Goldberg, R.B., DeGreef, W. and Leemans, J. (1992) Nature 357, 384–387.
85 Haselhoff, J. and Gerlach, W.L. (1988) Nature 334, 585–591.
86 Hiatt, A., Cafferkey, R. and Bowdish, K. (1989) Nature 342, 76–78.
87 Düring, K., Hippe, S., Kreuzaler, F. and Schell, J. (1990) Plant Mol. Biol. 15, 281–293.
88 Chaudhary, V.K., Batra, J.K., Gallo, M.G., Willingham, M.C., Fitzgerald, D.J. and Pastan, I. (1990) Proc. Nat. Acad. Sci. U.S.A. 87, 1066–1070.
89 Benfey, P.N. and Chua, N.H. (1989) Science 244, 174–181.
90 Battraw, M.J. and Hall, T.C. (1990) Plant Mol. Biol. 15, 527–538.
91 McElroy, D., Zang, W., Cao, J. and Wu, R. (1990) Plant Cell 2, 163–171.
92 Van der Zaal, E.J., Droog, F.N.J., Boot, C.J.M., Hensgens, L.A.M., Hoge, J.H.C., Schilperoort, R.A. and Libbenga, K.R. (1991) Plant Mol. Biol. 16, 983–998.
93 Skriver, K., Olsen, F.L., Rogers, J.C. and Mundy, J. (1991) Proc. Nat. Acad. Sci. U.S.A. 88, 7266–7270.
94 Meyer, R.C., Goldsbrough, P.B. and Woodson, W.R. (1991) Plant Mol. Biol. 17, 277–281.
95 Baumann, G., Raschke, E., Bevan, M. and Schoffl, F. (1987) EMBO J. 6, 1161–1166.
96 Kuhlemeier, C., Strittmatter, G., Ward, K. and Chua, N.H. (1989) Plant Cell 1, 471–478.
97 Keil, M., Sanchez-Serrano, J., Schell, J. and Willmitzer, L. (1990) Plant Cell 2, 61–70.
98 Schmid, J., Doerner, P.W., Clouse, S.D., Dixon, R.A. and Lamb, C.J. (1990) Plant Cell 2, 619–631.
99 Kertbundit, S., de Greve, H., Deboeck, F. and van Montagu, M. (1991) Proc. Nat. Acad. Sci. U.S.A. 88, 5212–5216.
100 Comai, L., Moran, P. and Maslyar, D. (1990) Plant Mol. Biol. 15, 373–381.

101 Last, D.I., Brettell, R.I.S., Chamberlain, D.A., Chaudhury, A.M., Larkin, P.J., Marsh, E.L., Peacock, W.J. and Dennis, E.S. (1991) Theor. Appl. Genet. 81, 581–588.
102 Dean, C., Tamaki, S., Dunsmuir, P., Favreau, M., Katayama, C., Dooner, H. and Bedbrook, J. (1986) Nucl. Acids Res. 14, 2229–2240.
103 Joshi, C.P. (1987) Nucl. Acids Res. 15, 9627–9640.
104 Mogen, B.D., Macdonald, M.H., Graybosch, R. and Hunt, A.G. (1990) Plant Cell 2, 1261–1272.
105 Ingelbrecht, I.L.W., Herman, L.M.F., Dekeyser, R.A., Van Montagu, M.C. and Depicker, A.G. (1989) Plant Cell 1, 671–680.
106 Gallie, D.R., Lucas, W.J. and Walbot, V. (1989) Plant Cell 1, 301–311.
107 Joshi, C.P. (1987) Nucl. Acids Res. 15, 6643–6653.
108 Heidecker, G. and Messing, J. (1986) Annu. Rev. Plant Physiol. 37, 439–466.
109 Kozak, M. (1989) J. Cell Biol. 108, 229–241.
110 Gallie, D.R., Sleat, D.E., Watts, J.W., Turner, P.C. and Wilson, T.M.A. (1987) Nucl. Acids Res. 15, 3257–3273.
111 Jobling, S.A. and Gehrke, L. (1987) Nature 325, 622–625.
112 Callis, J., Fromm, M. and Walbot, V. (1987) Genes Devel. 1, 1183–1200.
113 Tanaka, A., Mita, S., Ohta, S., Kyozuka, J., Shimamoto, K. and Nakamura, K. (1990) Nucl. Acis Res. 18, 6767–6770.
114 Cao, J., Zhang, W., McElroy, D. and Wu, R. (1991) in Rice Biotechnology (Khush, G.S. and Toenniessen, G.H., eds.), pp. 175–198, CAB International, Wallingford, U.K.
115 Maas, C., Laufs, J., Grant, S., Korfhage, C. and Werr, W. (1991) Plant Mol. Biol. 16, 199–207.
116 Jefferson, R.A., Burgess, S.M. and Hirsch, D. (1986) Proc. Nat. Acad. Sci. U.S.A. 83, 8447–8451.
117 Kavanagh, T.A., Jefferson, R.A. and Bevan, M.W. (1988) Mol. Gen. Genet. 215, 38–45.
118 Schmitz, U.K. and Lonsdale, D.M. (1989) Plant Cell 1, 783–791.
119 Iturriaga, G., Jefferson, R.A. and Bevan, M.W. (1989) Plant Cell 1, 381–390.
120 Farrell, L.B. and Beachy, R.N. (1990) Plant Mol. Biol. 15, 821–825.
121 Murray, E.E., Rocheleau, T., Eberle, M., Stock, C., Sekar, V. and Adang, M. (1991) Plant Mol. Biol. 16, 1035–1050.
122 Boutry, M., Nagy, F., Poulsen, C., Aoyagi, K. and Chua, N.H. (1987) Nature 328, 340–342.
123 Von Heijne, G., Steppuhn, J. and Herrmann, R.G. (1989) Eur. J. Biochem. 180, 535–545.
124 Lassner, M.W., Jones, A., Daubert, S. and Comai, L. (1991) Plant Mol. Biol. 17, 229–234.
125 Keegstra, K., Olsen, L.J. and Theg, S.M. (1989) Annu. Rev. Plant Physiol. Plant Mol. Biol. 40, 471–501.
126 Kalderon, D., Roberts, B.L., Richardson, W.D. and Smith, A.E. (1984) Cell 39, 499–509.
127 Van der Krol, A.R. and Chua, N.H. (1991) Plant Cell 3, 667–675.
128 Burr, F.A. and Burr, B. (1982) J. Cell Biol. 94, 201–206.

129 Von Heijne, G. (1985) J. Mol. Biol. 184, 99–105.
130 Torrent, M., Poca, E., Campos, N., Ludevid, M.D. and Palau, J. (1986) Plant Mol. Biol. 7, 393–403.
131 DellaPenna, D. and Bennett, A.B. (1988) Plant Physiol. 86, 1057–1063.
132 Denecke, J., Botterman, J. and Deblaere, R. (1990) Plant Cell 2, 51–59.
133 Chrispeels, M.J. (1991) Annu. Rev. Plant Physiol. Plant Mol. Biol. 42, 21–53.
134 Dorel, C., Voelker, T.A., Herman, E.M. and Chrispeels, M.J. (1989) J. Cell Biol. 108, 327–337.
135 Saalbach, G., Jung, R., Kunze, G., Saalbach, I., Adler, K. and Muntz, K. (1991) Plant Cell 3, 695–708.
136 Bednarek, S.Y. and Raikhel, N.V. (1991) Plant Cell 3, 1195–1206.
137 Mullet, J.E. (1988) Annu. Rev. Plant Physiol. Plant Mol. Biol. 39, 475–502.
138 Newton, K.J. (1988) Annu. Rev. Plant Physiol. Plant Mol. Biol. 39, 503–532.
139 Heslop-Harrison, J.S. and Bennett, M.D. (1990) Trends Genet. 6, 401–405.
140 Slatter, R.E., Dupree, P. and Gray, J.C. (1991) Plant Cell 3, 1239–1250.
141 Stief, A., Winter, D., Straling, W.H. and Sippel, A. (1989) Nature 341, 343–345.
142 Phi-Van, L., von Kries, J.P., Ostertag, W. and Straling, W.H. (1990) Mol. Cell Biol. 10, 2302–2307.
143 Kessler, D.A., Taylor, M.R., Maryanski, J.H., Flamm, E.L. and Kahl, L.S. (1992) Science 256, 1747–1749,1832.
144 Gould, F. (1988) BioScience 38, 26–33.

MOLECULAR BIOLOGY AND GENETICS OF PROTECTIVE FUNGAL ENDOPHYTES OF GRASSES

Christopher L. Schardl and Zhiqiang An*

Department of Plant Pathology
S-305 Agriculture Science-North
University of Kentucky
Lexington, KY 40546-0091

*Present address: Department of Physiological Chemistry
University of Wisconsin, Madison, WI 53706

INTRODUCTION

Clavicipitaceous endophytes, a group of seedborne fungal symbionts of C3 grasses, may be the most widespread agents of biological plant protection in agriculture, and the unwitting use of them in pasture grasses has continued for many centuries. Endophytes have been recognized in species of the Pooideae (the cool-season grasses) for nearly a century (1), but the agricultural and ecological importance of these associations has been appreciated only in recent decades. Those with evolutionary affinities to the teleomorph (sexual form) *Epichloë typhina* (Pers:Fr.) Tulasne (family Clavicipitaceae) are referred to as "clavicipitaceous endophytes" (here abbreviated c-endophytes) and can be particularly important for the fitness, competitiveness and persistence of their grass hosts under conditions of biotic or abiotic stress (2–4). The most intensely studied of these interactions have been the tall fescue (*Festuca arundinacea* Schreb.) and perennial ryegrass (*Lolium perenne L.*) symbiota. The multitude of fitness enhancements conferred by the c-endophytes (Table 1) are, in no small measure, contributory to the excellent persistence of these grasses in temperature pastures. However, since it was determined that they were associated with livestock toxicosis (5,6), the endophytes have not been

Genetic Engineering, Vol. 15, Edited by J.K. Setlow
Plenum Press, New York, 1993

Table 1
References Indicating Host Benefits or Possible Host Benefits of Grass Endophytes

Property	c-endophytes	p-endophytes
Anti-mammalian activity	5,6	
Anti-insect activity	7-11	
Anti-nematode activity	13-15	
Antifungal activity in culture	16	16
Enhancement of disease resistance	8,106	
Allelopathy against other plants	8	
Enhancement of drought tolerance	18	
Enhancement of tillering growth	17	
Greater seed set	19	

considered desirable by pastoralists. The most immediate goal of much of the present endophyte research is to identify or generate strains that cause minimal toxicosis to livestock, but still confer most of their protective benefits upon the host grasses.

Other than livestock toxicosis, the most intensively studied effect of the endophytes has been protection against insect herbivory (7–12). Anti-insect activity apparently is crucial for the survival of perennial ryegrass in New Zealand, where the Argentine stem weevil (*Listronotus bonariensis*) is a problem (2). Activities against nematodes (13–15) and fungal pathogens (8,16) have also been demonstrated. Improved growth characteristics (17) and protection from drought (18) are established fitness enhancements in tall fescue-endophyte symbiota. Other benefits include increased seed set (19), as well as allelopathy against other plants (8).

Many reviews of the c-endophytes have been published in recent years (2,12,20–30). The intention of this review is to focus on the technical aspects of these systems, with emphasis on the molecular biology of the endophytes and recently developed methods for DNA-mediated transformation. There are two main objectives of genetic transformation of these organisms: first, to eliminate, or at least to greatly ameliorate, problems of animal toxicosis while retaining most of the other protective benefits (31); second, to introduce new genetic capabilities to the endophytes in order to make them still more effective as biological control agents (32). For these aims to become reality much basic biology must be investigated. Pathways for indole alkaloid biosynthesis must be better elucidated, with key enzymes being identified and purified. Much more information on endophyte gene control is needed if new capabilities are to be introduced. The mechanisms underlying compatibility and incompatibility with host species and cultivars should be understood both as models for plant-fungus interaction, and to facilitate development of new, stable symbiota as agricultural cultivars.

TAXONOMIC RELATIONSHIPS OF CLAVICIPITACEOUS ENDOPHYTES

The c-endophytes are anamorphic (asexual) forms of the ascomycetous grass choke pathogen, *Epichloë typhina* (anamorph = *Sphacelia typhina* Sacc. = *Acremonium typhinum* G. Morgan-Jones *et* W. Gams). This fungus and other plant-associated Clavicipitaceae (genera *Claviceps, Balansia, Balansiopsis, Atkinsonella* and *Myriogenospora*) are biotrophs associated with grasses or sedges. Some are known to produce ergot alkaloids (33,34) and tremorgenic indole alkaloids (35). For example, the high levels of ergotamine in *Claviceps sclerotia* (or ergots; resting stages formed in host florets) make them dangerous contaminants of grains (36).

There are numerous similarities in the biology and host relationships of the endophytes and *E. typhina.* They share the endophytic life cycle of growth within aerial tissues (leaf sheaths and apical meristems) without causing any demonstrable pathology. Growth of fungal hyphae within host embryos (37) results in a high incidence (98% or more) of seed transmission of the endophytes (38). Most *E. typhina* strains can likewise be seed transmitted when, for reasons that are yet unclear, they do not exhibit their "choke" stage (23,39). Chokes are the sporogenous stromata which are produced in association with host inflorescences, and which arrest the development of the florets and prevent seed production. At this stage, *E. typhina* can undergo sexual crosses, producing ascospores which are, presumably, capable of infecting new host plants (though this has not yet been demonstrated) (40). It is this spore-forming stage which the asexual endophytes lack. However, the maternal line transmission due to seed dissemination of the endophytes is a highly efficient process, obviating the need for spore dissemination. Thus, the evolution of endophytes from *E. typhina* does not require the acquisition of new capabilities. All that is required is the loss of the pathogenic stage.

The c-endophyte and *E. typhina* group form a continuum of interactions from antagonistic to mutualistic (23). Some *E. typhina* strains are not capable of seed transmission (unpublished data), but most utilize this process as an alternative life cycle. Because the expression of the teleomorph is associated with sterilization of the host inflorescence, each grass-*E. typhina* association under any particular set of environmental conditions exhibits its own balance of stroma production versus seed set. The associations are grouped into type-I, type-II and type-III based on this balance (39). Type-I associations are antagonistic. They exhibit complete or nearly complete host sterilization and the fungus is not seed transmitted. Instead, it is propagated by spores and by mycelial growth in tillers and rhizomes. Type-II associations exhibit both seed transmission and choking on the same plant (sometimes even on the same panicle), and the association may be antagonistic or mutualistic depending on the relative importance of seed dissemination and vegetative propagation of the host (20). Type-III associations never, or very rarely, exhibit expression of external stromata. In these, the endophytes are entirely propagated by mycelial growth in host seed, tillers and rhizomes, and not by spores. They have lost all identity as individual organisms and have become only maternally inherited components of grass-fungus symbiota (41).

Mutualistic c-endophytes have been characterized and distinguished from one another at various levels. Morphological criteria, primarily focussing on the conidiophore and conidia (asexual spores) have traditionally been used to differentiate anamorphic species (42–45), but molecular phylogenetic methods have sometimes conflicted with this classification system (46,47). Isozyme and sequence analysis has indicated that related endophytes tend to be found in related hosts, indicating that host species should be one of the principal factors in endophyte taxonomy (48). However, the converse is not true. Sometimes a single host species may have at least two dissimilar endophytes (49). This observation suggests that the endophytes may be important contributors to the genetic diversity of their hosts. Such diversity is reflected in the variation in profiles of protective secondary metabolites (50).

PROTECTIVE ALKALOIDS

Several biological active secondary metabolites are produced in the grass-endophyte symbiota, but not in uninfected host grasses (10,51,52). These include four major alkaloids, although no single symbiotum is known to produce more than three. Livestock toxicosis is thought to be principally the result of fungus-specific indole-alkaloids: the vasoconstrictive ergovaline, and the tremorgenic neurotoxin, lolitrem B (53,54). Although indole alkaloids might contribute to other benefits, much of the anti-insect activity is attributable to other alkaloids: peramine (55) and saturated aminopyrrolizidines (loline and derivatives) (12). Peramine is believed to be an essential component in biological control of the Argentine stem weevil (2,55). Although peramine and aminopyrrolizidines are unrelated to the indole alkaloids, it is conceivable that they may also have some level of anti-mammalian activity. For example, immunosuppressive effects of the saturated aminopyrrolizidines have been noted (56). It is also conceivable (but not yet demonstrated) that the major anti-mammalian alkaloids contribute to other enhancements, such as protection from insects (51), nematodes, fungi and abiotic stresses. Nevertheless, it is postulated that the suitability of tall fescue and perennial ryegrass for livestock will be improved if compatible endophyte biotypes can be identified or developed that do not produce lolitrem B or ergovaline, but do produce the other protective alkaloids. As the biosynthetic pathways to the indole alkaloids continue to be elucidated (57,58), prospects improve for identification of genes for key enzyme setups, cloning and *in vitro* mutagenesis of these genes, and gene replacement (59,60) to eliminate the pathways.

The use of natural variation in endophyte alkaloid profiles offers another approach to cultivar improvement. In assessing alkaloid profiles, analysis of symbiota is more reliable than studies on cultured endophytes. Although paxilline and peramine have been identified in culture (61,62), identification of ergovaline in culture is unreliable (63). Saturated aminopyrrolizidines (loline and derivatives) have never been reported in culture, even though levels of over 1% on a dry-weight basis have sometimes been observed in vegetative tissues of certain symbiota (64, L.P. Bush, personal communication). Nevertheless, the firm association of lolines with only certain endophyte biotypes in natural, as

Table 2
Characteristics of c- and p-endophytes in Host Grasses

Property	c-endophyte	p-endophyte
Mycelium in seed	dense	sparse
Mycelium in leaf sheath	thicker, convoluted	thinner, highly branching
Sporulation in host tissue	no	yes
Rate of seed dissemination	>98%	~88%
Disease symptoms	none reported	none reported
Production of lolines, peramine and ergovaline	yes	no
Reacts with anti-*A. coenophialum* antiserum	yes	no
Reacts with anti-p-endophyte antiserum	no	yes

well as artificial host-endophyte combinations (10), strongly suggests that the endophytes participate directly in loline production. Possibly, they are responsible for converting a polyamine plant product into the saturated aminopyrrolizidines.

OTHER SEEDBORNE ENDOPHYTES

Non-clavicipitaceous seedborne endophytes have been reported in several of the same grass species that harbor c-endophytes. Often they occur together with c-endophytes in individual plants (43). Molecular data suggest that many of these, including the "*Phialophora* -like" and "*Gliocladium* -like" endophytes, are very closely related, perhaps constituting a single species (designated p-endophytes), and that they are not related to the Clavicipitaceae (65). The p-endophytes appear to be widely distributed throughout the world, but are not generally as common as the clavicipitaceous endophytes. They exhibit variable levels of seed dissemination, and produce potentially infective spores on senescent host tissues (43,66) (Table 2). It is possible that they are also mutualistic symbionts, since no pathology has been attributed to their presence, and they produce antifungal activity in culture (16). However, it is also conceivable that they are antagonistic parasites of their grass hosts. Their effects on host fitness, and on grazing animals, need to be elucidated. The p-endophytes may present complications in the study of the c-endophytes because the two types often occur together in seeds and leaves, and in the fungal cultures derived from these tissues.

Other endophytes have been reported in grass species not known to harbor c-endophytes. *Pseudocercosporella trichachnicola* in *Trichachne insularis* (67) and *Acremonium chilensis* in *Dactylis glomerata* (68) appear, on grounds of morphology and host interactions, to be non-clavicipitaceous. It seems likely that other species of seedborne fungi will be identified in future. Therefore, it is important to be able to identify and unambiguously distinguish c-endophytes.

DETECTION AND IDENTIFICATION OF CLAVICIPITACEOUS ENDOPHYTES

Methods most commonly used for detection of the fungal endophytes in grasses are tissue staining and serology. Staining is not technically demanding, and relatively fast if a small number of samples is being checked. Serology provides a definitive test for endophyte presence and type (c- or p-endophyte), and is more efficient for large numbers of samples.

Staining for Endophytes

Staining methods (69) can be used for rapid identification of c-endophytes and, sometimes, p-endophytes in seeds and vegetative plant tissues. Seeds are soaked overnight at room temperature in 5% aqueous sodium hydroxide, washed in tap water to remove the alkali, placed in Garner's solution (0.325 grams aniline blue, 100 ml water, 500 ml 85% lactic acid) and boiled for 3 to 5 minutes. They are deglumed and mounted on a slide with a drop of stain solution, overlaid with a coverslip, squashed and examined. Mycelium of the c-endophyte is darkly stained and densely packed around the aleurone cells.

In the vegetative tissues, the c-endophyte mycelia are most abundant in the outer and intermediate layers of the whorl (pseudostem). The inside epidermis is peeled from the leaf sheath with a razor blade and tweezers, and the tissue is placed on a slide with cotton blue in modified Garner's solution (aniline blue in lactic acid:glycerol:water, 1:2:1) (M.J. Christensen, personal communication). A coverslip is applied, and the slide is heated over an alcohol burner. Under microscopic examination, the c-endophyte mycelia are darkly stained and highly convoluted. They grow intercellularly, never penetrating the host cell wall, and run along the long axis of host cells.

The p-endophytes are easily distinguished from c-endophytes (66). Generally, the p-endophyte mycelia do not stain as darkly. In seed, the p-endophyte is highly sepatated and much sparser than c-endophyte. In leaf sheaths it can be as, or more abundant than c-endophyte. Mycelia are thinner, not as convoluted, and more highly branched than those of c-endophytes (Table 2).

Serology

Serological techniques provide a definitive identification of the endophyte types in plant tissues and culture. Fungal tissue is prepared for antiserum production by a modification of the method of Reddick and Collins (65,70). Protein extracted from 300 ml fungal culture is sufficient for each rabbit to be inoculated. Antisera against proteins extracted from c-endophytes are specific

```
                                .........•.........•.........•.........•.........•.........•.........•.........•.........•.......100
e19 (Festuca arundinacea)       GCATCGATGAAGAACGCAGCGAAATGCGATAAGTAATGTGAATTGCAGAATTCAGTGAATCATCGAATCTTTGAACGCACATTGCGCCCGCCAGTATTCT
Ety32 (F. rubra commutata)                                       TAATGTGAATTGCAGAATTCAGTGAATCATCGAATCTTTGAACGCACATTGCGCCCGCCAGTATTCT
Ety56 (Elymus canadensis)                                        TAATGTGAATTGCAGAATTCAGTGAATCATCGAATCTTTGAACGCACATTGCGCCCGCCAGTATTCT
Atkinsonella hypoxylon)                                          TAATGTGAATTGCAGAATTCAGTGAATCATCGAATCTTTGAACGCACATTGCGCCCGCCAGTATCCT
Claviceps purpurea                                               TAATGTGAATTGCAGAATTCAGTGAATCATCGAATCTTTGAACGCACATTGCGCCCGCCAGTATTCT
Neurospora crassa                                                TAATGTGAATTGCAGAATTCAGTGAATCATCGAATCTTTGAACGCACATTGCGCTCGCCAGTATTCT
p-endophyte p26                                                  TAATGTGAATTGCAGAATTCAGTGAATCATCGAATCTTTGAACGCATATTGCGCCCTCTGGTATTCC

                                .........•.........•.........•.........•.........•.........•.........•.........•.........•.......200
e19 (Festuca arundinacea)       GGCGGGCATGCCTGTTCGAGCGTCATTTCAACC-CTCAAGCCCGCTGCGCGCTTGCTGTTGGGGACCGGCTCA--CCCGCCT-------CGGCGGCGGCG
Ety32 (F. rubra commutata)      GGCGGGCATGCCTGTTCGAGCGTCATTTCAACC-CTCAAGCCCGCTGCGCGCTTGCTGTTGGGGACCGGCTCA--CCCGCCT-------CGGCGGCGGCG
Ety56 (Elymus canadensis)       GGCGGGCATGCCTGTTCGAGCGTCATTTCAACC-CTCAAGCCCGCTGCGCGCTTGGTGTTGGGGACCGGCCGG--CCCGCCT-------CGGCGGCGGCG
Atkinsonella hypoxylon)         GGCGGGCATGCCTGTTCGAGCGTCATTTCAACC-CTCAGAGCCTC----CCTCTGGTGTTGGGGATCGGCGCGCCCCCGTCTGCCCTCGCGCAGGCGGCC
Claviceps purpurea              GGCGGGCATGCCTGTTCGAGCGTCATTTCAACC-CTCAAGCCC------TGCTTGGTGTTGGGGACCGGCTCAGCGGGTGCGGGCTTCGGCCCGCCCCGT
Neurospora crassa               GGCGAGCATGCCTGTTCGAGCGTCATTTCAACCA-TCAAGCTC------TGCTT-GCGTTGGGGATCCG---------------------------CG
p-endophyte p26                 GGGGGGCATGCCTGTTCGAGCGTCATTATAACCACTCAAGCTCT-----CGCTTGGTATTGGGG-----------------------TTCGCGGTCTC

                                .........•.........•.........•.........•.........•.........•.........•.........•.........•.......300
e19 (Festuca arundinacea)       GCCGCCCCCGAAATGAATC---GGCGGT-CTCGTCGCAAGCC-TCCTTTGCGTAGTAGCA-----CACCACCTCGCAACCGGGAGCGCGGCGCGGCCACT
Ety32 (F. rubra commutata)      GCCGCCCCCGAAATGAATC---GGCGGT-CTCGTCGCA-GCC-TCCTTTGCGTAGTAACA-----CACCACCTCGCAACCGGGAGCGCGGCGCGGCCACT
Ety56 (Elymus canadensis)       GCCGCCCCTGAAATGAATT---GGCGGT-CTCGTCGCA-GCC-TCCCTTGCGTAGTAACA-----TACCACCTCGCAACCGGGAGCGCGGCGCGGCCACT
Atkinsonella hypoxylon)         GCCGTCCCCGAAATGAATT---GGCGGT-CTCGTCGCA-GCCTTCCTTTGCGTAGTAAGACTTTTTACCACCTCGCAACCAGGAGCGCAGCGCGGCCACT
Claviceps purpurea              GCCGCCCCCGAAATGGATC---GGCGGT-CTCGTCGCA-GCC-TTCTTTGCGTAGTAACA-----TACCACCTCGCAAC-AGGAGCGCGGCGCGGCCACT
Neurospora crassa               GCTGTCCGCTCAA--AATCAGTGGCGGG-CTCGTCA---GTCACACCGAGCGTGGTGACT-------CTACATCGCTAT-GGTCGTGCGGCG-GGTTCTT
p-endophyte p26                 GCGGCCCCTAAAATCAG----TGGCGGTGCCTGTCG---GCT---CTACGCGTAGTAATACTCCTCGC--------GTCTGGGTCCGACAGGTC-TACTT

                                .........•.........•.........•.........•.........•.........•.........•.........•.....388
e19 (Festuca arundinacea)       GCCGTAAAACGCCCAACTTTCTCCAAGAGTTGACCTCGAATCAGGTAGG-ACTACCCGCCGAACTTAAGCATATCAATAAGCGGAGGA
E32 (F. rubra commutata)        GCCGTAAAACGCCCAACTTTCTCCAAGAGTTGACCTCGAATCAGGTAGG-ACTACCCGCTGAACTTAA
E56 (Elymus canadensis)         GCCGTAAAACGCCCAACT-TCTCCAAGAGTTGACCTCGAATCAGGTAGG-ACTACCCGCTGAACTTAA
Atkinsonella hypoxylon          GCCGTAAAACGCCCAACT-TCTC-AAGAGTTGACCTCGAATCAGGTAGG-AGTACCCGCTGAACTTAA
Claviceps purpurea              GCCGTAAAACGCCCAACTTTTT---AGAGTTGACCTCGAATCAGGTAGG-AATACCCGCTGAACTTAA
Neurospora crassa               GCCGTAAAACCCCCCATT-TCT---AAGGTTGACCTCGGATCAGGTAGG-AATACCCGCTGAACTTAA
p-endophyte p26                 GCCA-ACAACCCCCAATTTTTT--ACAGGTTGACCTCGGATCAGGTAAGGATACCCGCTGAACTTAA
```

Figure 1. Sequence alignment of the internal transcribed spacer region 2 of nuclear rRNA genes (*rrnITS2*) of the tall fescue c-endophyte (e19), p-endophyte (p26), and teleomorphic fungi *Epichloë typhina* (E32 and E56), *Atkinsonella hypoxylon, Claviceps purpurea* (Clavicipitaceae) and *Neurospora crassa* (Sordariaceae). Plant hosts of e19 and the *E. typhina* isolates are given in parentheses. Underlined sequences correspond to oligonucleotide primers used for amplification of the DNA segments and subsequent sequence determination. Boxed sequences are recognition sites for diagnostic restriction endonucleases (see text). The *rrnITS2* sequence extends from position 138 to 327 in this alignment.

Table 3
Patristic Distance Matrix of Fungal *rrnITS*2 Sequences[1]

Species:	Species 1	2	3	4	5	6	7
1 *Acremonium coenophialum*	-	2	9	18	24	47	64
2 *Epichloë typhina* 32	2	-	7	16	22	45	62
3 *Epichloë typhina* 56	9	7	-	15	21	44	61
4 *Atkinsonella hypoxylon*	18	16	15	-	24	47	64
5 *Claviceps purpurea*	20	18	21	24	-	43	60
6 *Neurospora crassa*	41	39	40	43	43	-	55
7 p-endophyte	54	52	55	62	58	55	-

[1]Patristic distances are indicated above the diagonal; adjusted character distances are indicated below the diagonal.

for them, and do not react with p-endophytes (65), and the anti-p-endophyte antisera do not reach with c-endophytes. The antisera have also been tested against a number of grass pathogens (71), and found not to cross-react with them. Although there is some cross-reaction of both antisera to the saprophyte *Penicillium chrysogenum,* this does not detract from the usefulness of these antisera for routine screening for endophyte in seeds and leaf sheaths.

For serological detection, the procedures most commonly used have been direct and indirect enzyme linked immunosorbent assay (ELISA) (70,71). These protocols, although time consuming, give accurate and highly sensitive indications of the presence of the fungus of interest. The dilution of the antiserum used for serology depends upon the quality and specificity of the serum available, which should be determined empirically.

For rapid screening of individual plants or seed, the recently developed tissue immunoblot procedure (72) is preferable to ELISA. In this method, duplicate stem sections (1 mm long) or half-seeds are pressed onto nitrocellulose, then probed with the diluted antiserum, as for the indirect ELISA. A second probing with alkaline phosphatase-linked protein A, followed by an enzyme-catalyzed color reaction, are used to identify positive tissues. Color reactions should be scored with a dissecting microscope. The blots are best if read while still wet, but can be rinsed and dried, and read by re-wetting with water. It is important to incorporate both positive and negative controls on each filter, since a slight pink color is usually seen on the negative controls. Positives should be bright red.

DNA Sequence Analysis

Sequences of the variable internal transcribed spacers of the nuclear rRNA genes (*rrnITS*) can be used to identify c-endophytes, and to indicate relationships between them and strains of *E. typhina.* An alignment of the sequences is shown in Figure 1. As reflected in the patristic distance matrices (Table 3), calculated with the use of the PAUP software package (73), these sequences show a high degree of similarity between c-endophyte and *E. typhina* over their entire length and less, but clearly discernible, similarity with those of two other Clavicipitaceae, *Atkinsonella hypoxylon* and *Claviceps purpurea.* Much greater distances are indicated with *Neurospora crassa.* The p-endophyte sequence is still more divergent.

Sequence analysis is presently too expensive and time consuming to be of use in routine screening. A more practical approach is amplification of specified regions of the genome (74), followed by cleavage of the amplified DNA with various diagnostic restriction endonucleases. For example, there is a *Not* I site in the c-endophyte sequences which is absent in p-endophyte sequences. Conversely, there is an *Mlu* I site conserved in the p-endophyte sequence and absent in the c-endophyte sequence (Figure 1). Amplification of *rrnITS2* with the primers indicated in Figure 1 and cleavage with these two enzymes have been useful as a rapid identification of these two endophyte types (65).

ISOLATION, CULTURE AND MAINTENANCE

In nature the Clavicipitacease are biotrophic, requiring their hosts for completion of the life cycle. In the case of the strictly seedborne endophytes there is no stage of external growth. Nevertheless, most can be cultured, and many sporulate in culture. Complex or defined media may be used for culture. Many strains can be repeatedly subcultured over the course of a year or more without losing their properties or their host compatibility. However, other strains may die, or may lose their ability to reestablish stable associations with their hosts (M.R. Siegel and M.J. Christensen, personal communications). Furthermore, strains that sporulate poorly, or even *A. coenophialum* isolates which sporulate moderately well, do not generally survive long-term cold storage or freeze-drying. In such cases it is best to maintain the strains in the host plants in greenhouse conditions, and to obtain fresh cultures when required. Alternatively, they may be maintained in seed, but storage conditions (temperature and humidity) need to be carefully controlled because endophyte viability will be reduced more rapidly than will the viability of the seed (75).

The endophytes are isolated from surface sterilized leaf or seed tissues. Seeds are placed in 50% sulfuric acid, 20 min, rinsed three times with sterile water, placed into 50% Chlorox bleach (i.e., 1.5% sodium hypochlorite) for 20 to 30 min, then rinsed again at least three times. Treated seeds are placed into agar plates with complex media (see below), penicillin and streptomycin (both 100 mg/l). The seed may either be split to speed emergence of the endophyte, or cultured whole in order to obtain viable seedlings. Sufficient mycelial tissue to subculture should be obtained in two to six weeks, depending on the endophyte.

For isolation from vegetative tissues, three cm sections of pseudostems are cut from the crown level, washed in water and placed into 1.5% sodium hypochlorite for 30 min. They are washed three times in sterile water, blotted onto sterile filter paper, split, and placed onto medium with antibiotics, as described above.

Although the endophytes are biotrophs in nature, they may nevertheless be cultured on a minimal defined medium. This may be because the endophytes have no nutrient-absorbing, haustorial structures in their normal host associations, so they must obtain all their nutrients from the apoplast (76–78). Intensive study of the tall fescue endophyte, *A. coenophialum,* has been undertaken (79). In addition to minimal salts and fixed carbon and nitrogen, the fungus requires thiamine for growth. As a nitrogen source, it utilizes ammonium more efficiently than nitrate. It can use a wide variety of simple carbon sources, although catabolism of complex carbohydrates available in the apoplast has not been systematically investigated. Sugar alcohols are preferred over sugars. Carbon utilization is variable among the endophytes and teleomorphic *E. typhina* strains. Interestingly, reducing sugars such as fructose and arabinose are often inhibitory for growth of anamorphic endophytes, but teleomorphs tend to grow well on them (80).

Standard media such as potato-dextrose agar or cornmeal malt agar are adequate for growth. Based on the nutritional study (79), a proline-glutamate agar (PGA) and broth have been employed for routine culture of endophytes and teleomorphs. The agar medium consists of 3% glucose, 0.5% L-proline, 0.5% L-glutamic acid, 5% D-sorbitol, 0.1% KH_2PO_4, 0.03% $MgSO_47H_2O$, 0.2% yeast extract, 2% agar, pH 5.85. Specialized media have also been developed to enhance production of the ergot alkaloids (21).

A minimal medium is advantageous for genetic or molecular genetic studies because auxotrophic mutants may be characterized and complemented by, for example, transforming with the appropriate cloned gene. To prepare one such medium (79), autoclave 212 ml water with 4.75 g agar, then add 25 ml 10 X basal salts stock, 12.5 ml 500 mM D-mannitol, 0.148 ml 1 mM thiamine and 0.25 ml 1000 X trace elements stock (final volume 250 ml). The 10 X basal salts consist of 10 g ammonium sulfate, 15 g KH_2PO_4, 10 g $K_2HPO_4{\cdot}3H_2O$, 2.5 g $MgSO_4{\cdot}7H_2O$, pH adjusted to 6.4 with the appropriate potassium phosphate (monobasic dibasic), and made up to 500 ml with water. For 1000 X trace elements, add 55 mg boric acid, 37.5 mg cupric sulfate, 30 mg potassium iodide, 75 mg ferric sodium EDTA, 32.5 mg manganese sulfate, 27.5 mg sodium molybdate and 27.5 mg zinc sulfate, and bring to 250 ml. After autoclaving, some precipitation may occur.

The system used by D.B. Scott and coworkers (personal communications) for long-term endophyte storage involves freezing in a 50% glycerol solution. Mycelia (often containing conidia) are coarsely ground under sterile conditions, equilibrated with 50% glycerol on ice for 15 min, then divided into 2 ml cryogenic storage tubes. These are frozen at -80° C. It is preferable, though not essential, to have the temperature lowered gradually (~ 1° C per min). This method worked well for various endophyte and *E. typhina* isolates, but a preliminary test (unpublished data) suggests that *A. coenophialum* may not survive ultra-cold storage.

If greenhouse space or other growth facilities are available, then it is preferable to maintain endophytes in their host plants. An artificial inoculation procedure developed by Latch et al. (81) works well for most endophytes. Because of the high degree of host specificity exhibited by many strains it is essential that the natural host species be inoculated. There have been some preliminary indications that different cultivars may exhibit different levels of compatibility with the conspecific endophyte isolates (M.R. Siegel and M.J. Christensen, personal communications), so it is advisable to use a variety of host genotypes and ecotypes. Typical success rates for compatible interactions are 20% infection of surviving seedlings (over 50% of the seedlings should survive) (M.R. Siegel, personal communication). Probably the most efficient method for screening inoculated plants is by serology, as described earlier. This should be done after the plants have produced multiple tillers, again three months later, and periodically each year.

It should never be assumed, and always be established ahead of time, that the seed lot to be inoculated is virtually endophyte free. Even then, it is advisable to use whatever means available including isozyme analysis, DNA polymorphisms, genetically marked strains, mating type, morphology and alkaloid profiles, to characterize the symbionts following artificial introductions.

ISOLATION OF NUCLEIC ACIDS

Standard methods for DNA and RNA isolation from fungi require slight modifications (82,83). Endophytes are cultured by plating ground mycelia, or, when they occur in abundance, conidia, onto cellophane overlays on nutrient agar plates. They are allowed to grow to confluence, but not far beyond this stage, before being harvested for nucleic acids. The yield is approximately 0.5 g from each 82 mm cellophane disk. It is important that the mycelia not be so old that they begin to collapse. Otherwise the yield and quality of the nucleic acids, particularly RNA, is compromised. Prior to DNA isolation, the mycelia can be harvested and freeze-dried, stored at -20° C or -80° C. For RNA isolation it is best to use fresh mycelia, or to flash freeze fresh mycelia and store at -80° C. To isolate nuclear or mitochondrial DNA fractions, organelles can first be separated from fresh, ground mycelia by differential centrifugation techniques (83).

To isolate total DNA, five g fresh mycelia or an equivalent amount of freeze-dried material is ground in a mortar, with liquid nitrogen, then suspended in 150 mM EDTA, 50 mM Tris-HCl, pH 8.0, 1% sodium lauroyl sarcosine, 2 mg/ml proteinase K. Cellular debris is removed by centrifugation (2000 x g, 4° C, 10 min). The supernatant is incubated, 37° C, 20 min, then extracted with high-purity phenol, phenol/chloroform/isoamyl alcohol (25:24:1), and chloroform/isoamyl alcohol (24:1), always with the upper aqueous phase saved, but material at the interphase interface avoided. A subsequent high-speed centrifugation (25,000 x g, 4° C, 20 min) removes insoluble carbohydrates. DNA is precipitated with an equal volume of cold isopropanol and pelleted at 8,000 x g, 4° C, 10 min, then redissolved in TE buffer (10 mM Tris-

HCl, 1 mM EDTA, pH 8.0). An optional CsCl-ethidium bromide buoyant density gradient (84) can then be used to remove RNA and polysaccharides.

As a source of RNA it is preferable to use fresh cellophane cultures, flash frozen in liquid nitrogen and, if so desired, stored at -80° C. The material is ground to a fine powder in a mortar cooled on dry ice. Grinding thick tissue, such as endophyte-infected grass crowns, is difficult. Instead, these tissues are flash frozen, mixed with dry ice and milled into a precooled mortar. RNA is then isolated by a modification of the method of Chirgwin et al. (85), as previously described (86). A polyadenylated fraction is obtained by chromatography on oligo-deoxythymidine cellulose (87).

CONSTRUCTION OF CLONE LIBRARIES

Epichloë typhina and *A. coenophialum* genomic clone libraries have been constructed in phage lambda FIX II (Stratagene Cloning Systems, La Jolla, CA), which takes inserts of 9 to 23 kb in length (30). Endophytes have genome sizes approximately similar to those of other filamentous fungi. An estimate of 47 Mb was obtained by electrophoretic karyotyping of endophyte 187 BB from *L. perenne* (32). *Acremonium coenophialum* appears to have approximately the same genome size, whereas those of *E. typhina* MP-I and MP-II are smaller (unpublished data). Therefore, genomic libraries in lambda FIX II require an estimated 15,000 independent clones for a ≥ 95% probability of obtaining any individual single-copy sequence (88). Cosmid libraries with inserts ≥ 30 kb will require at least 5,000 clones.

Prior to cloning genomic libraries (82), the DNAs are partially cleaved with *Mbo* I, then size fractionated on rate-zonal CsCl density gradients. Fragments of ≥ 30 kb are separated from those of ~ 8 to 25 kb. The larger size fraction is for cosmid clone libraries, and the smaller for lambda phage libraries. A partial end-fill system is used to maximize recombinants with the vectors. The phage vector, as prepared by the manufacturer, has been cleaved by *Xho* I, and the 5′ overhanging ends have been treated with DNA polymerase I large fragment, dGTP and dATP. Size fractionated fragments are partially filled with dCTP and dTTP in order to generate ends that are compatible with the vector ends. These are ligated to the lambda phage arms, packaged into lambda coat proteins, and introduced by transfection into *Escherichia coli* for amplification. Phage are stored at -80° C in SM buffer (88) with 7% dimethylsulfoxide.

Libraries of cDNA, made from *A. coenophialum* and *E. typhina* mRNA (30), were also prepared in lambda ZAP and ZAP II (Stratagene Cloning Systems) (89). These vectors incorporate the plasmid (phagemid) pBluescript. Once cDNA is cloned into these vectors, subclones can be generated by *in vivo* excision with the Ff helper phage. Libraries were also prepared from the *A. coenophialum-F. arundinacea* symbiotum, stromatal tissue of *E. typhina* on *L. perenne,* and the corresponding host tissues (30).

TRANSFORMATION

Endophyte transformation has been performed by variations on techniques that have become standard for filamentous ascomycetes (31,32). The endophytes serve as good models for molecular manipulation of many naturally occurring and ecologically important fungi which may not be as tractable as *Emericella nidulans* and *Neurospora crassa*. Clavicipitaceous endophytes grow slowly, and do not generally produce abundant conidiospores. Different endophyte isolates, even from the same grass species, may exhibit significantly different growth characteristics and physiologies (50,90). Therefore, transformation protocols developed for one isolate do not necessarily work well if applied, unmodified, to another. It is important to be fully familiar with the characteristics of the endophyte strain, and to do considerable preliminary experimentation. Among the critical parameters to use when developing the transformation system are protoplast viability before and after treatment. Culture conditions and the conditions during cell wall digestion are important for high protoplast viability.

The choice of a selectable marker can also be important. The two isolates transformed to date are both sensitive to hygromycin B. This is by no means a general condition of the c-endophytes. Also, all *E. typhina* strains which have been screened are resistant to this antibiotic at concentrations up to 500 mg/l or more (unpublished data). Other markers, such as genes for phleomycin or bialaphos resistance (91–93), need to be explored for such strains. It appears that many fungal promoters can be used to drive expression of the selectable markers, and that they do not necessarily need to be derived from endophytes or closely related fungi. However, different promoters, and even different DNA sequences flanking the promoter, give different levels of transformation in controlled comparisons (31). It seems advisable to choose constructs that give high-level transformation in a number of fungal systems. Among the vectors used in endophyte transformation are pAN7-1 which utilizes the *gpd* promoter from *Em. nidulans* (94), pCSN43 with a *trpC* promoter from *Em. nidulans* (95, C.A. Staben, personal communication), and pKAES105 employing the *tub2* promoter from *E. typhina* (31,82). The inclusion of a 3′ flanking DNA segment containing a fungal transcription terminator sometimes improves the transformation efficiency (31). These vectors are maintained as integrated DNA in transformed fungal genomes (31,32).

Two endophyte isolates have been transformed with the use of different protocols. Isolate 187BB (PN2000) differs from the more common perennial ryegrass endophytes (*Acremonium lolii* Latch, Christensen *et* Samuels) in that plants with 187BB do not accumulate detectable levels of the neurotoxic agent lolitrem B (32). To prepare mycelia of 187BB, mycelia and associated conidia are ground into a suspension, and inoculated into potato-dextrose broth. The cultures are grown for 5 to 6 d, 22° C with gentle shaking, then harvested by filtration. Mycelia grown in this fashion yield abundant viable protoplast from this strain, but not from *A. coenophialum* (unpublished data). For the latter, plate-grown mycelia are used. The plate cultures are prepared by finely grinding a small amount of culture in sterile water, then spreading the suspension onto a cellophane layer on PGA. After approximately one week, the mycelia are ready for protoplasting. The appropriate time is determined empirically to follow, within 24 hr, the peak accumulation of conidia on the plate.

For both endophytes, preparation of protoplasts is as described by Yelton et al. (96) except that, in the case of plate-grown *A. coenophialum,* a vacuum infiltration step is used to wet the tissue thoroughly with the enzyme suspension used for cell wall digestion. Following the preparation of protoplasts, different protocols have been employed for transformation of the different strains. Electroporation was used for e19, whereas a polyethylene glycol (PEG) treatment was employed for transformation of 187BB. In the PEG procedure, inclusion of sorbitol osmoticum in each step is of benefit (32). Electroporation is a faster procedure, but the transformation frequency obtained by electroporation of e19 was not as high as those obtained by PEG transformation of 187BB. However, it is unknown whether this was due to the different methods or to other factors, such as strain differences. A direct comparison of the two procedures on a single protoplast preparation has not yet been undertaken.

For electroporation, 80 μl protoplast suspensions, with 10^9 protoplasts per ml in cold STC buffer (1.2 M D-sorbitol, 10 mM Tris-HCl, pH 7.5, 10 mM $CaCl_2$), are incubated with 5 to 10 μg vector DNA, placed into cuvettes with 0.1 cm electrode gaps, then subjected to an exponential pulse delivered by the GenePulser™ apparatus (BioRad, Richmond, CA). Settings were optimized at 25 μF, 0.4 to 0.5 kV (i.e., 4,000 to 5,000 v/cm), 100 to 200 Ω serial resistance. Optimum settings will differ if other buffer conditions are used. Following electroporation, the treated protoplasts are resuspended in 7 ml PGAtop (similar to PGA, but with 1.2 M D-sorbitol and 0.7% low gelling temperature agarose), poured onto 88 mm diameter PGA plates containing hygromycin B (42 μg/ml in ~ 20 ml medium each) and incubated, inverted, at least 3 weeks at 21° C. Many small colonies are obtained in the DNA treatments, but not in the no-DNA controls. Most of these do not appear to be stable transformants. The transformation frequencies are scored by counting colonies that exceed 2 mm diameter within the three week incubation period. Up to 20 transformants have been obtained per μg linearized DNA, per 1.25 x 10^7 protoplasts, i.e., up to 120 transformants per pulse. The transformation frequencies vary between protoplast preparations. There was also considerable dependence on the vector used, and on its conformation. Linearized vector generally performed better than supercoiled DNA, and the highest frequencies obtained were with pCSN43 and pKAES105 (31), both with modified *E. coli hph* genes. The former used the *Em. nidulans trpC* promoter and terminator, whereas the latter employed the *E. typhina tub2* promoter and *Em. nidulans trpC* terminator.

The transformation method used by Murray et al. (32) is a modification of the method of Vollmer and Yanofsky (97). Protoplasts are diluted to 1.25 x 10^8 ml in STC, then 20 μl PEG solution (40% PEG 4,000, 50 mM $CaCl_2$, 1 M D-sorbitol, 50 mM Tris-HCl, pH 8.0) is added to 80 μl aliquots, to which are also added 2 μl 50 mM spermidine, 5 μl heparin (5 mg/ml) and 5 μg DNA. After mixing gently, the suspension is incubated on ice, then 900 μl PEG solution are added. The suspension is incubated at room temperature, 15 to 20 min, diluted and plated in CM (98) top agar onto CM plates with 200 μg/ml hygromycin B. Transformation frequencies obtained by this procedure have been up to 200 per μg DNA per 2.5 x 10^7 protoplasts with circular vector, and up to 500 stable transformants with linearized vector.

Murray et al. (32) also investigated the possibility of cotransformation by employing a second marker, the *E. coli uidA* gene (encoding β-glucuronidase) on plasmid pNOM-1 (99). In one test, 80% of the hygromycin B-resistant transformants also had the *uidA* marker. β-glucuronidase proved a useful histochemical marker for the transformants in culture and *in planta.* Future investigations are planned to determine the frequencies of homologous integration following transformation with vectors containing various cloned segments of endophyte DNA (D.B. Scott, personal communication). Homologous recombination is crucial for gene disruption strategies. For some filamentous ascomycetes it is very common, but for others it can be quite rare. The observation of frequent homologous recombination following transformation of the related fungus, *Claviceps purpurea* (60), is encouraging.

MENDELIAN GENETICS OF *Epichloë typhina*

Epichloë typhina has been characterized as heterothallic with two mating types (bipolar) in studies involving two biotypes: one in *Agrostis hiemalis,* and another in *Elymus canadensis* (100). Crosses have been initiated by rubbing stromata together to transfer spermatia. Conidia produced in culture can also be employed as spermatia. So far, the female parent must produce a stroma on its host, and the crosses must be done on the plant. For the complete sexual cycle to be completed, the two parents must be of opposite mating type, and within the same mating population (see below). First, a dense white mycelial growth occurs over the surface, followed by development of raised white bumps. These are perithecia (fruiting bodies), and change to a tan-orange color as they mature. They are darkest at the center near the developing ostioles, the pores at the top of the perithecia from which ascospores are ultimately discharged. The ascospores are filamentous, 500 to 700 μm long, arranged eight in parallel in each ascus. The whole process from inoculation to ascospore discharge occurs over a span of 2 to 4 weeks.

Epichloë typhina, as presently described, apparently comprises several cryptic biological species. For example, several isolates from *Festuca* subgenus *Festuca* are compatible with one another, and crosses between any two with opposite mating types yield abundant viable ascospores (30). They constitute a single mating population, designated MP-I. The MP-I strains are not sexually compatible with known teleomorphs that have been identified in other grasses including *Ag. hiemalis, Agrostis stolonifera, D. glomerata, El. canadensis, L. perenne* and *Sphenopholis obtusata.* Close sequence homology of variable genomic regions (*rrnITS* or *tub2* introns) appears to be a better predictor of sexual compatibility than do relationships of host grass species (unpublished data).

Incompatibility in crosses between opposite mating types of different mating populations is characterized by proliferation of a white mycelial mat overlaying the inoculated portion of the stroma, and ultimately the formation of structures similar to perithecia. Microscopic inspection indicates that these structures are barren. Superficially they look similar to those from compatible crosses, but they are sparser, develop more slowly, become less intensely colored, and do not contain asci (30). In contrast, crosses between strains with the same mating type result in no noticeable change in development of the stromata, whether such attempts involve the same or different mating populations.

A modification of the method of Bacon and Hinton (40) has been used to obtain single ascospore isolates. Perithecia are first monitored for maturation by microscopy. First asci, then the filamentous ascospores, are evident. At this point, stem sections containing stromata are removed and taped to the lids of petri plates. The plates, which contain water agar overlaid with sterile cellophane, are stored—overturned to minimize contamination—at ambient temperature under fluorescent room lights. Ascospores are forcibly ejected upward from the perithecia. The plates are monitored periodically under 400 X magnification for the presence of ascospores or part-spores on the surface of the cellophane layer. Monitoring may be done hourly or every several seconds depending on how frequently the spores are being ejected. Ascospores may be harvested immediately, or the plate may be stored at 4° C to arrest germination for several days.

If ascospores are to be taken for genetic analysis, it must be noted that they tend to break into part-spores. Taking multiple part-spores from the same meiotic product should be avoided by taking spores from different groups on the plate. Another problem which can arise is that some pairs or groups of spores from single asci may not separate. Therefore, single conidia should be isolated from the germinating ascospores. The ascospores germinate on water or nutrient agar to form multiple phialides perpendicular to the axis of the spore (40,101). At the end of each phialide is a single, slipper-shaped conidium. These can be taken directly, or older colonies from picked ascospores can be streaked for single conidia. Ascospores can be lifted by touching the blade of a scalpel to the spore, and transferred by brushing the blade across or through the agar on a nutrient plate, then monitored for growth. Viable ascopores should germinate within 48 to 78 hr, although it is possible that some strains are somewhat slower growing. Periodic microscopic examination should reveal growth and the presence of conidia, which tend to be abundant at the early growth stage.

The sexual cycle has not yet been used for analysis of genetic linkage groups, but segregation of length polymorphism of the nontranscribed spacer (intergenic or IGS region) has indicated that the nuclear rRNA genes are at a single locus in *E. typhina* MP-I (30). This test has also confirmed the heterothallic nature of the cross. However, very rare homothallic ascospores have been obtained in another study involving the type-I *E. typhina* from *L. perenne* (MP-II) (29). Although the mating systems of type -II (100) and type-I (unpublished data) *E. typhina* strains have been elucidated, no success has been achieved in attempted crosses with type-III endophytes inoculated onto stromata of known teleomorphs (29,30).

The Mendelian genetic system should prove useful, in conjunction with various molecular techniques, to identify alleles that otherwise give no phenotypic indications of their presence: techniques such as isozyme analysis (46,48), or amplified fragment length polymorphism (102,103), Alternatively, auxotrophic or antibiotic resistant mutants may be generated, or foreign genes may be introduced into the genomes by transformation, then their linkage to other markers assessed in crosses. In *N. crassa*, ectopic repeats of genes generated by transformation will frequently become inactivated by extensive repeat-induced point mutations (RIP) during sexual crosses (104,105). It will be of interest and practical value to determine if a similar phenomenon occurs in *E. typhina*.

CONCLUSIONS

Molecular and classical genetic investigations of the c-endophytes and the related teleomorphic species, *Epichloë typhina*, are in their very early stages. Because of the slow growth of the endophytes, this system is certainly not the preferred model for basic molecular biology of filamentous fungi. Other systems lend themselves to investigations of cell cycles, developmental regulation, physiological adaptation and other aspects of organismal biology. However, there are compelling reasons for transferring the techniques and knowledge developed with the more tractable systems to the endophytic fungi. Most importantly, these organisms exist in a unique symbiosis with their hosts. Taken together with *E. typhina*, the symbioses range from antagonistic to mutualistic. Even in the antagonistic associations, most of the life cycle of the fungus is spent in vegetative plant tissues without eliciting any symptoms of disease. In the mutualistic associations, the c-endophytes have no known external infective stage, and have therefore lost all separate identities as organisms. The grass-endophyte symbiota have become individual units, with the heritable traits of each unit determined by the combination of host and fungal genotypes. In those systems which have been most thoroughly investigated—tall fescue and perennial ryegrass symbiota—the c-endophytes have been demonstrated to confer a wide range of protective effects and other fitness enhancements. For this reason, they have unwittingly been used for centuries as natural agents of biological protection of pasture grasses.

Endophytes exhibit considerable diversity in host specificity and biosynthesis of protective secondary metabolites. These metabolites provide important protection from various pests, but some also can cause toxicosis to grazing livestock. The problem of livestock toxicosis provides most of the present focus for efforts to genetically modify the c-endophytes. Future applications will make use both of existing endophyte diversity and of molecular genetic modifications. In addition, it is likely that the genetic transformation systems developed for the c-endophytes will be used to introduce additional genes into their genomes, in order to enhance the genetic capabilities of the symbiota.

Acknowledgments: This work was supported by USDA CRGO grant 88-37151-3860, USDA NRICGP grant 90-00694, NSF New Zealand Program grant INT-912083, and by the McKnight Foundation. This is publication number 92-11-216 of the Kentucky Agricultural Experiment Station, and is published with approval of the director.

REFERENCES

1 Freeman, E.M. (1904) Phil. Trans. Royal Soc. London Ser. B. 196, 1–27.
2 Siegel, M.R., Latch, G.C.M. and Johnson, M.C. (1987) Annu. Rev. Phytopathol. 25, 293–315.
3 Clay K. (1990) Annu. Rev. Ecol. Syst. 21, 275–295.
4 Hill, N.S., Belesky, D.P. and Stringer, W.C. (1991) Crop Sci. 31, 185–190.
5 Bacon, C.W., Porter, J.K., Robbins, J.D. and Luttrell, E.S. (1977) Appl. Environ. Microbiol. 34, 576–581.
6 Gallagher, R.T. and Hawkes, A.D. (1985) N. Zeal. J. Agric. Res. 28, 427–431.
7 Funk, C.R., Halisky, P.M. Ahmad, S. and Hurley, R.H. (1985) in Proceedings of the 5th International Turf Conference, Avignon (Lemaire, F., ed.), pp. 137–145, INRA, Versailles.
8 Ford, V.L. and Kirkpatrick, T.L. (1989) Proceedings of the Arkansas Fescue Toxicosis Conference, Arkansas Agricultural Experiment Station, University of Arkansas.
9 Mathias, J.K., Ratcliffe, R.H. and Hellman, J.L. (1990) J. Econ. Entomol. 83, 1640–1646.
10 Siegel, M.R., Latch, G.C.M., Bush, L.P., Fannin, F.F., Rowan, D.D., Tapper, B.A., Bacon, C.W. and Johnson, M.C. (1990) J. Chem. Ecology 16, 3301–3315.
11 Clay, K. (1991) in Microbial Mediation of Plant-Herbivore Interactions (Barbosa, P., Kirschik, L. and Jones, E., eds.), pp. 199–226, John Wiley and Sons, New York, NY.
12 Dahlman, D.L., Eichenseer, H. and Siegel, M.R. (1991) in Microbial Mediation of Plant-Herbivore Interactions (Barbosa, P., Kirschik, L. and Jones, E., eds.), pp. 227–252, John Wiley and Sons, New York, NY.
13 Pederson, J.F., Rodriguez-Kabana, R. and Shelby, R.A. (1988) Agronomy J. 80, 811–814.
14 West, C.P., Izekor, E., Oosterhuis, D.M. and Robbins, R.T. (1988) Plant and Soil 112, 3–6.
15 Kimmons, C.A., Gwinn, K.D. and Bernard, E.C. (1990) Plant Dis. 74, 757–761.
16 Siegel, M.R. and Latch, G.C.M. (1991) Mycologia 83, 525–537.
17 Hill, N.S., Stringer, W.C., Rottinghaus, G.E., Belesky, D.P., Parrot, W.A. and Pope, D.D. (1990) Crop Sci. 30, 156–161.
18 Arechavaleta, M., Bacon, W.C., Hoveland, C.S. and Radcliffe, D.E. (1989) Agronomy J. 81, 83–90.
19 Rice, J.S., Pinkerton, B.W., Stringer, W.C. and Undersander, D.J. (1990) Crop Sci. 30, 1303–1305.
20 Bacon, C.W. and De Battista, J. (1990) in Soil and Plants (Avora, D.K., Rai, B., Mukerji, K.G. and Knudsen, G.R., eds.), Vol. 1, pp. 231–256, Marcel Dekker, Inc., New York, NY and Basel.

21 Bacon, C.W. (1990) in Isolation of Biotechnological Organisms from Nature (Labeda, D.P., ed.), pp. 259–282, McGraw-Hill, New York, NY.
22 Bacon, C.W. and Siegel, M.R. (1988) J. Prod. Agric. 1, 45–55.
23 Clay, K. (1988) in Coevolution of Fungi with Plants and Animals (Pirozynski, K.A. and Hawksworth, D., eds.), pp. 79–105, Academic Press, London.
24 Clay, K. (1988) Ecology 69, 10–16.
25 Clay, K. (1986) in Microbiology of the Phyllosphere (Fokkema, N.J. and van den Heuvel, J., eds.), pp. 190–201, Cambridge University Press, Cambridge.
26 Carroll, C.G. (1988) Ecology 69, 2–9.
27 Siegel, M.R., Latch, G.C.M. and Johnson, M.C. (1985) Plant Dis. 69, 179–183.
28 Siegel, M.R., Dahlman, D.L. and Bush, L.P. (1989) in Integrated Pest Management for Turfgrass and Ornamentals (Leslie, A.R. and Metcalf, R.L., eds.), pp. 169–186, United States Environmental Protection Agency, Washington, DC.
29 Schardl, C.L. and Siegel, M.R. (1993) Agr. Ecosyst. Environ. (in press).
30 Schardl, C.L. and Tsai, H.-F. (1993) Natural Toxins 1, 1–14.
31 Tsai, H.-F., Siegel, M.R. and Schardl, C.L. (1992) Curr. Genet. 22, 399–406.
32. Murray, F.R., Latch, G.C.M. and Scott, D.B. (1992) Mol. Gen. Genet. 233, 1–9.
33 Stadler, P.A. and Stutz, P. (1975) The Alkaloids 15, 1–40.
34 Rèháček, Z. (1991) Folia Microbiologica 36, 323–342.
35 Cole, R.J., Dorner, J.W., Lansden, J.A., Cox, R.H., Pape, C., Cunfer, B., Nicholson, S.S. and Bedell, D.M. (1977) J. Agric. Food Chem. 25, 1197–1201.
36 Kavaler, L. (1965) Mushrooms, Molds, and Miracles: The Strange Realm of Fungi, The John Day Company, Ltd., New York, NY.
37 Philipson, M.N. and Christey, M.C. (1986) N. Zeal. J. Bot. 24, 125–134.
38 Siegel, M.R., Johnson, M.C., Varney, D.R., Nesmith, W.C., Buckner, R.C., Bush, L.P., Burrus, P.B., II., Jones, T.A. and Boling, J.A. (1984) Phytopathology 74, 932–937.
39 White, J.F. (1988) Mycologia 80, 442–446.
40 Bacon, C.W. and Hinton, D.M. (1988) Trans. Brit. Mycol. Soc. 90, 563–569.
41 Siegel, M.R. and Schardl, C.L. (1991) in Microbial Ecology of Leaves (Andrew, J.H. and Hirano, S.S., eds.), pp. 198–221, Springer Verlag, Berlin.
42 Morgan-Jones, G. and Gams, W. (1982) Mycotaxon 15, 311–318.
43 Latch, G.C.M., Christensen, M.J. and Samuels, G.J. (1984) Mycotaxon 20, 535–550.
44 White, J.F., Jr. and Morgan-Jones, G. (1987) Mycotaxon 30, 87–95.
45 Gams, W., Petrini, O. and Schmidt, D. (1990) Mycotaxon 37, 67–71.
46 Leuchtmann, A. and Clay, K. (1990) Phytopathology 80, 1133–1139.
47 Schardl, C.L., Liu, J.-S., White, J.F., Finkel, R.A., An, Z. and Siegel, M.R. (1991) Pl. Syst. Evol. 178, 27–41.

48 Leuchtmann, A. (1993) Natural Toxins (in press).
49 An, Z.-Q., Liu, J.-S., Siegel, M.R., Bunge, G. and Schardl, C.L. (1992) Theor. Appl. Genet. 85, 366–371.
50 Christensen, M.J., Latch, G.C.M. and Tapper, B.A. (1991) Mycolog. Res. 95, 918–923.
51 Clay, K. and Cheplick, G.P. (1989) J. Chem. Ecol. 15, 169–182.
52 Jackson, J.A., Yates, S.G., Powell, R.G., Hemden, R.W., Bush, L.P., Boling, J.A., Zavos, P.M. and Siegel, M.R. (1989) Drug Clin. Toxicol. 12, 147–164.
53 Lacey, J. (1991) in Mycotoxins and Animal Foods (Smith, J.E. and Henderson, R.S., eds.), pp. 363–414, CRC Press, Boca Raton, FL.
54 Raisbeck, M.F., Rottinghaus, G.E. and Kendall, J.D. (1991) in Mycotoxins and Animal Foods (Smith, J.E. and Henderson, R.S., eds.), pp. 647–677, CRC Press, Boca Raton, FL.
55 Rowan, D.D. and Gaynor, D.L. (1986) J. Chem. Ecol. 12 647–658.
56 Dew, R.K., Boissonneault, G.A., Gay, N., Boling, J.A., Cross, R.J. and Cohen, D.A. (1990) Vet. Immunol. Immunopathol. 26, 285–295.
57 Krupinski, V.M., Robbers, J.E. and Floss, H.G. (1976) J. Bacteriol. 125, 158–165.
58 Laws, I. and Mantle, P.G. (1989) J. Gen. Microbiol. 135, 2679–2692.
59 Barnes, D.A. and Thorner, J. (1985) in Gene Manipulations in Fungi (Bennett, J.W. and Lasure, L.L., eds.) pp. 197–226, Academic Press, New York, NY.
60 Smit, R. and Tudzynski, P. (1992) Mol. Gen. Genet. 234, 297–305.
61 Weedon, C.M. and Mantle, P.G. (1987) Phytochemistry 26, 969–971.
62 Fannin, F.F., Bush, L.P., Siegel, M.R. and Rowan, D.D. (1990) J. Chromatography 503, 288–292.
63 Bacon, C.W. (1988) Appl. Env. Microbiol. 54, 2615–2618.
64 Kennedy, C.W. and Bush, L.P. (1983) Crop Sci. 23, 547–552.
65 An, Z.-Q., Tsai, H.-F., Siegel, M.R., Hollin, W., Schmidt, D. and Schardl, C.L. (1993) Appl. Envir. Microbiol. (in press).
66 Philipson, M.N. (1989) N. Zeal. J. Bot. 27, 513–519.
67 White, J.F.J., Morrow, A.C. and Morgan-Jones, G. (1990) Mycologia 82, 218–226.
68 Morgan-Jones, G., White, J.F., Jr. and Pointelli, E.L. (1990) Mycotaxon 39, 441–454.
69 Clark, E.M., White, J.F. and Patterson, R.M. (1983) J. Microbiol. Meth. 1, 149–155.
70 Reddick, B.B. and Collins, M.H. (1988) Phytopathology 78, 418–420.
71 Johnson, M.C., Pirone, T.P., Siegel, M.R. and Varney, D.R. (1982) Phytopathology 72, 647–650.
72 Gwinn, K.D., Shepard-Collins, M.H. and Reddick, B.B. (1991) Phytopathology 81, 747–748.
73 Swofford, D.L. (1991) PAUP: Phylogenetic Analysis Using Parsimony, Version 3.0q, Illinois Natural History Survey, Champaign, IL.
74 White, T.J., Bruns, T.D., Lee, S. and Taylor, J. (1990) in PCR Protocols: A Guide to Methods and Applications (Innis, M.A., Gelfand, D.H.,

Sninsky, J.J. and White, T.J., eds.), pp. 315–322, Academic Press, San Diego, CA.
75 Rolston, M.P., Hare, M.D., Moore, K.K. and Christensen, M.J. (1986) N. Zeal. J. Agric. Res. 14, 297–300.
76 Hinton, D.M. and Bacon, C.W. (1985) Can. J. Bot. 63, 36–42.
77 Rykard, D.M. and Bacon, C.W. and Luttrell, E.S. (1985) Phytopathology 75, 950–956.
78 Siegel, M.R., Jarlfors, U., Latch, G.C.M. and Johnson, M.C. (1987) Can. J. Bot. 65, 2357–2367.
79 Kulkarni, R.K. and Nielsen, B.D. (1986) Mycologia 78, 781–786.
80 White, J.F., Jr., Morrow, A.C., Morgan-Jones, G. and Chambless, D.A. (1991) Mycologia 83, 72–81.
81 Latch, G.C.M. and Christensen, M.J. (1985) Ann. Appl. Biol. 107, 17–24.
82 Byrd, A.D., Schardl, C.L., Songlin, P.J., Mogen, K.L. and Siegel, M.R. (1990) Curr. Genet. 18, 347–354.
83 Mogen, K.L., Siegel, M.R. and Schardl, C.L. (1991) Curr. Genet. 20, 519–526.
84 Sambrook, J., Fritsch, E.F., and Maniatis, T. (1989) Molecular Cloning: A Laboratory Manual, Cold Spring Harbor Laboratory Press, Cold Spring Harbor, NY.
85 Chirgwin, J.M., Przybyla, A.E., MacDonald, R.J. and Rutter, W.J. (1979) Biochemistry 18, 5294–5299.
86 Choi, G.H., Marek, E.T., Schardl, C.L., Richey, M.G., Chang, S. and Smith, D.A. (1990) J. Bacteriol. 172, 4522–4528.
87 Aviv, H. and Leder, P. (1972) Proc. Nat. Acad. Sci. U.S.A. 69, 1408–1412.
88 Ausubel, F.M., Brent, R., Kingson, R.E., Moore, D.D., Seidman, J.G., Smith, J.A. and Struhl, K. (eds.) (1987–1992) Current Protocols in Molecular Biology, John Wiley and Sons, New York, NY.
89 Short, J.M., Fernandez, J.M., Sorge, J.A. and Huse, W.D. (1988) Nucl. Acids Res. 16, 7583–7599.
90 Christensen, M.J. and Latch, G.C.M. (1991) Mycolog. Res. 95, 1123–1126.
91 Gatignol, A., Baron, M. and Tiraby, G. (1987) Mol. Gen. Genet. 207, 342–348.
92 Avalos, J., Geever, R.F. and Case, M.E. (1989) Curr. Genet. 16, 369–372.
93 van Engelenburg, F., Smit, R., Goosen, T., van den Broek, H. and Tudzynski, P. (1989) Appl. Microbiol. Biotcchnol. 30, 364–370.
94 Punt, P.J., Oliver, R.P., Dingemanse, M.A., Pouwels, P.H. and van den Hondel, C.A.M.J.J. (1987) Gene 56, 117–124.
95 Cullen, D., Leong, S.A., Wilson, L.J. and Henner, D.J. (1987) Gene 57, 21–26.
96 Yelton, M.M., Hamer, J.E. and Timberlake, W.E. (1984) Proc. Nat. Acad. Sci. U.S.A. 81, 1470–1474.
97 Vollmer, S.J. and Yanofsky, C. (1986) Proc. Nat. Acad. Sci. U.S.A. 83, 4869–4873.
98 Oliver, R.P., Roberts, I.N., Harling, R., Kenyon, L., Punt, P.J., Dingemanse, M.A. and van den Hondel, C.A.M.J.J. (1987) Curr. Genet. 12, 231–233.

99 Roberts, I.N., Oliver, R.P., Punt, P.J. and van den Hondel, C.A.M.J.J. (1989) Curr. Genet. 15, 177–180.
100 White, J.F. and Bultman, T.L. (1987) Amer. J. Bot. 74, 1716–1721.
101 Bacon, C.W. and Hinton, D.M. (1991) Mycologia 83, 743–751.
102 Kohn, L.M. (1992) Mycologia 84, 139–153.
103 Smith, M.L., Bruhn, J.N. and Anderson, J.B. (1992) Nature 356, 428–431.
104 Cambareri, E.B., Jensen, B.C., Schabtach, E. and Selker, E.U. (1989) Science 244, 1571–1575.
105 Marathe, S., Connerton, I.F. and Fincham, J.R.S. (1990) Mol. Cell. Biol. 10, 2638–2644.
106 Gwinn, K.D. and Gavin, A.M. (1992) Plant Dis. 76, 911–914.

PROSPECTS FOR HUMAN GENE THERAPY

A.B. Moseley[1†] and C.T. Caskey[1,2]

[1]Institute for Molecular Genetics
[2]Howard Hughes Medical Institute
Baylor College of Medicine
Houston, TX 77030

[†]Present address: Applied Immune Sciences
5301 Patrick Henry
Santa Clara, CA 95052

INTRODUCTION

The introduction of the adenosine deaminase (ADA) gene into mature lymphocytes of a patient with ADA deficiency offers only a temporary treatment of the immunodeficiency but has paved the way for potential cure of this and other inherited diseases through the use of somatic gene therapy. Rapid progress in the development of molecular technology has resulted in the identification of a number of disease genes, the subsequent cloning of their normal counterparts, and has enabled scientists to address correction of these inherited disorders. Large amounts of gene products can now be produced by recombinant technology and initial therapeutic efforts focused on the *in vivo* administration of such purified recombinant proteins. This therapeutic approach has been ineffective in the treatment of genetic diseases in which the activity of the enzyme is located within the cell, with the notable exceptions of ADA deficiency and Gaucher disease (1). Even when addition of the deficient enzyme has been found to correct the biochemical defect *in vitro, in vivo* administration of the enzyme has generally not proven to be adequate therapy for correction of the defect due to either accelerated metabolism of the protein or an enhanced immune response to the protein. Research efforts in the correction of inherited defects have thus been directed to the transfer of normal genes to autologous target cells and the reintroduction of these cells into the patient. The use of various strategies to transfer genetic sequences into somatic

Genetic Engineering, Vol. 15, Edited by J.K. Setlow
Plenum Press, New York, 1993

cells from patients and their readministration to the patient for clinical benefits constitutes somatic gene therapy (2). The requirements for a successful gene therapy protocol include: (i) identification and availability of the defective gene (cloned) (ii) knowledge of the biology and regulation of the gene in the target tissue and (iii) the ability to transduce sufficient numbers of target cells as well as ensuring proper gene regulation. Although most commonly considered in the context of inherited single gene defects because of the extensive molecular knowledge of the defective gene and target cell involved, gene therapy plays an increasingly greater role in the therapy of acquired diseases, as the genes and target cells involved in these disorders are identified and new methods of gene transfer are developed.

The goal of somatic gene therapy is to institute a lifelong treatment of the disease. The transferred gene must be maintained indefinitely and continue to function in the target cells. To achieve permanent correction of the genetic defect, the target cells would need to persist indefinitely. As most target cells have a short half-life, the issue of the maintenance of the transfected gene in the target population has become one of great importance in the design of gene therapy protocols. For certain acquired diseases, only short-term expression of the gene product might be required and this would allow the use of alternative gene transfer strategies.

Almost 5,000 genetic disorders have been described and disease genes have been identified for fewer than 10% of these disorders (3). About 30% of the disorders are recessive or X-linked and are often the result of the lack of a protein product, making them particularly attractive for gene therapy (4). For the successful correction of a number of inherited diseases, the genetic defect must be corrected in the specific cells in which the disease gene manifests itself and therefore, gene transfer must take place in the target cell or its progenitor. Examples of such disorders include beta-thalassemia, sickle cell anemia, cystic fibrosis and muscular dystrophy. The lack of expression of a gene product that is not critical for the function of the cell in which it is synthesized, but rather is critical for the function of associated cells or tissues, is seen in inherited disorders such as hemophilia A and B and hormone deficiencies, as well as any disorders which result from inadequate circulating protein products or insufficient detoxification of specific harmful products (2). In these disorders, the targeted cell for gene transfer would not necessarily be the same cell that normally synthesizes the gene product, thereby allowing greater potential for gene transfer strategies. Attempts to inhibit the expression of a gene product might prove useful for certain autosomal dominant diseases.

TRANSPLANTATION AS A FORERUNNER FOR GENE THERAPY

Bone marrow transplantation from histocompatible normal donors was first performed in 1968 and led to the correction of two lethal genetic disorders, severe combined immunodeficiency (SCID) and Wiskott-Aldrich syndrome (5). The successful marrow transplantation for Wiskott-Aldrich and osteopetrosis with the use of a preparative regimen without radiation has led to the increased interest in the potential of marrow transplants as a curative approach for a variety of genetic diseases of the hematopoietic system and as a treatment modality for certain lethal disorders of metabolism (6). For patients with certain inherited diseases, such as cystinosis, hyperoxaluria and alpha-1-antitrypsin deficiency, which produce pathology in solid organs such as kidney and liver, transplantation of these organs from normal donors has resulted in increased long-term survival for these patients (7). The transplants have not, however, altered the extrarenal or extrahepatic manifestations of the primary metabolic defects. Even after successful transplantation, patients require lifelong treatment with immunosuppressive agents which have significant morbidity.

Allogeneic bone marrow transplant has become the treatment of choice for several lethal immunodeficiencies, particularly the different variants of SCID as well as congenital deficiency of interleukin-1 receptor, congenital deficiencies of T-cell activation and interleukin-2 receptor, the Bare lymphocyte syndrome (8). Thalassemia major has been the genetic disorder for which the most HLA-matched bone marrow transplants have been performed worldwide. The major benefit of transplant is seen in younger patients who have not developed hepatic dysfunction due to hemosiderosis; however, with a 10 to 15% mortality associated with transplantation, the risks are considered too great in a child who could survive into late teens or early twenties with vigorous transfusion support and chelation therapy with deferoxamine (9). Recently, transplants have been suggested for homozygous sickle cell anemia patients who develop destructive vaso-occlusive events (10). Other rare disorders are treated by bone marrow transplantation (for review, see Ref. 11). Stem cell transfusions from HLA matched cord blood transplants have been reported to be successful for certain of these rare disorders which are inherited, such as Fanconi's anemia (12). Osteopetrosis, an autosomal recessive disorder characterized by deficiencies of osteoclasts resulting in an inability to resorb and reform bone, has been successfully treated with marrow transplantation resulting in both increased survival as well as resorption of sclerotic bone and normalization of bone growth (13).

Following allogeneic bone marrow transplantation, donor-derived tissue-based macrophages can be found, including Langerhans cells in the skin, Kupfer cells in the liver, alveolar macrophages in the lung, osteoclasts in bone and microglial cells in the brain (14). After it was demonstrated that certain enzymatic defects were corrected after co-culture with normal cells, marrow transplantation for the correction of storage was attempted. Enzyme levels were detectable in tissues after allogeneic transplantation for the mucopolysaccharidoses I, II, III, IV and VI, mannosidosis and Gaucher disease. No effect on enzyme activity was seen after transplantation for Pompe's, Type II glycogen storage disease. Stabilization of disease has been reported in two patients with arylsulfatase A deficiency (metachromatic leukodystrophy), as well as patients with Krabbe's disease, Nieman-Pick, adrenoleukodystrophy, Hurler's (MPS-I) and Sanfillipo, type B (MPS-III). No benefit was seen in Hunter's (MPS-II), Sanfillipo type A and Lesch-Nyhan patients who failed to arrest neurologic progression (15).

AUTOLOGOUS GENE TRANSFER

While HLA matched bone marrow transplantation has been suggested to be a potential cure for many of the disorders mentioned above, fewer than 30% of patients with these diseases have an HLA matched sibling. Even with a matched marrow, a 10 to 15% mortality is associated with transplantation and surviving patients require lifelong immunosuppressive therapy. A therapeutic protocol with the patient's own genetically manipulated bone marrow would alleviate the need for a matched donor and would result in decreased morbidity and mortality. All hematopoietic lineages are thought to be derived from a primitive progenitor which has an extensive self-renewal capacity (16). Gene transfer targeted to the primitive stem cell would ensure that the gene would be present in all replenishing populations of cells and therefore continued expression of the transduced gene would be maintained. The transfer of genes to the hematopoietic stem cells has been accomplished in a number of systems as determined by long-term expression of transduced genes in transplanted animals (for review, see Ref. 17).

Adenosine deaminase (ADA) deficiency has replaced beta-thalassemia as the most likely candidate for bone marrow gene therapy. ADA is a rare autosomal recessive disorder accounting for approximately 15% of the SCID cases. ADA catalyzes the irreversible deamination of adenosine and deoxyadenosine (dAdo) to inosine and deoxyinosine, respectively. In the absence of ADA, dAdo accumulates in many tissues, especially those of the lymphoid system which express high amounts of the enzyme. It has been suggested that the accumulation of dAdo and its metabolites, particularly deoxyATP, inhibits DNA synthesis resulting in profound T-cell and subsequent B-cell dysfunction. The severe immunodeficiency leads to recurrent opportunistic infections, failure to thrive and, if untreated, will invariably be fatal within the first few years of life (5). HLA identical bone marrow transplantation has been curative for ADA deficiency, and is still the treatment of choice. While T-cell depleted haploidentical marrow transplantation has been quite successful in non-ADA deficient SCID, the results have not been encouraging in ADA deficient patients. Enzyme replacement is an attractive mode of therapy because dAdo is freely diffusable, and therefore the ADA protein does not need to enter the target cells to be partially effective. Early therapeutic efforts with irradiated red blood cell infusions demonstrated transient clearing of dAdo, but transfusion-related risks outweigh the benefit in this form of treatment. Injections of bovine ADA protein conjugated to polyethylene glycol (PEG-ADA) have been shown to reverse the biochemical abnormalities, and restore some T-lymphocyte function; however, total immune reconstitution has not been demonstrated (1). Anti-PEG-ADA antibodies have been demonstrated in some patients. Recently, a trial has begun to treat ADA deficient SCID patients who are on PEG-ADA treatment with *ex vivo* -expanded autologous peripheral T-cells after retroviral-mediated gene transfer of the human ADA gene (18). Both PEG-ADA injections and administration of retroviral vector modified peripheral T-cells result in only temporary treatment of the ADA enzyme defect. Transfer of the human ADA gene into the hematopoietic progenitor cell pool would lead to permanent correction of the defect. Our group has recently demonstrated the transfer of the ADA gene with retroviral supernatant into normal and ADA-deficient human CD34+ bone marrow cells, the population of cells thought to contain the hematopoietic stem cells, and maintenance of expression after 6 weeks in long-term culture (19), supporting previous results of long-term expression obtained after co-cultivation with retrovirus-producing cells (20).

Gaucher disease is an autosomal recessively inherited lipid storage disorder resulting from the deficiency of the lysosomal enzyme glucocerebrosidase. Glycolipid accumulates in large amounts in the reticuloendothelial system causing hepatosplenomegaly, pancytopenia, lytic bony lesions (types I, II and III) and neurologic symptoms (types II and III) (21). Allogenic bone marrow transplantations have been performed for approximately two dozen patients and five year follow-up has demonstrated favorable outcome with respect to the hepatosplenomegaly and the skeletal degeneration in most of the patients, but the procedure has had high morbidity and mortality due to immunologic complications (22). Enzyme replacement with purified human glucocerebrosidase has shown benefit in early clinical trials but the therapy requires monthly intravenous infusions (23) and the treatment is costly. Gene transfer of the glucocerebrosidase gene into a purified population of CD 34+ cells from bone marrow from a Gaucher patient resulted in normalization of enzyme activity *in vitro* (24). Studies examining the expression of the human glucocerebrosidase gene in long-term reconstituted mice have demonstrated that hematopoietic stem cells could be transduced efficiently and maintained after transplantation (25).

Although the defect in beta-thalassemia is impaired synthesis of the beta-globin polypeptide, the pathology seen in the patient, severe anemia with resulting erythroid proliferation, results from the precipitation of excess alpha-globin chains in erythroid cells. The erythroid proliferation leads to hepatosplenomegaly and congestive heart failure (26). Current treatment is largely supportive and consists primarily of red cell transfusions to decrease the anemia and the resultant erythroid response. Iron overload is a potentially lethal side effect from transfusion therapy (27). Few centers have had promising results with HLA-matched bone marrow transplantation (28). Although a stem cell gene transfer strategy, such as that proposed for ADA deficiency, could be effective for the treatment of beta-thalassemia, the requirements for gene transfer are more stringent because beta-globin is expressed only in erythroid cells. The balanced synthesis of alpha- and beta-globin chains in the targeted cells after beta-globin gene transfer is essential, because too little beta-globin would be of no benefit and too much would result in an alpha-thalassemia phenotype. The discovery of the DNase hypersensitive regions (tissue-specific consensus sequences remote from the expressed genes which are required for regulated beta-globin synthesis), has resulted in the construction of vectors that contain both sequences and promise more regulated gene expression in the target cells (29).

Many of the inborn errors of metabolism are due to defective enzymes in the liver. Gene therapy of these disorders must take into account the need to reconstitute a complete metabolic pathway, not merely the missing gene product. Diseases amenable to liver gene therapy include urea cycle defects, organic acid disorders, phenylketonuria, galactosemia, homocystinuria and maple syrup urine disease, glycogen storage diseases and familial hypercholesterolemia (30). Liver transplantation has resulted in a cure for a number of these diseases, although the availability of the organ, the high morbidity and mortality associated with the procedure and the lifelong need for immunosuppression have identified a need for gene therapy targeted to hepatocytes. An *ex vivo* method has been postulated for the introduction of the vector into the hepatic tissue whereby hepatocytes are removed from the body, cultured with the vector and returned to the liver *via* an intravenous route (31). Studies with retroviral vectors and the genes for OTC (ornithine transcarbamylase), alpha-1-antitrypsin and LDL receptor have demonstrated successful *ex vivo* transduction of hepatocytes (32–34). Two human protocols with this method have passed the Recombinant DNA Committee of the NIH, one for hepatocyte marking (35) and one for treatment of hypercholesterolemia (36).

Familial hypercholesterolemia is an autosomal dominantly inherited disease due to a deficiency of low density lipoprotein (LDL) receptors, resulting in increased serum cholesterol and premature development of coronary artery disease. Patients with one abnormal allele have moderate elevation of serum cholesterol and premature coronary artery disease. The prevalence of this disease is thought to be 1 in 500 and these patients represent about 5% of all patients under 45 who have had a myocardial infarction. Patients who are homozygous for the disease will die in the first decade of life of severe coronary artery disease. Greater than 90% of the LDL receptors are found in the liver. Drugs stimulating the expression of the normal allele have been partially beneficial in heterozygotes; however, in nearly all homozygotes and in those patients with less than 2% LDL receptor these drugs have not been beneficial. Liver transplant from donors with normal LDL receptors has been successful in correcting the hyperlipidemia, suggesting that correction of hepatic LDL receptor expression would be sufficient to correct the disease (37). An animal model, the Watanabe heritable hyperlipidemic (WHHL) rabbit, has been available for development of gene transfer strategies. *Ex vivo* transduction of defective hepatocytes with the normal LDL receptor gene has resulted in transiently decreased levels of serum cholesterol and thus short-term correction of the defect (38). Recently, a study reporting the use of a hepatocyte-targeted vector for *in vivo* gene transfer demonstrated transient improvement of the hyperlipidemia in WHHL rabbits (39).

X-linked OTC deficiency is the most common urea cycle disorder in man. The severe form of the disease presents with hyperammonemia in the newborn period, leading to coma and frequently to irreversible brain damage. Considerable progress has been made in the dietary and pharmacologic treatment of the disease. These therapies are beneficial in mildly affected males and carrier females, but are unable to prevent episodes of hyperammonemia and their sequelae in affected males (40). Nevertheless, these treatments have made it possible to stabilize patients long enough to be eligible for liver transplantation, which has been shown to be curative for the metabolic defect. There are two mouse models of OTC deficiency, one, the *spf* (sparse fur), which represents a mild deficiency with 10 to 15% of wild-type enzyme activity, and the second, *spf-ash* (sparse fur/abnormal skin and hair), with a more severe defect and only 5% of wild-type activity. A recent study has demonstrated the correction of the defect in both these mice with a retroviral vector containing the normal OTC gene (32). Current gene transfer strategies must be targeted to the hepatic tissues because the substrate, carbamyl-phosphate, is synthesized only in the liver and the intestine. Previous experiments utilizing an adenoviral vector with the OTC gene have shown transduction of the gene in hepatic tissue (41). The development of *in vivo* transfer techniques for transduction of hepatocytes with the OTC gene are currently under way in our group (Morsy, unpublished data).

NEED FOR ALTERNATIVE GENE TRANSFER STRATEGIES

Certain tissues such as muscle and neurons do not replicate after development and the integration of retroviral vectors becomes a significant obstacle to gene therapy for diseases of these target tissues. Duchenne muscular dystrophy is an X-linked disease of progressive muscle weakness and atrophy, particularly of the limb and girdle muscles. The disease has a prevalence of 1 in 3,000 and many of the patients are wheelchair bound by childhood (42). A mouse model for muscular dystrophy, the *mdx* mouse, has been corrected by the transgenic administration of the normal dystrophin sequence (43). The gene encoding the dystrophin protein has been cloned, and the large size of the gene, 14 kb, makes it difficult to develop retroviral vectors that can deliver the entire DNA sequence. Recent research efforts by a number of investigators have concentrated on alternative strategies for gene transfer of the dystrophin molecule into muscle tissue (for review see 44). Cultured myoblasts have been transduced with the dystrophin gene and experiments are under way to examine their potential for gene therapy. Techniques aimed at the direct transfer of the dystrophin gene into muscle are being developed. Becker muscular dystrophy is a milder form of muscular dystrophy and "minigenes" have been developed which have gene deletions seen in the Becker patients. Development of vectors that can accommodate the large size of the dystrophin cDNA are currently under development in our group.

The cloning of the disease genes for such inherited disorders as cystic fibrosis (1), fragile X syndrome (45) and myotonic dystrophy (46) has opened up the door to the early diagnosis of these disorders. Further research focused on the structure and function of the gene product as well as the molecular pathogenesis of the disease is necessary to determine the respective target tissues and the appropriate gene transfer strategy.

GENE THERAPY FOR ACQUIRED DISEASES

Acquired diseases such as cancer and acquired immunodeficiency syndrome represent conditions in which gene therapy may play a potential role. The pathology at the molecular level is not known for many of the acquired diseases, and the role of specific gene products may not be completely understood. A major focus in the treatment of cancer has been the immunotherapy of cancer. The studies have focused on enhancing the natural cytolytic response of the natural killer and cytotoxic T-cells in the immunologic reaction against tumor cells (for review see 47). Lymphocyte-activated killer (LAK) cell therapy is defined as the *ex vivo* expansion of natural killer cells with interleukin-2 and the return of these cells to the patient. This protocol has been effective in a small number of metastatic cancers. A related protocol, *ex vivo* expansion of tumor-infiltrating lymphocytes (TIL therapy), has been demonstrated to be effective against metastatic melanoma (48). By utilizing gene transfer of cytokine genes, such as tumor necrosis factor, into the TIL, it is hoped that the TIL will have an even greater anti-tumor effect (49). Modification of the tumor by transfer of a gene product that would confer greater antigenicity upon the tumor could enhance the body's cancer surveillance against metastatic disease. The transfer of drug resistance genes into a stem cell population would allow for greater doses of chemotherapy to be given to the tumor cells while the normal progenitor cells would be resistant (50).

For acquired immunodeficiency syndrome, gene transfer techniques, termed "suicide therapy," have focused on the selective killing of those cells harboring the human immunodeficiency virus (HIV). CD4+ T-lymphocytes were transfected with herpes simplex virus type 1 thymidine kinase (HSV1-TK) under the control of the HIV promoter and transactivation response element sequences, and upon HIV infection, these sequences were transactivated. The HSV1-TK activation resulted in the phosphorylation and toxic accumulation of the drug acyclovir and resulted in the subsequent death of the HIV infected cells (51). The induction of specific cytotoxic T-lymphocyte responses by gene transfer of selected HIV gene products has been proposed to function as an enhancement to the natural immune response to the infection (52). These methods represent alternative strategies of gene therapy based on the increasing knowledge of the pathophysiology of this acquired disease.

CONCLUSION

The development of even the most successful prenatal diagnosis at the molecular level cannot prevent the occurrence of rare recessive diseases and many of the X-linked and autosomal diseases which demonstrate a high new mutation rate. Gene therapy for the treatment of somatic diseases has become a possibility for a few select diseases. Gene transfer technology offers the potential for permanent correction of the genetic or acquired disorders without the associated morbidity and mortality of allogenic organ transplantation, the most successful therapy for these diseases to date. Scientific research efforts should focus not only on the identification and regulation of the disease genes, but also on the biology of potential target tissues and gene transfer technologies into these tissues.

Acknowledgments: CTC is an Investigator with the Howard Hughes Medical Institute.

REFERENCES

1 Levy, Y., Hershfield, M.S., Fernandez, M.C., Polmar, S.H., Scudiery, D., Berger, M. and Sorensen, R.U. (1988) J. Pediat. 113, 312–317.
2 Mulligan, R. (1991) in Etiology of Human Disease at the DNA Level (J. Lindsten and U. Pettersson, eds.), pp. 143–189, Raven Press, New York, NY.
3 Collins, F.S. (1991) Hospital Practice 26, 93–98.
4 Belmont, J. and Caskey, C.T. (1986) in Gene Therapy (R. Kurcherlapati, ed.), pp. 411–420, Plenum Press, New York, NY.
5 Bach, F.H., Albertini, R.J. and Joo, P. (1968) Lancet 2, 1364–1366.
6 O'Reilly, R.J., Brochstein, J., Dinsmore, R. and Kirkpatrick, D. (1984) Semin. Hematol. 21, 188–221.
7 Langlois, R.P., O'Regan, S., Pelletier, M. and Robitaille, P. (1981) Nephron 28, 273–278.
8 O'Reilly, R.J., Keever, C.A., Small, T.N. and Brochstein, J. (1989) Immunodef. Rev. 1, 273–309.
9 Luccarelli, G., Galimberti, M., Polchi, P. Angelucci, E., Baronnoni, D., Giardini, C., Politi, F., Durazzi, S.M., Morello, P. and Albertini, T. (1990) N. Engl. J. Med. 322, 417–421.
10 Vermylen, C., Fernandez-Robles, E., Ninane, J. and Cornu, G. (1988) Lancet 21, 1427–1428.
11 O'Reilly, R.J. (1991) in Etiology of Human Disease at the DNA Level (J. Lindsten and U. Pettersson, eds.), pp. 261–289, Raven Press, New York, NY.
12 Broxmeyer, H.E., Douglas, G.W., Hangoc, G., Cooper, S., Bard, J., English, D., Arny, M., Thomas, L. and Boyse, E.A. (1989) Proc. Nat. Acad. Sci. U.S.A. 86, 3828–3832.
13 Gerritsen, E.J.A., Van Loo, I.H.G. and Fisher, A. (1990) Bone Marrow Transplantation 5 (Supp. 2), 12.
14 Thomas, E.D., Ramberg, R.F. and Sale, G.E. (1976) Science 192, 1016–1018.
15 Nyhan, W.L., Page, T., Gruber, H.E. and Parkman, R. (1986) Birth Defects 22, 41–53.
16 Sutherland, H.J., Lansdorp, P.M., Henkelman, D.H., Eaves, A.C. and Eaves, C.J. (1990) Proc. Nat. Acad. Sci. U.S.A. 87, 3584–3588.
17 Cournoyer, D., Scarpa, M., Moore, K.A., Fletcher, F.A., MacGregor, G.R., Belmont, J.W. and Caskey, C.T. (1991) in Etiology of Human Disease at the DNA Level (J. Lindsten and U. Pettersson, eds.), pp. 199–220, Raven Press, New York, NY.
18 Culver, K.W., Anderson, W.F. and Blaese, R.M. (1991) Human Gene Therapy 2, 107–110.
19 Moseley, A., Mitani, K., Deisseroth, A. and Caskey, C.T. (1992) J. Cell. Biochem. (Suppl. 16F), 50.
20 Cournoyer, D., Scarpa, M., Mitani, K., Moore, K.A., Markowitz, D., Bank, A., Belmont, J.W. and Caskey, C.T. (1991) Human Gene Therapy 2, 203–213.

21 Barranger, J.A. and Ginns, E.I. (1989) in the Metabolic Basis of Inherited Disease, 6th Edition (C.R. Scriver, A.L. Beaudet, W.S. Sly and D. Valle, eds.), pp. 1677–1698, McGraw-Hill, New York, NY.
22 Ringdon, O., Groth, C.-D., Erickson, A., Backman, L., Granqvist, S., Mansson, J.-E. and Svennerholm, L. (1988) Transplantation 46, 66–70.
23 Barton, N.W., Furbish, F.S., Murray, G.J., Garfield, M. and Brady, R.O. (1990) Proc. Nat. Acad. Sci. U.S.A. 87, 1913–1916.
24 Kohn, D.B., Nolta, J.A., Wienthal, J., Bahner, I., Yu, X.J., Lilley, J. and Crooks, J.M. (1991) Human Gene Therapy 2, 101–106.
25 Correll, P.H., Kew, Y., Perry, L.K., Brady, R.O., Fink, J.K. and Karlsson, S. (1990) Human Gene Therapy 1, 277–288.
26 Weatherall, D.J., Clegg, J.B., Higgs, D.R. and Wood, W.G. (1989) in the Metabolic Basis of Inherited Disease (C.R. Scriver, A.L. Beaudet, W.S. Sly and D. Valle, eds.), pp. 2305–2339, McGraw-Hill, New York, NY.
27 de Montalembert, M., Llados, A., Hannadouche, T. and Girot, R. (1989) Arch. Fr. Pediatr. 46, 99–103.
28 Fosburg, M.T. and Nathan, D.G. (1990) Blood 76, 435–444.
29 Novak, U., Harris, E.A.S., Forrester, W., Groudine, M. and Gelinas, R. (1990) Proc. Nat. Acad. Sci. U.S.A. 87, 3386–3390.
30 Ledley, F.D. (1987) J. of Ped. 110, 167–174.
31 Ponder, K.P., Dunbar, R.P., Wilson, D.R., Darlington, G.J. and Woo, S.J.L. (1991) Human Gene Therapy 2, 41–52.
32 Grompe, M., Jones, S.N., Loulseged, H. and Caskey, C.T. (1992) Human Gene Therapy 3, 35–44.
33 Kay, M.A., Baley, P., Rothenberg, S., Leland, F., Fleming, L., Ponder, K.P., Liu, T.-J., Finegold, M., Darlington, G., Pokorny, W. and Woo, S.L.C. (1992) Proc. Nat. Acad. Sci. U.S.A. 89, 89–93.
34 Wilson, J.M., Chowdhury, N.R., Grossman, M., Wajsman, R., Epstein, A., Mulligan, R.C. and Chosdhury, J.R. (1990) Proc. Nat. Acad. Sci. U.S.A. 87, 8437–8441.
35 Ledley, F.D. (1991) Human Gene Therapy 2, 331–358.
36 Wilson, J.M. (1991) Federal Register Announcement, Human Gene Therapy 2, 378.
37 Goldstein, J.L. and Brown, M.S. (1989) in The Metabolic Basis of Inherited Disease, 6th Edition (C.R. Scriver, A.L. Beaudet, W.S. Sly and D. Valle, eds.), pp. 1215–1250, McGraw-Hill, New York, NY.
38 Grossman, M., Raper, S.E. and Wilson, J.M. (1991) Somatic Cell. and Mol. Genetics 17, 601–607.
39 Wilson, J.M., Grossman, M., Wu, C.H., Chowdhury, N.R., Wu, G.Y. and Chowdhury, J.R. (1992) J. Biol. Chem. 267, 963–967.
40 Brusilow, S.W. and Horwich, A.L. (1989) in The Metabolic Basis of Inherited Disease, 6th Edition (C.R. Scriver, A.L. Beaudet, W.S. Sly and D. Valle, eds.), pp. 629–670, McGraw-Hill, New York, NY.
41 Stratford-Perricaudet, L.D., Levero, M., Chasse, J.-F., Perricaudet, M. and Briand, P. (1990) Human Gene Therapy 1, 241–256.
42 Harper, P.S. (1989) in The Metabolic Basis of Inherited Disease, 6th Edition (C.R. Scriver, A.L. Beaudet, W.S. Sly and D. Valle, eds.), pp. 2869–2902, McGraw-Hill, New York, NY.

43 Leee, C.C., Pearlman, J.A., Chamberlain, J.S. and Caskey, C.T. (1991) Nature 349, 334–336.
44 Rossiter, B.J.F., Stirpe, N.S. and Caskey, C.T. (1992) Neurology 42, 1413–1418.
45 Verkerk, A.M.J.H., Pieretti, M., Sutcliffe, S., Fu, Y.-H., Kuhl, D.P.A., Pizzuti, A., Reiner, O., Richards, S., Victoria, M.F., Zhang, F., Eussen, B.E., Ommen, G.-J., Blonden, L.A.J., Riggins, G.J., Chastain, J.L., Kunst, C.B., Caskey, C.T., Nelson, D.L., Oostra, B.A. and Warren, S.A. (1991) Cell 65, 905–914.
46 Fu, Y.-H., Pizzuti, A., Fenwick, Jr., R.G., King, J., Rajnarayan, S., Dunne, P.W., Dubel, J., Nasser, G.A., Ashizawa, T., de Jong, P., Wieringa, B., Korneluk, R., Perryman, M.B., Epstein, H.F. and Caskey, C.T. (1992) Science 255, 1256–1258.
47 Rosenberg, S.A. (1992) J. Clin. Oncology 10, 180–199.
48 Rosenberg, S.A. (1990) Human Gene Therapy 1, 463–471.
49 Sorrentino, B.P., Brandt, S., Gottesman, M., Pastan, I., Bodine, D. and Nienhuis, A.W. (1991) Blood 78 (Suppl. 1), 191.
50 Caruso, M. and Klatzman, D. (1992) Proc. Nat. Acad. Sci. U.S.A. 89, 182–186.
51 Jolly, D.J. (1991) Human Gene Therapy 2, 111–112.
52 Jolly, D.J. (1992) Abstracts, 1992 Meeting on Gene Therapy (W.F. Anderson, T. Friedmann and R. Mulligan, eds.) Cold Spring Harbor Press, New York, NY.

THE USE OF MICROPARTICLE INJECTION TO INTRODUCE GENES INTO ANIMAL CELLS *IN VITRO* AND *IN VIVO*

Stephen A. Johnston and De-Chu Tang

Internal Medicine and Biochemistry
University of Texas Southwestern Medical Center
Dallas, TX 75235-8573

INTRODUCTION

The ability to introduce DNA into cells has been a *sine-qua-non* for the developments in molecular biology. One of the most unusual and unexpected technological breakthroughs in this arena is biolistic transformation. This technique goes under several names, including microparticle injection, gene gun, particle bombardment, etc. The basic idea, as originally conceived of by Sanford and coworkers (1), is disarmingly simple. DNA or other biological material is coated onto heavy metal microprojectiles and the microprojectiles are shot into the target cells, carrying the DNA in with them.

This technology has already had a tremendous impact on plant research. It has proved more effective than other techniques for transformation of several plant species, including ones of economic importance such as soybeans (2). The most significant breakthrough was the recent accomplishment of transformation of corn (3). The biolistic technology remains the only way to transform corn and has opened this crop to recombinant-DNA improvement. The technique has also proved useful for basic research in plant biology. For example, DNA can be introduced directly into the cells in intact plant tissue allowing gene constructs to be assayed directly in the tissue of interest in the authentic environment. One striking demonstration of this capability was the testing of the interaction of pigmentation regulatory genes by co-shooting them into the endosperm and assaying for pigmented cells (4). The applications of the biolistic technique to plant science have been reviewed (5).

Microprojectile bombardment has also been used for important breakthroughs in organellar and microbial transformation. The first demonstration of mitochondria transformation employed the biolistic technique (6) and it remains the only way to transform mitochondria (7). It is also the only

Genetic Engineering, Vol. 15, Edited by J.K. Setlow
Plenum Press, New York, 1993

reproducible method for transforming chloroplast DNA, and has been used to transform both *Chlamydomonas* and higher plant chloroplasts. It remains to be seen whether this technique can be extended to the mitochondria of higher plants or animals. The first demonstration of the effectiveness of this technology for the nuclear transformation of microbial targets was with *Saccharomyces cerevisiae* (8). The usefulness of microprojectile bombardment for microbial transformation probably lies in its ability to introduce DNA into hard-to-transform species. For example, the bacterium, *Bacillus megaterium* (9), and the virulent form of the pathogenic fungi, *Cryptococcus neoformans* (S.A. Johnston and J. Perfect, unpublished data), are only transformed by this technique. The robustness of this technology for plant, microbial and organellar transformation apparently rests in its insensitivity to the state of the cell membranes and walls, therefore not requiring any special pretreatment.

By comparison to plants, there has been relatively little application of the biolistic technology to animal cells. However, there has been enough experimentation to provide some indication of the potential uses and limitations of this technique for the introduction of DNA into animal cells. In this chapter we will review the current state of the technology for the devices. The research to date will be summarized with particular emphasis on the potential applications to gene therapy.

PRINCIPLES OF OPERATION

Since the first description of a biolistic device there have been many new forms reported. However, all have the same principles of operation. DNA or other biological material is coated onto microprojectiles of 0.5 to 5 μm in diameter. These "BBs" are of a heavy metal, usually gold or tungsten. Material of lesser density has proven inefficient in penetrating the target cells, even those without cell walls. The DNA-coated BBs are placed on a carrier macroprojectile. In the original form of the device this was a plastic bullet. Current versions use a Mylar or Kapton disc. A high velocity gas blast or shock wave is generated to propel the macroprojectile against a stopping screen which restrains the macroprojectile but allows the BBs to keep going into the target cells. The target cells are placed in a vacuum chamber for the bombardment so that the BBs can travel enough distance to disperse while not being stopped by the air. The different devices differ principally in the method used to accelerate the macroprojectile.

The original device of Sanford and co-workers was literally a short-barreled 22-caliber shotgun. The propellant was a gunpowder cartridge. This form was available commercially and most of the reports in the literature to date employed this technology. However, recently one of us (SAJ) in collaboration with the Sanford laboratory has designed a new helium-driven version of the biolistic device which is substantially superior to the gunpowder form (10). The only other gene gun design which has a proven record is the one developed by McCabe and coworkers. This system relies on an electric discharge to create the impelling shock wave. We are not familiar with the details of the operation of the electrical system, but from comparisons of

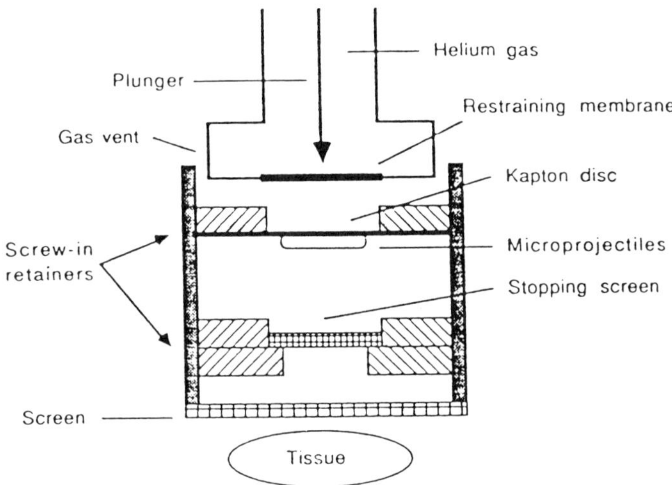

Figure 1. Diagram of the helium-driven gene gun. DNA-coated microprojectiles are spread onto a macrocarrier disc. A membrane restraining the helium gas is ruptured by a lance and the disc is propelled toward the target tissue by the resulting shock wave. The macrocarrier is stopped by a screen while the microprojectiles continue on to penetrate the tissue.

published data this device seems to produce results similar to those of the helium-driven system.

The helium-driven design resulted from efforts to adapt the biolistic technique for the direct introduction of DNA into cells in the living animal. We wanted to use a more controllable propellant than gunpowder and to produce less trauma to the target tissue. The gunpowder bombardment tended to produce substantial lethality in the center of the blast area. The intent was to create a hand-held unit that could be used for animal gene transfer and somatic cell modification. A schematic of the current design is depicted in Figure 1. The propelling shock wave is generated by the release of a small volume of helium under high pressure, between 800 and 2400 psi depending on the application. The gas can be released either by spontaneously bursting the restraining membranes (Kapton) or by piercing the membrane with an electrically controlled lance. These two forms of release produce distinctive shock waves that may have different biological effects.

As the shock wave speeds ahead of the expanding gas it launches the macrocarrier. The macrocarrier, a Kapton or Mylar disc, is slammed onto a restraining screen, impelling the DNA-coated BBs spread on its surface toward the target. The critical dimensions are the distance between the end of the shocktube and the macrocarrier, the dimensions of the macrocarrier and the distance the macrocarrier flies. These operating variables were determined empirically to maximize the velocity of the microprojectiles while minimizing the assault to the target tissue. An essential feature of the design is that the macrocarrier lands on the restraining screen in such a way as to deflect the gas blast away from the target cells. Figure 2 consists of photographs of the device currently in use in our laboratory. It is configured so that a vacuum can be applied, or not, over the target, or the chamber can be flushed with helium. The details of the construction have been published (10). Ironically, though this system was designed for animal applications, when it was placed onto the in-chamber, gunpowder unit it gave dramatically better results (10). The in-chamber device now commercially available employs the helium-driven system. As yet, the hand-held version is not commercially produced.

The biolistic technology is still under development and more improvements are likely. Two areas in particular limit the applications of this technique. First, there are several different designs for the apparatus and it is not clear which is best for different applications. With direct comparisons of the different units it is likely that new hybrid designs will be developed which will extend the utility of this technology. A second important limitation is the quality of the microprojectiles available. Most reports have been of using the commercial sources of BBs which vary widely in their characteristics and quality. Often microprojectiles of particular diameters will not coat with DNA or they aggregate. Once there is available a wide range in diameters of microprojectiles which coat well with DNA and other biological material, then the effectiveness of this technology should be dramatically improved.

TRANSFECTION OF CULTURED CELLS AND TISSUES

The in-chamber device can be used to transfect cultured animal cells with efficiencies comparable to those of conventional protocols such as calcium phosphate, electroporation and lipofection. Zelenin et al. (11) first reported using this technique to stably transfect mouse NIH-3T3 cells using a variation of the gunpowder device. The biolistic technique has since been used for both stable and transient transfection on a number of cell types. The procedure for adherent cells involves first removing most of the medium over the cells. The culture dish is then placed in the chamber and a full vacuum is drawn on the chamber (30 inches Hg). The cells are bombarded and then the medium is replaced. None of the cell types we have tested has experienced a loss in viability during exposure to the vacuum. With the earlier gunpowder version there was significant cell death in the center of the bombardment area. There is little or none in the much gentler helium system. Non-adherent cells can also be bombarded by filtering them onto a support or artificially attaching them to the culture dish.

For standard cell types that are readily transfected by conventional means the biolistic protocol does not offer any advantage in efficiency. It may be a useful alternative when DNA or RNA is at a premium as 100 ng or less is required to transfect cells in a 100-mm dish. As with microbial transformation, the advantage of the biolistic technology resides in its ability to transfect cells which are difficult or impossible to be transfected by other protocols. As examples we have found that the biolistic technique is effective for transfecting several difficult cell types including macrophage-derived cells and primary

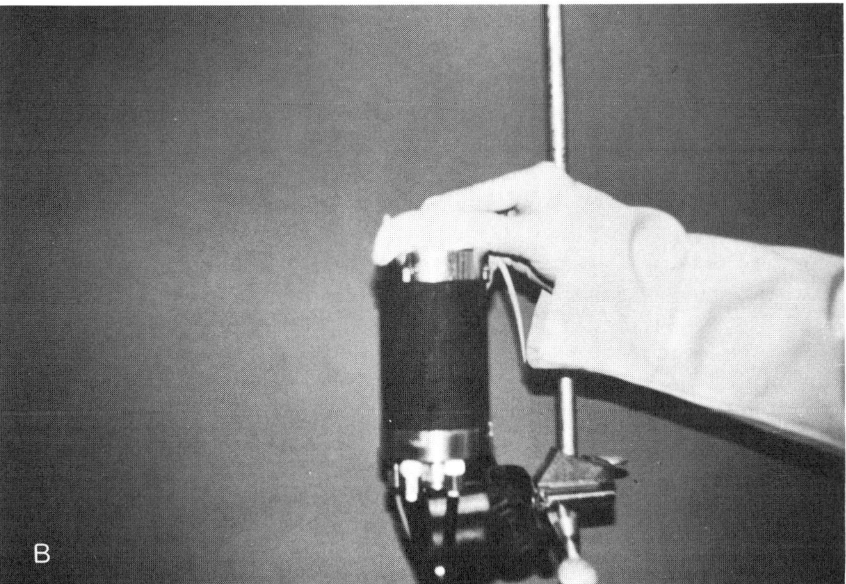

Figure 2. The gene gun setup in SAJ's lab. (A) The unit consists of a helium tank, a vacuum pump, a computer box, a compressor, the in-chamber device, and the hand-held wand. The vacuum pump is not shown. (B) The wand is launching DNA-coated gold microprojectiles into the liver of a living mouse.

Table 1
Expression of Firefly Luciferase Gene Following Transfection by Microparticle Bombardment of Post-Mitotic Skeletal Myotubes

	Luciferase Activity (pg/50 mm culture dish)
Mock transfection	0
pG-LUC	11
pG-MCK-LUC	619
pHβ-LUC	2,876
pHβ-LUC + ara-C	2,759
pHβ-LUC (modified apparatus)	49,232

Primary cultures of skeletal myotubes from chick embryos were bombarded with DNA-coated microprojectiles in a commercially available biolistic device, or a newly developed apparatus designed to minimize ballistic trauma and to produce a more even distribution of the particle beam (**modified apparatus**). Some cultures were also treated with cytosine arabinoside (**ara-C**) to inhibit growth of residual undifferentiated myoblasts or non-myogenic cells. Plasmid constructions containing the firefly luciferase gene were driven either by a minimal promoter containing only a TATA element (**pG-LUC**), a minimal promoter linked to a muscle-specific enhancer from the mouse M-creatine kinase gene (**pG-MCK-LUC**), or the human β-actin promoter (**pHβ-LUC**).

cultures of myotubes. Data on the transfection of primary cells of chick skeletal myotubes are presented below.

Though the primary myoblasts are readily transfected by lipofection or calcium phosphate methods, the differentiated myotubes, which are mitotically inactive, are very inefficiently transfected. At least a 20-fold higher transfection efficiency was obtained with the biolistic technique. A luciferase gene introduced under the transcriptional control of the constitutive, human β-actin promoter (pHβ-LUC) produces high level activity (Table 1). That this transient activity is produced by post-mitotic cells is supported by the observation that comparable activities are observed in the presence of a DNA-synthesis inhibitor (pHβ-LUC+ara-C, Table 1).

Two experiments done in collaboration with R.S. Williams support the contention that myotubes are transfected in this protocol. First, transfection with a reporter gene and histochemical staining indicates that greater than 75% of the stained cells are myotubes as opposed to uninuclear cells (Figure 3). In this experiment the transgene was the *Drosophila* ADH (alcohol dehydrogenase) gene under control of the RSV (Rous sarcoma virus) promoter. With the helium system approximately 12% of the myotubes on the dish express the transgene. In contrast, when myotubes were transfected with the same reporter with the use of lipofection or calcium phosphate, most of the expressing cells were mononuclear. A second experimental result supporting the idea that

Figure 3. Transgene expression in skeletal myotubes. Cells were fixed and stained for ADH activity 24 hr following microprojectile bombardment and photographed under bright field (50X). Some myotubes are densely stained in a diffuse manner, while staining is limited to a perinuclear distribution in other cells. Nontransformed myotubes and mononuclear cells show only faintly under bright field but are of normal appearance under phase contrast (not shown).

myotubes are transfected is that when the transfection is done with the luciferase gene under control of the myotube-specific promoter, M-creatine kinase, a 55-fold higher activity is observed than when an identical plasmid without the enhancer is used (Table 1). Also evident in Table 1 is the relative improvement in the helium system over the gunpowder one. The pHβ-LUC vector produces approximately 17-fold higher activity when fired in the helium unit (modified apparatus).

Thc plasmid introduced into the myotubes seems to be maintained and expressed for a period of time, even though these cells are not dividing. This was ascertained by transfecting with a luciferase reporter under control of a heat-shock promoter, HSP70. As indicated in Figure 4, the transfected reporter gene was still responsive to heat induction a week after bombardment. This indicates that the plasmid introduced is transcriptionally active even in non-dividing cells.

The in-chamber device can also be effective in transfecting cells in explanted organs or tissues. Yang et al. reported using the electrical biolistic system to transfect isolated ductal segments from rat and human mammary glands (12). We have used the technique to successfully transfect explanted human and mouse tumors, epithelial cells in pig eye and limb buds from newts. This technology should work on most explanted tissues.

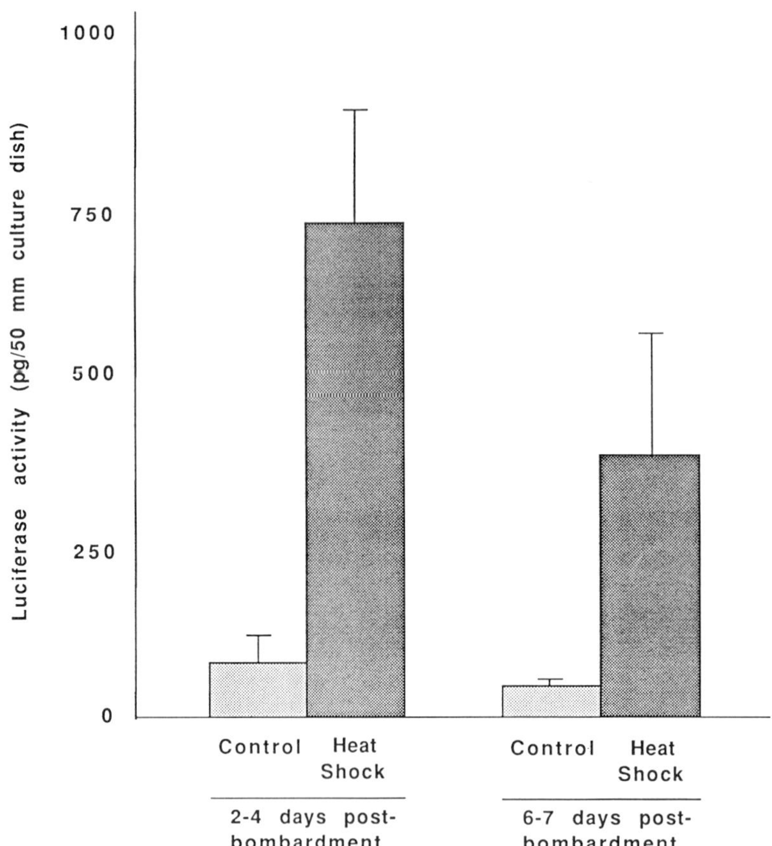

Figure 4. Heat-inducibility of the human HSP70 promoter at early and late time points following transformation of skeletal myotubes by microprojectile bombardment. Luciferase activity was measured in sister cultures that either were maintained at 37° C (Control) or placed at 45° C for 90 min, followed by recovery at 37° C for 3 hr (Heat Shock). Bars represent mean values (± SEM) from four separate cultures. The cultures maintained for 6 to 7 days following bombardment were re-fed at two-day intervals with conditioned media (depleted of mitogenic growth factors) from nontransfected myotube cultures. Both basal and heat-inducible luciferase activity tended to decline over time, but substantial heat-inducible activity remained even one week after transfection.

In summary, the in-chamber biolistic system may be useful for the transfection of: 1) cultured cells that are difficult or impossible to be transfected by conventional protocols, 2) post-mitotic cells, 3) primary cells, 4) explanted tissues or organs.

IN VIVO ANIMAL TRANSFECTION

The in-chamber device is not amenable to bombarding tissues in the living animal. We have reported an adaptation to the in-chamber unit which permits bombarding tissues *in situ* in the mouse (13), but this system is awkward and cannot be used on animals larger than a mouse. The hand-held helium unit we described above allows one to address any tissue easily in a surgical setting. McCabe and coworkers have also developed a hand-held version of the electrical discharge unit (K. Barton, personal communication). Zelenin and coworkers have also reported using a gunpowder version to transfect liver cells *in vivo* (14).

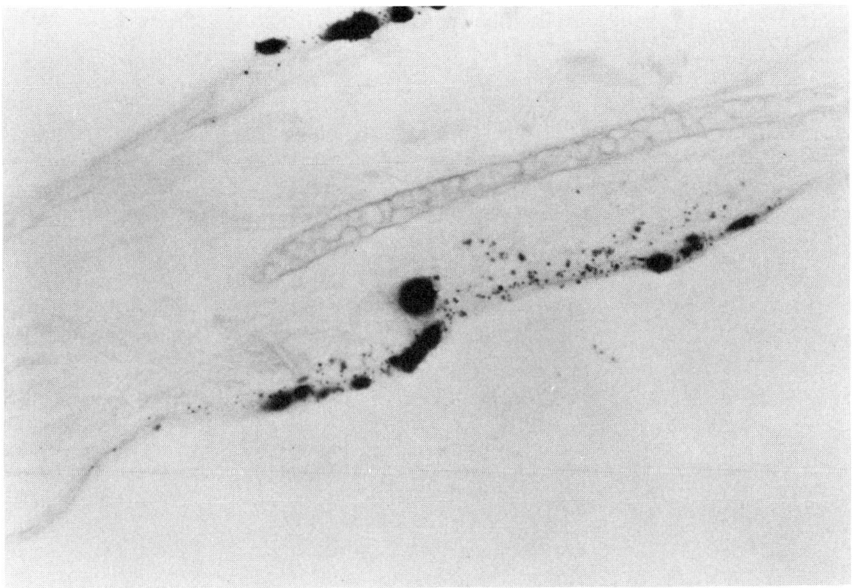

Figure 5. Cross-section of a bombarded mouse ear displaying the distribution of microprojectiles and transgene-expressing cells. The ear was removed, fixed, and stained for β-galactosidase activity one day after inoculating the ear with microprojectiles coated with the SV40 large T-antigen nuclear localization signal/*E. coli* β-galactosidase fusion gene. The stained ear was subsequently embedded, sectioned, and photographed under bright field (100X). Microprojectiles appear as little black dots. Cells expressing the transgene appear as cells with dark nuclei.

With the hand-held helium system transfection of the skin involves first removing the hair with a depilatory. A vacuum of 20 inches Hg is drawn over the skin during the bombardment. As we have documented (15), the BBs penetrate well into the dermis and DNA is carried with them. However, when the skin is bombarded with a β-galactosidase reporter gene most of the expressing cells are detected in the epidermis (Figure 5). Whether this restricted expression is due to differences in metabolic activity or that biologically active DNA is only left in the superficial layers is not clear.

While transfection of skin is reproducible and yields high levels of transgene expression, bombardment of internal soft tissue is more difficult. Liver, muscle, spleen, intestine and blood vessels have been transfected. We have had the most experience in the bombardment of liver. A small incision is made in an anesthetized mouse to expose the liver. It is bombarded without applying a vacuum to the tissue, with the end of the hand-held device against the organ. Penetration of the BBs can be quite deep (20 or more cell layers, as shown in Figure 6), but is quite variable. As in skin the DNA is carried deep into the tissue. The expression levels are also erratic. Within the same experiment individual livers can express high to undetectable levels of activity when bombarded with the same DNA-BB preparation. At this point we do not know the source of this high variability. It probably restricts the usefulness of the biolistic technique for internal organs to qualitative questions.

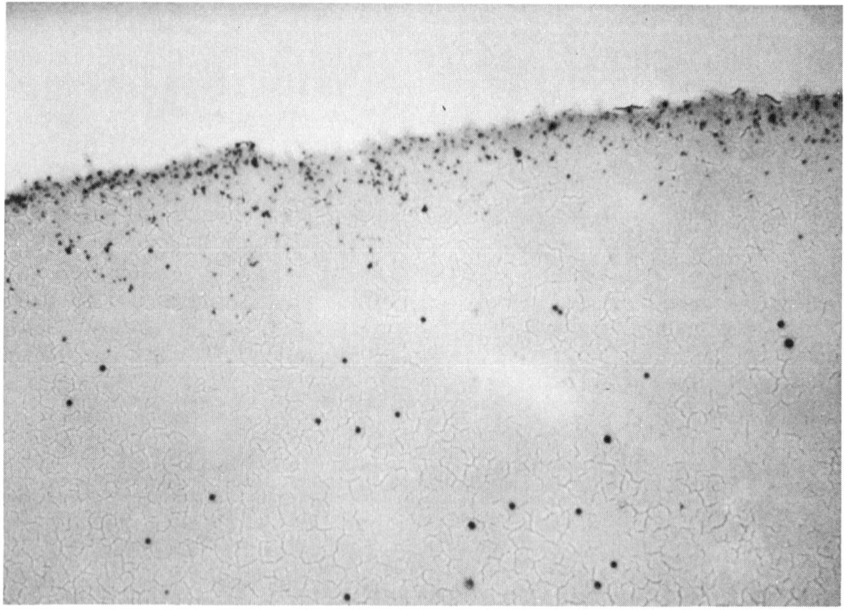

Figure 6. Cross-section of a bombarded mouse liver displaying the penetration of microprojectiles. One day after the inoculation, the liver was removed, embedded, sectioned, stained, and photographed (100X).

PRACTICAL APPLICATIONS

Given the technology as it exists today, are there some potential useful applications? One consideration is that only a restricted number of cells can be transfected *in vivo*. With this in mind and the fact that skin bombardment worked well, we explored the possibility of using the technology for "genetic immunization." The basic idea is that DNA could be directly introduced into the cells in the skin and that these cells would express the foreign gene, eliciting an immune response. This idea has been borne out (16). Simple plasmid constructs have been inoculated directly into the skin of mice and as a consequence a humoral response was induced to human growth hormone, human α1-antitrypsin, luciferase or the gp120 protein of HIV. If on further testing this technique proves solid, it may have two uses. One would be to produce antibodies to proteins of interest which are difficult to purify. The second is as a mode of vaccination. Of particular interest in this regard is whether an effective cytotoxic T-lymphocyte response is produced. This remains to be tested.

A second possible application which relies on a catalytic response is the cytokine-induced immunity against tumor cells. There is some evidence that induced expression of particular cytokines in tumor cells can elicit an immune attack even on non-expressing tumor cells of the same kind (17–19). In collaboration with J. Minna and D. Carbone we have shown that tumor cells can be transfected directly in the mouse with the hand-held device. Whether this protocol can actually produce an anti-tumor immune response in an effective manner remains to be seen.

As concerns gene replacement therapy the biolistic approach has several limitations. One is that the depth of penetration of the BBs may limit the cells which can be addressed. The other is simply the limit on the total number of cells that can be bombarded. As the guns are designed now it would require multiple bombardments superficially to cover a human liver, for example, with microprojectiles. A third consideration is the fate of the introduced DNA. Experience with cultured cells indicates that only a small fraction of the transfected cells stably integrate the transgene. This may not be an important impediment when the target cells are non-dividing, but in dividing cells effective gene therapy will require integration into the chromosome. Until a method to effect high-frequency integration is devised the biolistic protocol will probably not be a useful gene therapy tool when dividing cells are the target. It may prove useful, however, when only limited or local expression is required.

SUMMARY

The biolistic technology has had a tremendous impact on plant and microbial research and development. The experience to date indicates that it should also have important uses with animal applications. The in-chamber system should prove increasingly valuable for transfecting non-dividing cells, primary cells, and other difficult-to-transfect targets. The hand-held version is currently useful for *in situ* transfection of cells in skin, but has limited effectiveness for internal organ transfection. This limitation may be lessened with further improvements in gun design and microprojectile coating. For gene therapy the most promising applications of this technology appear to be for genetic immunization and protocols which elicit an anti-tumor response.

REFERENCES

1 Sanford, J.C., Klein, T.M., Wolf, E.D. and Allen, N. (1987) Particulate Sci. Technol. 5, 27–37.

2 McCabe, D.E., Swain, W.F., Marinell, B.J. and Christou, P. (1988) Bio/Technology 6, 923–926.

3 Gordon-Kamm, W.J., Spencer, T.M., Mangano, M.L., Adams, T.R., Daines, R.J., Start, W.G., O'Brien, J.V., Chambers, S.A., Adams, W.R., Willetts, N.G., Rice, T.B., Mackey, C.J., Krueger, R.W., Kausch, A.P. and Lemaux, P.G. (1990) Plant Cell 2, 603–618.

4 Klein, T.M., Roth, B.A. and Fromm, M.E. (1989) Proc. Nat. Acad. Sci. U.S.A. 86, 6681–6685.

5 Sanford, J.C. (1990) Physiologia Plantarum 79, 206–209.

6 Johnston, S.A., Anziano, P.Q., Shark, K., Sanford, J.C. and Butow, R.A. (1988) Science 240, 1538–1541.

7 Klein, T.M., Arentzen, R., Lewis, P.A. and Fitzpatrick-McElligott, S. (1992) Bio/Technology 10, 286–291.

8 Armaleo, D., Ye, G.-N., Klein, T.M., Shark, K.B., Sanford, J.C. and Johnston, S.A. (1990) Curr. Genet. 17, 97–103.

9 Shark, K.B., Smith, F.D., Harpending, P.R., Rasmussen, J.L. and Sanford, J.C. (1991) Appl. Environ. Microbiol. 57, 480–485.

10 Sanford, J.C., DeVit, M.J., Russell, J.A., Smith, F.D., Harpending, P.R., Roy, M.K. and Johnston, S.A. (1991) Technique 3, 3–16.

11 Zelenin, A.V., Titomirov, A.V. and Kolesnikov, V.A. (1989) FEBS Lett. 244, 65–67.

12 Yang, N-S., Burkholder, J., Roberts, B., Martinell, B. and McCabe, D. (1990) Proc. Nat. Acad. Sci. U.S.A. 87, 9568–9572.

13 Johnston, S.A., Riedy, M., DeVit, M.J., Sanford, J.C., McElligott, S. and Williams R.S. (1991) In Vitro Cell. Dev. Biol. 27, 11–14.

14 Zelenin, A.V., Alimov, A.A., Titomirov, A.V., Kazansky, A.V., Gorodetsky, S.I. and Kolesnikov, V.A. (1991) FEBS Lett. 280, 94–96.

15 Williams, R.S., Johnston, S.A., Riedy, M., DeVit, M.J., McElligott, S.G. and Sanford, J.C. (1991) Proc. Nat. Acad. Sci. U.S.A. 88, 2726–2730.

16 Tang, D., DeVit, M.J. and Johnston, S.A. (1992) Nature 356, 152–154.

17 Tepper, R.I., Pattengale, P.K. and Leder, P. (1989) Cell 57, 503–512.

18 Golumbeck, P.T., Lazenby, A.J., Levitsky, H.I., Jaffee, L.M., Karasuyama, H., Baker, M. and Pardoll, D.M. (1991) Science 254, 713–716.

19 Porgador, A., Tzehoval, E., Katz, A., Vadai, E., Revel, M., Feldman, M. and Eisenbach, L. (1992) Cancer Res. 52, 3679–3686.

INDEX